ACTIN COMPUTATION
Unlocking the Potential of Actin Filaments for Revolutionary Computing Systems
Volume 3

WSPC Book Series in Unconventional Computing

Print ISSN: 2737-5218
Online ISSN: 2737-520X

Published

Vol. 3   *Actin Computation: Unlocking the Potential of Actin Filaments for Revolutionary Computing Systems*
edited by Andrew Adamatzky

Vol. 2   *Unconventional Computing, Arts, Philosophy*
edited by Andrew Adamatzky

Vol. 1   *Handbook of Unconventional Computing*
(In 2 Volumes)
Volume 1: Theory
Volume 2: Implementation
edited by Andrew Adamatzky

WSPC BOOK SERIES IN UNCONVENTIONAL COMPUTING   *Volume 3*

ACTIN COMPUTATION

Unlocking the Potential of Actin Filaments for Revolutionary Computing Systems

Editor **Andrew Adamatzky**

University of the West of England, UK

World Scientific

NEW JERSEY · LONDON · SINGAPORE · BEIJING · SHANGHAI · HONG KONG · TAIPEI · CHENNAI · TOKYO

Published by

World Scientific Publishing Co. Pte. Ltd.

5 Toh Tuck Link, Singapore 596224

USA office: 27 Warren Street, Suite 401-402, Hackensack, NJ 07601

UK office: 57 Shelton Street, Covent Garden, London WC2H 9HE

Library of Congress Cataloging-in-Publication Data

Names: Adamatzky, Andrew, editor.

Title: Actin computation : unlocking the potential of actin filaments for revolutionary
   computing systems / editor Andrew Adamatzky, University of West of England, UK.

Description: New Jersey : World scientific, [2024] | Series: WSPC book series in unconventional
   computing, 2737-5218 ; vol. 3 | Includes bibliographical references and index.

Identifiers: LCCN 2023040941 | ISBN 9789811285066 (hardcover) |
   ISBN 9789811285073 (ebook for institutions) | ISBN 9789811285080 (ebook for individuals)

Subjects: LCSH: Biocomputers--Technological innovations. | Biocomputers--Materials. |
   Actin. | Information technology--Technological innovations.

Classification: LCC QA76.884 .A28 2024 | DDC 006.3/842--dc23/eng/20231114

LC record available at https://lccn.loc.gov/2023040941

British Library Cataloguing-in-Publication Data

A catalogue record for this book is available from the British Library.

For any available supplementary material, please visit
https://www.worldscientific.com/worldscibooks/10.1142/13643#t=suppl

Desk Editors: Balasubramanian Shanmugam/Julio Hong

Typeset by Stallion Press
Email: enquiries@stallionpress.com

Preface

The book establishes a solid theoretical and modelling foundation for the future experimental prototyping of actin-based cytoskeleton computers in laboratory settings. Actin, a globular protein abundantly found in the cells of various eukaryotic organisms, serves as a fundamental building block of intracellular cytoskeletal filaments. These actin filaments, along with tubulin microtubule filaments, play vital roles in facilitating coordinated cellular movements and functioning as the "nervous system" within cells. They are involved in information processing and contribute to cellular learning processes. By delving into the mechanisms of signal propagation and perturbations on actin filaments, the book aims to uncover the underlying principles of information processing at the sub-cellular level. Through this understanding, novel principles for information processing can be developed, potentially leading to the design of self-sustaining and evolving actin-based computers. These unconventional computing systems would harness the unique properties of actin filaments and could revolutionise the field of computational biology.

The book's theoretical and modelling framework serves as a crucial stepping stone towards the experimental realisation of actin-based cytoskeleton computers in the laboratory. By bridging the gap between theory and practice, researchers and scientists can leverage this knowledge to conduct experiments, validate hypotheses, and explore the practical implications of actin-based computing systems. Ultimately, the goal is to unlock the full potential of actin filaments as a versatile and powerful platform for information processing, opening up new avenues in both biological and computational sciences.

The book is structured as follows.

In the opening chapter, we are provided with a concise introduction to actin and the computation performed by the cytoskeleton.

The second chapter presents the initial formalism of actin automata, marking an important historical milestone. In our research, we depict actin filaments as two chains of one-dimensional semi-totalistic automaton arrays with binary states. This representation enables us to simulate hypothetical signalling events occurring within actin filaments. Each node in the actin automaton is assigned a state of either '0' (resting) or '1' (excited). The states of the nodes are updated discretely over time based on the states of their neighbouring nodes.

Our analysis focuses on exploring the entire rule space of actin automata. We investigate integral characteristics of the space–time configurations generated by these rules and identify state transition rules that facilitate the movement and localisation of signals. We propose methods for selecting rules that promote localisation based on global characteristics. By considering Shannon entropy, activity levels, and the degree of incoherence in excitation between the polymer chains, we find that certain properties of actin automata rules can be predicted.

We demonstrate that it is possible to determine whether a given rule supports travelling or stationary localisations by examining the ratios of excited neighbours, which play a crucial role in the generation of localisations. By applying bio-molecular hypotheses to this model, we explore the implications of our findings in the context of cell signalling and the emergence of complex behaviour in cellular computation.

The third chapter presents an exhaustive analysis of Rule 22 cellular automaton, which is the abstract version of actin automata. Rule 22 elementary cellular automaton (ECA) has a 3-cell neighbourhood, binary cell states, where a cell takes state '1' if there is exactly one neighbour, including the cell itself, in state '1'. In Boolean terms, the cell-state transition is an XOR function of three cell states. In physico-chemical terms, the rule can be interpreted as describing the propagation of self-inhibiting quantities/species. The space–time dynamics of Rule 22 demonstrate non-trivial patterns and quasi-chaotic behaviour. We characterise the phenomena observed in this rule using mean field theory, attractors, de Bruijn diagrams, subset diagrams, filters, fractals, and memory.

The fourth chapter introduces innovative concepts in actin automata by incorporating a memory function. Our proposed model represents actin filaments as a dual chain of finite state machines called nodes. These nodes can exist in states '0' or '1', representing the absence or presence of a sub-threshold charge on the corresponding actin units. The state of each node is discretely updated over time, with all nodes simultaneously changing their states. In contrast to previous actin automata models that only considered instantaneous state transitions, our enhanced model introduces the notion of memory. We assume that node states depend not only on the current states of neighbouring nodes but also on their past states. This inclusion enables us to explore the impact of memory on the dynamics of actin automata.

Through computational experiments, we elucidate the effects of memory on the behaviour of actin automata. We discover that memory decelerates the spread of perturbations within the system. As a result, perturbations require more time to propagate throughout the filament due to the influence of past states. The introduction of memory reduces the entropy of the space–time patterns generated by the automata. Memory transforms moving activity patterns, known as travelling localisations, into stationary oscillators. Consequently, these patterns oscillate in a fixed position instead of propagating along the filament. Additionally, memory can convert stationary oscillations into static or motionless patterns with no observable movement or oscillation. The introduction of memory alters the propagation speed, reduces entropy, converts travelling localisations into stationary oscillators, and stationary oscillations into static patterns.

The fifth chapter focuses on the dynamic behaviour of excitation in graph automata models of actin filaments. We introduce a model called the F-actin automaton, which represents a filamentous actin molecule as a graph composed of finite state machines. Each node in the graph can exist in one of three states: resting, excited, or refractory. The states of all nodes are updated simultaneously and follow the same rule within discrete time steps. In our analysis, we explore two distinct rules: the threshold rule and the narrow excitation interval rule.

Under the threshold rule, a resting node transitions to an excited state if it has at least one excited neighbour. On the other hand, the narrow excitation interval rule only allows a resting node to become excited if it has exactly one excited neighbour. By studying the dynamics of the F-actin automaton, we investigate the distribution of transient periods and the lengths of limit cycles that emerge during its evolution. These limit cycles represent recurring patterns or oscillations within the system. Furthermore, we delve into the underlying mechanisms that drive the formation of these limit cycles. In addition to analysing the cyclic behaviour, we assess the density of information storage in F-actin automata. Through our investigation of the F-actin automaton's characteristics, we gain insights into the quantity and efficiency of information that can be stored within this model.

The remaining chapters of the study introduce several modelled prototypes of actin computers. These prototypes are designed to explore the potential applications and capabilities of actin-based computing systems. Each chapter presents a distinct model and examines its functionality within the context of actin-based computation.

The sixth chapter digs into the realm of electrical analogue computing circuits implemented on actin filaments. We consider actin bundles as conduits resembling electrical wires, where the density of filaments varies along the bundles. This unique perspective allows us to analyse and discuss the key electrical parameters of the system.

Our analysis revolves around solving a set of equations that describe the network, accounting for various initial conditions. We introduce input voltages, representing information bits, at specific locations within the system, and compute the resulting output voltages at other designated positions. To explore different scenarios, we examine two distinct situations. First, we explore an idealised scenario where point-like electrodes can be effortlessly inserted at any desired points along the actin bundles. In this ideal setting, we investigate the system's capability to implement logical gates and function as a finite state machine. Second, we explore a more realistic scenario where electrodes are placed on a surface and possess dimensions typical of those available in the industry. Despite the constraints imposed by this setting, we investigate the system's performance and its ability to emulate logical gates and operate as a finite state machine.

Actin filaments exhibit the ability to support travelling solitons, which can manifest as electrical, mechanical, and potentially optical solitons. In the seventh chapter, we present computational evidence suggesting that the interaction of solitons on junctions of actin filaments can enable the implementation of Boolean logical gates. Our investigation focuses on studying the propagation of solitons on one-dimensional arrays of nodes, where adjacent nodes interact via Morse interaction.

By employing numerical integration techniques, we showcase that solitons colliding at the junctions of two arrays can function as Boolean logic gates, specifically the AND, OR, and NOT gates. The specific logic gate implemented depends on the geometry of the junctions and the phase difference between the colliding solitons. By analysing the behaviour of solitons at these junctions, we identify the conditions that allow the solitons to execute the desired logic operations. Through this chapter, we provide computational evidence supporting the idea that solitons on actin filaments can serve as the building blocks for implementing Boolean logical gates. The ability to manipulate solitons at junctions opens up avenues for novel approaches to information processing, potentially expanding the capabilities of actin-based computing systems.

In the eighth chapter, we utilise a coupled nonlinear transmission line model to examine the dynamics of voltage pulses travelling along actin filaments. We assign the logical value of "True" to the presence of a voltage pulse and "False" to its absence, enabling us to represent digital information transmission along these filaments. The interactions between two pulses, which correspond to the Boolean values of input variables, can either facilitate or inhibit their propagation. Through an exploration of this phenomenon, we demonstrate the construction of Boolean logical gates and a one-bit half-adder using the interacting voltage pulses. By leveraging the interactions between pulses, we are able to manipulate the flow of information and realise basic computational operations.

The ninth chapter is about actin quantum automata. We adopt a model where actin filaments are represented as two chains of one-dimensional quantum automata arrays. This model allows us to explore hypothetical signalling events that propagate along these chains. Our focus is on investigating various functions of automaton state transitions and analysing their behaviour, particularly considering the role of superposition in the initial states. Through our analysis, we observe and study the propagation of localised particles along the actin chains. One significant finding of our study is the realisation that logical gates can be implemented through collisions between these travelling particles. We further leverage these particle collisions to successfully implement a binary adder, showcasing the computational capabilities of the system.

In the tenth chapter, we present a novel approach for realising three-valued logical gates in actin quantum automata. Our model represents actin as a helix consisting of two one-dimensional quantum automata arrays, extending our previous work and leveraging the quantum properties of the automaton, such as superposition. By selecting functions for automaton state transitions, we demonstrate the computation of actin automata evolution and successful implementation of three-valued logical gates within this framework. We provide illustrative examples showcasing the implementation of operators from various three-valued logical systems.

The eleventh chapter focuses on exploring the computational potential of a single actin molecule. We evaluate the molecule's computational capabilities by investigating the distributions of logical gates implemented through the propagation of excitation patterns. Our approach involves representing the actin molecule as an excitable network of automata, known as the F-actin automaton. In this model, each atom's state is updated based on the states of its neighbouring atoms, considering both chemical bonds (hard neighbours) and atoms in close proximity (soft neighbours). By analysing the behaviour of the F-actin automata, we demonstrate their effectiveness in implementing logical gates, including OR, AND, XOR, and AND-NOT, through the interaction of excitation patterns. Among these gates, the AND gate is the most commonly observed, while the XOR gate appears less frequently. Leveraging the identified gate architectures, we further construct a one-bit half-adder and controlled-not circuits using the F-actin automata.

In the twelfth chapter, we explore the mining of logical gates in a molecular automata model of actin filaments. Our approach involves utilising an automaton network to simulate an actin filament, where each atom can exist in either two or three states. The state updates occur simultaneously for all atoms, and these updates are based on the ratios of their neighbouring atoms in specific states. We explore two different state transition rules within this model. First, in the semi-totalistic Game of Life-like actin filament automaton, atoms have binary states (0' and 1'), and their state transitions depend on the ratios of neighbouring atoms in the state '1'. Second, in the excitable actin filament automaton, atoms possess three states: resting, excited, and refractory. A resting atom becomes excited if the ratio of its excited neighbours falls within a specified interval. Transitions from the excited state to the refractory state and from the refractory state to the resting state occur unconditionally.

To investigate the capabilities of these automata, we conduct computational experiments involving perturbation and excitation dynamics on the actin filament automata. Through a series of computational trials, we discover a range of eight-argument Boolean functions by mapping all possible configurations of eight-element binary strings to the excitation outputs of the input/output domains. By exploring these mappings, we uncover the logical gates and functions that can be effectively implemented within the actin filament automata framework.

In the thirteenth chapter, we shift our focus to the cytoskeleton scale of actin networks and examine the behaviour of bundles of actin filaments as conductive pathways. Our main objective is to investigate the propagation of excitation waves within these actin filament bundles. We employ computational experiments on a two-dimensional slice of an actin bundle network. Rather than controlling individual filaments, our approach explores the collective behaviour of the bundle as a whole. By arranging electrodes in an arbitrary configuration within the bundle, we aim to demonstrate the feasibility of implementing circuits that have two inputs and one output. Through our experiments, we show how the actin bundle network can be utilised to implement these two-inputs-one-output circuits. This approach allows us to tap into the potential computational capabilities offered by actin filaments while circumventing the challenges associated with controlling excitations at the level of individual filaments.

The book concludes with the fourteenth chapter, where we introduce the ultimate design of an actin computer capable of implementing mappings between sets of binary strings. This advanced computer model, known as the actin droplet machine, simulates a three-dimensional network of actin bundles within a droplet of physiological solution. The actin bundle network within the droplet facilitates the propagation of travelling impulses, which form the basis of information processing in this machine. To interact with the actin droplet machine, a selected set of k electrodes is employed. These electrodes are responsible for applying stimuli in the form of binary strings of length k, represented by impulses generated on the electrodes. The actin droplet machine records the responses to these stimuli in the form of impulses and subsequently converts them into corresponding binary strings. The state of the machine is represented by a binary string of length k, where a '1' is present at the ith position if an impulse is recorded on the ith electrode, and '0' otherwise. In this chapter, we present the detailed design of the actin droplet machine, including its architectural layout and operational principles. We analyse the state transition graphs of the machine to understand its computational dynamics and capabilities.

The book serves as a comprehensive treatise that delves into the theoretical analysis of actin computer architectures across multiple levels. These architectures encompass arrays of finite state machines, molecular automata on graphs, solitons, nonlinear electrical lines, and excitable substrates. By exploring these diverse approaches, the book offers a pioneering resource for physicists, chemists, biologists, computer scientists, mathematicians, and engineers who aspire to develop tangible prototypes of cytoskeleton-based computing devices. It presents an array of computer architectural designs, making it a valuable reference for researchers and practitioners in these fields, providing them with a foundation to embark on the development of physical prototypes in the realm of cytoskeleton-based computing.

Andrew Adamatzky
Bristol, UK
January 2024

About the Editor

Andrew Adamatzky is Professor of Unconventional Computing and Director of the Unconventional Computing Laboratory, Department of Computer Science, University of the West of England, Bristol, UK. He does research in molecular computing, reaction-diffusion computing, collision-based computing, cellular automata, slime mould computing, massive parallel computation, applied mathematics, complexity, nature-inspired optimisation, collective intelligence and robotics, bionics, computational psychology, non-linear science, novel hardware, and future and emergent computation.

He has authored seven books, mostly notable are *Reaction-Diffusion Computing*, *Dynamics of Crow Minds*, and *Physarum Machines*, and has edited 22 books in computing, most notable are *Collision-Based Computing*, *Game of Life Cellular Automata*, and *Memristor Networks*. He has also produced a series of influential artworks published in the atlas *Silence of Slime Mould*. He is Founding Editor-in-Chief of *Journal of Cellular Automata* and *Journal of Unconventional Computing* and Editor-in-Chief of *Journal Parallel, Emergent, Distributed Systems* and *Parallel Processing Letters*.

Contents

Chapter 1

Actin

Andrew Adamatzky*,§, Jörg Schnauß†,¶, and Jack Tuszyński‡,∥

*Unconventional Computing Lab, UWE Bristol, UK
†Soft Matter Physics Division, Peter Debye Institute for Soft Matter
Physics, Faculty of Physics and Earth Science, Leipzig University,
Germany and Fraunhofer Institute for Cell Therapy and Immunology
(IZI), DNA Nanodevices Group, Leipzig, Germany
‡Department of Oncology, University of Alberta, Edmonton,
AB T6G 1Z2, Canada; DIMEAS, Politecnico di Torino,
Corso Duca degli Abruzzi 24, 10129, TO, Turin, Italy
§andrew.adamatzky@uwe.ac.uk
¶joerg.schnauss@uni-leipzig.de
∥jack.tuszynski@gmail.com

Abstract
We present the concept of actin networks, which serve as intracellular
frameworks, actuator systems, and pathways for information transfer
and processing within cells.

The cytoskeletal protein actin is known in its filamentous form as
F-actin. Filaments are organised in a double helix structure consisting of polymerised globular actin monomers (G-actin) (Fig. 1.1).
Typical filaments have a length of roughly 10 μm, but also micro-
filaments with a length of less than 1 μm and a diameter of 8 nm
(80Å) [179] can be formed. These individual actin filaments can be
interconnected, cross-linked, and associated with the cytoskeleton
and other cellular structures through various actin-binding proteins
(ABPs). Examples of commonly encountered ABPs that fulfil these
functions include profilin, Arp2/3 complex, filamin, spectrin, and
α-actinin [272].

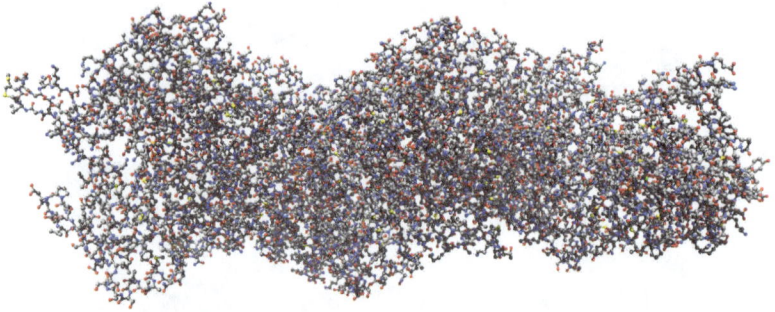

(a)

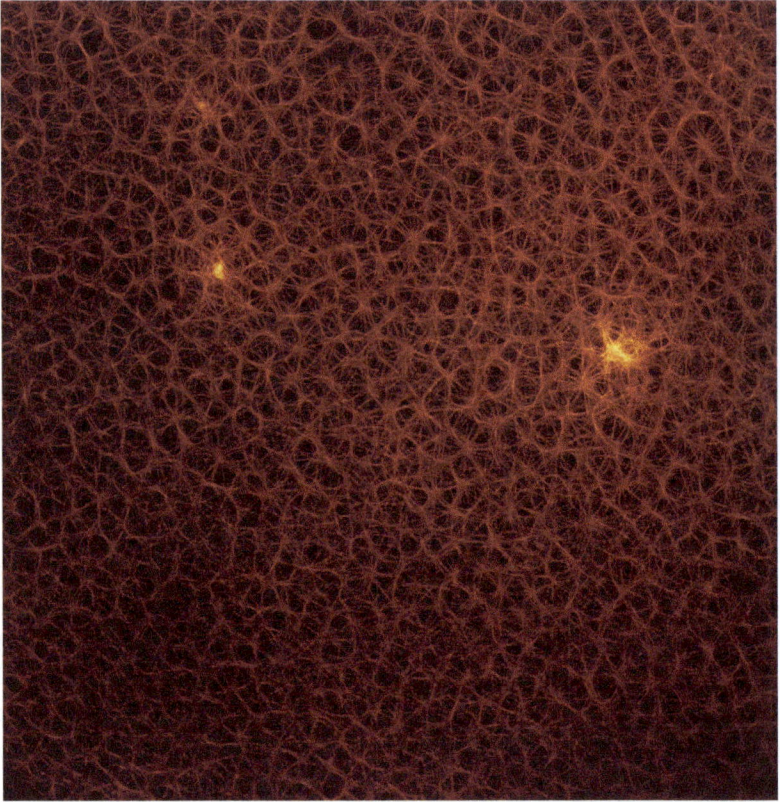

(b)

Fig. 1.1. Actin polymer. (a) A pseudo-atomic model of F-actin [93] in Corey–Pauling–Koltun colouring. (b) The actin network, produced in laboratory experiments on formation of regularly spaced bundle networks from homogeneous filament solutions [242].

Actin is a protein that is abundantly expressed in all eukaryotic cells [121]. It plays a crucial role in cellular functions by forming intracellular scaffolds, actuators, and pathways for information transfer and processing. These functions include maintaining the structural integrity of the cell, contributing to muscle contraction, and facilitating the movement of cytoplasmic molecules and organelles. The cytoskeleton, consisting of helical actin filaments, bundles of tubulin microtubules, and various intermediate filaments and cytoskeletal-binding proteins, collectively contributes to these cellular processes. While actin and tubulin are widely present and highly conserved, the specific intermediate filaments found in a cell vary depending on their role. For example, human epithelial cells contain significant amounts of cytokeratins, which enhance their structural integrity [239].

The cytoskeleton is widely recognised for its significant role in cell signalling events, encompassing processes such as signal transduction pathways, electrical potential, quantum events, and mechanical stress [136, 138, 247, 253]. Actin, a crucial component of the cytoskeleton, plays an essential role in cellular signalling by detecting and transmitting various environmental stimuli through its interactions with the cell membrane and membrane-bound proteins, including ion channels and receptors [136]. Notably, actin filaments are involved in the transmission of cell-to-cell signalling events and mechanical stress through the focal adhesion complex. This multi-protein structure, connected to the cell membrane, facilitates the transduction of mechanical signals along associated filaments via signalling cascades or direct transmission of mechanical force [127, 196].

From the perspective of natural computation, the cytoskeleton presents an intriguing model for understanding how information is transmitted and transformed within a cell. While the precise mechanisms underlying many of these processes are still under investigation, it has been proposed that 'data' travelling through actin filaments could undergo computation through collisions between mobile localisations representing the data [7, 8, 13, 30, 252, 282]. Moreover, the cytoskeleton may exhibit some degree of processing capability, potentially performing Boolean logical operations as signals pass through specific proteins or branches within the cytoskeletal network [151].

Actin filament networks also play a crucial role in modulating synaptic terminals by interacting with ion channels [57] and filtering noise in synapses [65, 80]. Through the modulation of dendritic ion channel activity, actin filaments govern neural information processing and enhance the computational capabilities of dendritic trees. They achieve this by facilitating ionic condensation and the propagation of ion clouds [204].

Actin and tubulin filaments have been implicated in information processing and learning mechanisms [66, 74, 111, 130, 157, 206, 207, 209, 253]. Dysfunction in actin assembly or its association with other intracellular components has been linked to psychiatric and neurological disorders [86, 144, 192, 226, 258, 259]. Therefore, by unravelling the mechanics of signal propagation and perturbation on actin filaments, we have the potential to develop new principles of information processing at the sub-cellular level. Furthermore, this knowledge may contribute to the development of nano medicine-based treatments for neurological disorders.

There is also evidence to suggest that actin may also transmit electrical potential/ionic waves [254] and quantum protein transitions [49, 247], in addition to mechanical force and signalling cascades. The actin electrical phenomena have been both experimentally observed and modelled as bio-wires capable of conducting ionic waves [202, 205, 216–218, 227, 254]. Actin filaments are polyelectrolytes surrounded by counterions, the filaments therefore possess the capacity of transmitting signals or sustaining ionic conductances [56, 254].

Under physiological conditions, actin filaments represent highly charged polyelectrolytes [56], and as originally predicted by Manning's condensation theory [160], polyelectrolytes may attract condensed ions to their surroundings. These ions are called counterions, and condense along the length of the polymer if a sufficiently high linear charge density is present on the solvent exposed surface of the polymer. Since actin filaments are highly negatively charged, according to the Manning theory, formation of this counterion layer is formed in the saline solution with co-ions in this solution being repelled from the filaments' vicinity. Functionally, actin polymers have been modelled as nonlinear inhomogeneous electrical transmission "cables" and shown to propagate nonlinear ionic solitary representing electrical signals that propagate with virtually no loss of energy along their path [153]. However, the underlying biophysical principles and molecular mechanisms that support the ionic conductance and transport along actin filaments are still poorly understood. Approximate theories using infinitely long cylindrical filament models have become a powerful tool to characterise the electrical conductivity properties of these polyelectrolytes. For instance, the high surface charge of the polyelectrolyte, usually present under physiological conditions, causes an inhomogeneous arrangement of counter- and co-ions forming an electrical double layer (EDL) around its surface. The accumulation of these ions builds up an ionic conductivity and capacitance layer intrinsic to the EDL [125, 254].

A recent study [188] has characterised AC conduction characteristics of both globular and polymerised actin and compared the obtained quantitative values to those predicted theoretically before. Actin filaments were previously demonstrated to act as conducting nanobiowires forming a signalling network capable of transmitting ionic waves in cells. These measurements revealed two relevant characteristics. First, the polymerised actin, arranged in filaments, has a lower impedance than its globular counterpart. Second, increasing the actin concentration leads to higher conductivities. It has also been hypothesised that the actin filaments represent electrical RLC circuits, where the resistive contribution is due to the viscous ion flows along the filaments, inductive contributions are due to the solenoidal flows along and around the helix-shaped filament, and the capacitive contribution is due to the counterion layer formed around each negatively charged filament.

Actin filaments, being rod-like polymers, are particularly likely to have counterions adsorbed to their surface at high ionic concentrations, such as those in the intra-neuronal environment, ions would be expected to densely adsorb to the surface of actin filaments due to complementary charges [207]. The actin filaments are also capable of supporting propagation of discrete breathers (nonlinear localised modes of excitation) as a consequence of nonlinearity in pure, translationally invariant systems of any dimensionality (similar to intrinsic localised modes in anharmonic crystals) [90, 141, 148]. The solitonic signals propagating on actin networks are capable of realising collision-based logical circuits [3, 8, 11]. These events may therefore also be considered as discrete data packets from a computational perspective.

Computational models of tubulin microtubules have been developed in 1990s and used to demonstrate that computation could be implemented in tubulin protofilaments by classical and quantum means [110, 113, 135, 209]. Less attention was paid to actin double helix filaments, despite the importance of the actin in learning and information preprocessing as might be hinted by the predominant presence of actin networks in synapses [65, 80, 87, 142].

Actin offers several advantages over other polymers when it comes to developing unconventional computers, as summarised in our publication [28]. The following are the key points:

- *High density charges*: Actin filaments exhibit a significantly higher charge density compared to DNA and microtubules, resulting in extensive charges in electric dipole momentum [153]. This characteristic allows for efficient signal transmission and processing.
- *Nonlinear dispersive wave propagation*: Actin filaments serve as nonlinear inhomogeneous transmission lines, supporting the propagation of nonlinear dispersive waves and solitons [205, 217, 256]. This property enables the manipulation and processing of signals in complex ways.
- *Self-renewal through polymerisation and depolymerisation*: Actin can undergo polymerisation and depolymerisation, processes that can be regulated and tuned using accessory proteins or bionic complexes [155, 246]. This dynamic behaviour allows for the adaptation and reconfiguration of actin structures, enhancing their versatility in computing applications.
- *Structural simplicity*: Compared to DNA, actin is less structurally complex on relevant length scales, making experimental prototyping easier to achieve. This simplicity facilitates the construction and manipulation of actin-based computing circuits.

- *Macro-molecular actuator*: Actin acts as a macro-molecular actuator, enabling the development of embedded controllers for molecular machinery [95, 124]. This characteristic expands the application domains of actin computing circuits, allowing for the integration of computational processes with physical actuation.
- *Self-assembly and stability*: Actin structures can self-assemble into energetically favourable configurations, leading to stable structures that can persist for days without additional treatment [123]. Even after experiencing mechanical deformations, actin structures can re-anneal quickly and regain their functional properties.

These advantages position actin as a promising candidate for developing unconventional computers, offering unique properties and capabilities for information processing and manipulation at the nanoscale.

Chapter 2

Actin Automata: Phenomenology and Localisations

Andrew Adamatzky[*,‡] and Richard Mayne[†,§]

*Unconventional Computing Lab, UWE, Bristol, UK
†Mayne Bio Analytics Ltd., Cinderford, UK
‡andrew.adamatzky@uwe.ac.uk
§director@mayneba.com

Abstract

We model actin filaments by representing them as two chains of one-dimensional binary-state semi-totalistic automaton arrays. This allows us to simulate hypothetical signalling events occurring within actin filaments. Each node in the actin automaton has a state of either '0' (resting) or '1' (excited). The state of each node is updated discretely over time based on the states of its neighbouring nodes. Our analysis focuses on exploring the entire rule space of actin automata. We examine integral characteristics of the space–time configurations generated by these rules and determine state transition rules that facilitate the movement and localisation of signals. We propose approaches for selecting rules that promote localisation based on global characteristics. We discover that certain properties of actin automata rules can be predicted by considering Shannon entropy, activity levels, and the degree of incoherence in excitation between the polymer chains. Additionally, we demonstrate that it is possible to determine whether a given rule supports travelling or stationary localisations by examining the ratios of excited neighbours, which are crucial for the generation of localisations. By applying biomolecular hypotheses to this model, we explore the implications of our findings in the context of cell signalling and the emergence of complex behaviour in cytoskeleton computation. This research sheds light on the significance of actin automata and their potential role in understanding cellular processes.

2.1 Introduction

We have developed a model to describe a generalised "information transmission" event that occurs within the actin-component cytoskeleton. This event involves the propagation of energy from one G-actin molecule to its neighbouring molecules through chemical bonds between actin molecules. When a globular actin (G-actin) molecule is stimulated, it enters an excited state, which corresponds to an increase in energy. This state is then transferred to the neighbouring molecules, causing the original molecule to return to a non-excited resting state. The model we have engineered is versatile and can be applied to various forms of data transmission within the actin-component cytoskeleton. This includes the conduction of electrical potential, quantum energy states, or physical waves/strand compression. The model suggests that actin filaments are stimulated at specific points where external data are perceived by the cell. These points can be articulated to the cell membrane or membrane-bound receptors. The associated "output" of this information transmission is a generalised system in which the data packet elicits a repeatable and predictable response. For example, it may lead to the activation of a signal transduction cascade at a target protein within an organelle. By studying this model, we aim to gain insights into how information is transmitted and processed within the actin-component cytoskeleton. This has implications for understanding cellular signalling pathways and how cells respond to environmental stimuli.

2.2 Automaton model

Each G-actin molecule (except those at the ends of F-actin strands) has four neighbours, as demonstrated in Fig. 2.1. An actin automaton consists of two chains, x and y, of semi-totalistic binary-state automata. Each automaton takes two states, '0' (resting) and '1' (excited). Automata update their states by the same rule in discrete time. Each automaton updates its state depending on the states of its immediate neighbours. The neighbourhood of an automaton x_i in chain x (Fig. 2.1) is $u(x_i) = \{x_{i-1}, x_{i+1}, y_i, y_{i-1}\}$ and neighbourhood

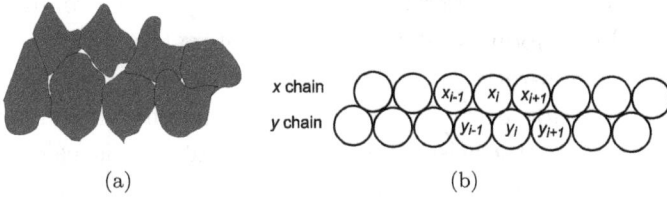

x chain

y chain

(a)   (b)

Fig. 2.1. Schematic diagram of F-actin strands. (a) Structure of actin detected by X-ray fibre diffraction. Adapted from Ref. [187]. (b) Actin automata.

of an automaton y_i in chain y is $u(y_i) = \{y_{i-1}, y_{i+1}, x_i, x_{i+1}\}$. States of automata x_i and y_i at time step t are x_i^t and y_i^t.

Let $\sigma_{x_i}^t = x_{i-1}^t + x_{i+1}^t + y_i^t + y_{i-1}^t$ and $\sigma_{y_i}^t = y_{i-1}^t + y_{i+1}^t + x_i^t + x_{i+1}^t$ be sums of excited, state '1', neighbours of automata x_i and y_i. The automaton state transition function is determined by a two-dimensional matrix $F = (f_{i,j})$, $0 \leq i \leq 1$ and $0 \leq j \leq 4$, $f_{i,j} \in \{0, 1\}$. Each automaton updates its states as the following: $x^{t+1} = f_{x_i^t, \sigma_{x_i}^t}$ and $y^{t+1} = f_{y_i^t, \sigma_{y_i}^t}$. There are 1024 states of F: each determines a unique automaton state transition rule. We will decode the rules via decimal representations of sub-matrices F_0 and F_1. For example, Rule (10, 4) encodes the transition when $F_0 = (01010)$ and $F_1 = (00100)$, or in words, a node in state '0' takes state '1', or is excited, if it has one or three neighbours in state '1' (otherwise, the cell remains in the resting state '0') and a cell in state '1' remains in the state '1', or remains excited, if it has two neighbours in state '1'; otherwise, the cells switches to state '0', or returns to the resting state.

Assuming there are n automata in each chain and the actin automaton evolves for τ time steps, the following integral measures are calculated on space–time configurations generated by actin automata rules. In experiments presented here, $n = 300$ and $\tau = 1000$. To compute the characteristics defined in what follows, we excited a resting actin automaton with a random configuration of '0' and '1' states, where each state is represented with probability 0.5. We evolved the automata t steps and collected linear configurations into a two-dimensional matrix $M = (m_{ij})$, where $1 \leq i \leq n$ and $1 \leq j \leq \tau$, and m_{ij} is a state of automaton i at time step j. The integral measures are as follows:

- Shannon entropy H. Let W be a set of all possible configurations of a 9-cell neighbourhood of entry (i, j) of matrix M. We calculate number of non-resting configurations as $\eta = \sum_{a \in M} \epsilon(a)$, where $\epsilon(a) = 0$ if for every resting a all its neighbours are resting, and $\epsilon(a) = 1$ otherwise. The Shannon entropy is calculated as $-\sum_{w \in W} (\nu(w)/\eta \cdot ln(\nu(w)/\eta))$, where $\nu(w)$ is a number of times the neighbourhood configuration w is found in matrix M.
- Simpson diversity index $D = 1 - \sum_{w \in W} (\nu(w)/\eta)^2$.
- Space filling P is a ratio of non-resting nodes in actin chains in space–time configuration of $n \times \tau$ of entities: $P = \sum_{1 \leq i \leq n, 1 \leq j \leq \tau} m_{ij}$.
- Morphological richness R is a ratio of different w found in M to the total number of possible w.
- Activity $A = (\sum_{1 \leq i \leq n, 1 \leq t \leq \tau} x_i^t) \cdot (n \cdot \tau)^{-1}$.
- Incoherence $I = |\sum_{1 \leq i \leq n, 1 \leq t \leq \tau} x_i^t - \sum_{1 \leq i \leq n, 1 \leq t \leq \tau} y_i^t| \cdot (n \cdot \tau)^{-1}$.
- Compressibility $Z = s^{-1}$, where s is a size of space-configuration M, in bytes, compressed by LZ algorithm using Java Zip library.

These measures are proven to be successful in characterising behaviour of one- and two-dimensional automaton networks, as demonstrated in our previous works [14, 15, 22, 186, 210].

The following approach to detect rules supporting localisations, i.e. compact stationary or travelling (gliders) patterns that include non-resting states, was used. We excited resting chains x and y with seeds $\boldsymbol{x} = (....000...0x_0x_1x_2x_3x_40...000...)$ and $\boldsymbol{y} = (....000...0y_0y_1y_2y_3y_40...000...)$ where $x_i, y_j \in \{0, 1\}$ for $0 \leq i \leq 4$. For each seed we started the evolution of actin automata from a seed $s = \langle \boldsymbol{x}, \boldsymbol{y} \rangle$, evolved for τ steps and calculated activity A. If $10 \leq A \leq 6 \cdot \tau$, then s is assumed to generate travelling or stationary localisation. For each rule, we counted a number T of seeds generating travelling localisations and a number S of seeds generating stationary localisations. Only localisations on otherwise resting background were taken into account, as we did not consider localisations travelling in the periodically changing backgrounds, as e.g. one shown in Fig. 2.2.

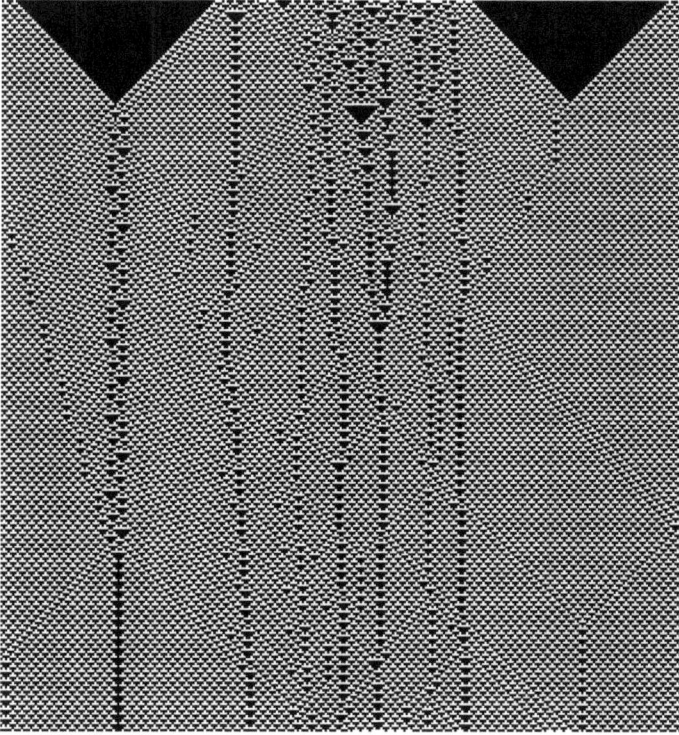

Fig. 2.2. Development of rule $(28, 17)$ actin automaton.

We illustrate the dynamics of actin automata only with space–time configurations of x chains. Morphological differences between configurations of x and y chains do not make a substantial contribution to characterisation of the automaton dynamics (Fig. 2.3). Patterns of incoherence indeed give us additional insights into the dynamics of actin automata, highlighting a potential topic for further study.

2.3 Global characteristics

A plot of Shannon entropy H versus Simpson diversity index D is shown in Fig. 2.4(a). An accurate approximation would be polynomial $D = 0.05 + 0.08*H + 0.62*H^2 - 0.38)*H^3 + 0.08*H^4 - 0.01*H^5$, although a simpler logarithmic approximation $D = 0.3\ln(H) + 0.6$

Fig. 2.3. Development of rule $(4, 25)$ actin automaton. (Left) Space–time configuration of x chain. (Centre) Space–time configuration of y chain. (Right) Pattern of incoherence between configurations of x and y chains.

-->

Fig. 2.4. Shannon entropy versus Simpson diversity index. (a) Plot H vs. D. (b–g) Space-time configurations developed of automaton x chain governed by exemplar rules, (b) rule $(10, 10)$, $H(10, 10) = 4.8$, (c) rule $(11, 6)$, $H(11, 6) = 4.5$, (d) rule $(7, 29)$, $H(7, 29) = 4$, (e) rule $(11, 14)$, $H(11, 14) = 3.5$, (f) rule $(14, 9)$, $H(14, 9) = 3$, (g) rule $(20, 13)$, $H(20, 13) = 2.5$. Time goes down. A node in state '1' is black pixel, and in state '0' is blank. Initially the automata are perturbed by a random configuration of 100 nodes, where each takes '0' or '1' with probability 0.5.

(a)

(b)

(c)

(d)

(e)

(f)

(g)

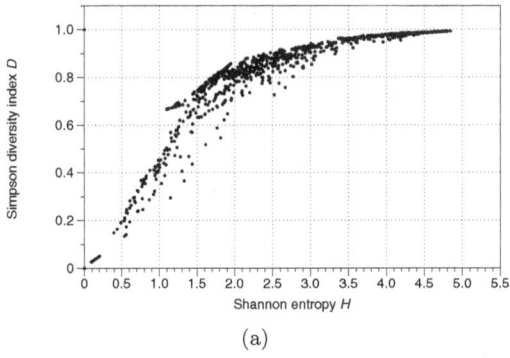

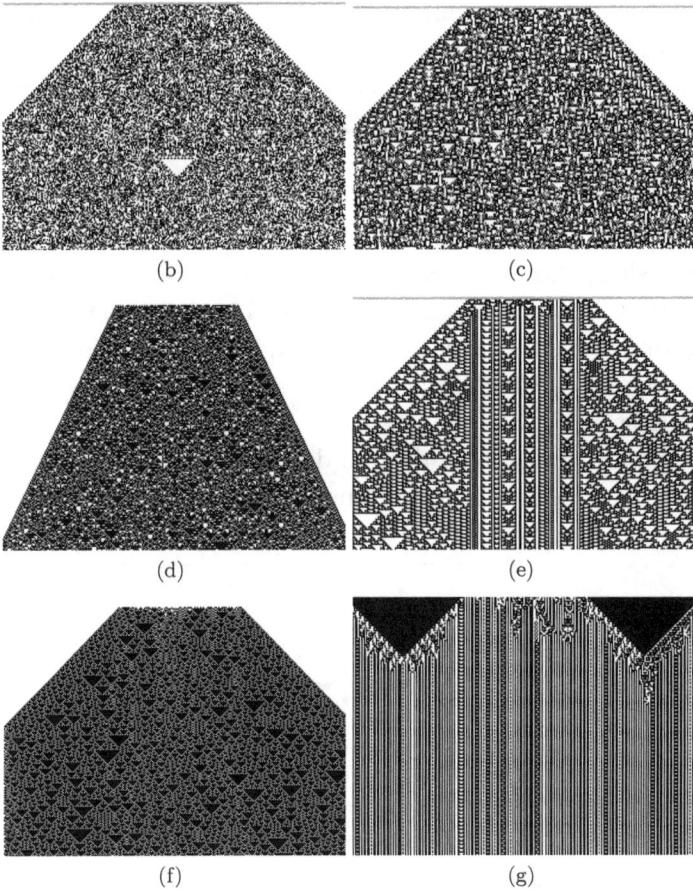

Fig. 2.4. (*Continued*)

provides a good result with coefficient of determination $R^2 = 0.905$.
Values of D are further provided for completeness, but our statements
will consequently be based on H. The distribution of rules verses H
values is shown in Fig. 2.5. The distribution is well approximated by
a polynomial with maxima at entropy values around 1.8 and 4.3. If
we arrange rules in ascending order of their values H, we find that for
$H < 1.8$, rules generate solid configurations, filled with either '0' or
'1' states, or patterns of still localisations in otherwise resting chains.
When H exceeds 1.8 space–time, the chains become filled with regular patterns of activity. Morphology of the space–time configuration
becomes less regular and more complex when H exceeds 4.5.

Rules with the highest H values exhibit disordered, quasi-chaotic
dynamics with no apparently visible structure, e.g. the space–time
configuration generated by rule $(10, 10)$ is shown in Fig. 2.4(b–g).
The emergence of local domains of regularly arranged states is
manifested as decrease of H, e.g. the configuration transitions seen in
Fig. 2.4(c) and (d): note how the triangles of solid '1'-state domains
get their sharp boundaries. Further decreases in entropy are due
to the formation of stationary domains of activity co-existing with
propagating wave-like patterns (Fig. 2.4(e, f)). Stationary localisations dominate the automata chains for low values of the Shannon
entropy (Fig. 2.4(g)).

Node state transition rules are described by vectors F_0 and F_1,
where $F_{0i} = 1$ means that a resting cell becomes excited if it has i

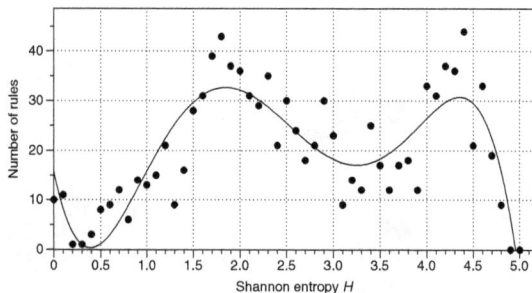

Fig. 2.5. Distribution of a number of rules on Shannon entropy H values of the
space–time configurations. The line is an approximation of the distribution by a
polynomial of degree five.

excited neighbours. Is there any particular value of i, $0 \le i \le 4$, that is responsible for space–time configurations having particular value of the Shannon entropy? To answer the question, we rank all rules on values of entropies of configurations generated by the functions and split rules in classes $L = \{L^1, \ldots, L^5 0\}$ of entropy values starting from 0, incrementing with 0.1 and ending at 5. There are between 7 to 30 rules in each of the 50 classes of entropy. Then we calculate the frequency vectors G_0^z and G_1^z as follows: $G_{ki}^z = |L^z|^{-1} \cdot \sum \{F_{0i} : F_0 \in L^z\}$, $k = 0, 1$, $|L^z|$ is a size of class L^z. That is the frequency vector G_k^z shows how often value '1' appears in vectors F_k, $k = 0, 1$ for classes of entropy, normalised by size of each class.

The frequency vectors as a function of Shannon entropy are shown in Fig. 2.6. Let us exemplify the construction. In Fig. 2.6(a), for $H = 0.9$ we see labels in the following order: empty square, empty triangle up, empty circle, solid triangle down, and solid disk. This means that rule-vector F_0 corresponding to the rule generating configuration with entropy $H = 0.9$ more often has 1 in Position 4, $F_{04} = 1$, less often in Position 2, less often in Position 0, less often in Position 3, and less often in Position 1. In words, if a rule generates space–time configurations with entropy $H = 0.9$, then it is more likely that a resting node is excited if it has four excited neighbours, less likely two excited neighbours, less likely zero neighbours, less likely three neighbours, and almost never one neighbour. To detect what number of neighbours exciting a node is most responsible for generating configurations of certain entropy, we select, for each class/interval of entropy, only positions that have maximum frequencies in the class/interval. Thus, for a transition from state '0' to state '1', the following numbers of excited neighbours are most critical, they are ordered by increasing H: 444443324234 344000000110 440222222241 1433141211. The corresponding sequence for a transition from state '1' to state '1' is 444444444434 400121330034 331110401244 0003442420. The sequences are intentionally split into four parts each, the parts correspond to four equal intervals of the entropy H. For each interval, we calculate the dominating position as follows. For $0 \le H \le 1.25$, transitions $0 \to 1$ and $1 \to 1$ happen more likely if

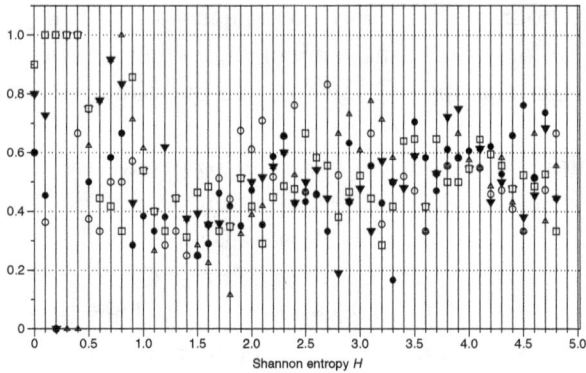

(a)

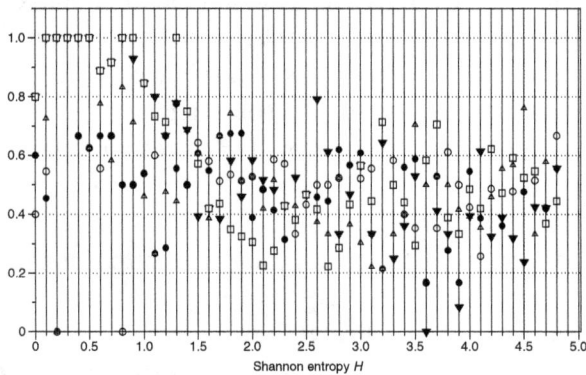

(b)

Fig. 2.6. Frequency vectors for rules corresponding Shannon entropy values H ranging from 0 to 1, with increment 0.1. (a) Vector G_0 reflecting conditions of node-state transition from state '0' to state '1', excitation; and (b) Vector G_1 reflecting conditions of node-state transition from state '1' to state '0', retaining excitation. Frequencies are marked as followed: G_{k0}: empty circle, G_{k1}: solid disc, G_{k2}: triangle up, G_{k3}: triangle down, G_{k4}: square, where $k = 0$ (a) or $k = 1$ (b).

four neighbours are excited. For entropy in the interval [1.25, 2.5], the node is excited or remains excited even if no neighbours are excited, a phenomenon called autonomous excitation. Autonomous excitations pose little interest in terms of communication of signals via travelling localisations, therefore, we discard dominating position 0 and look at the sub-sequences 344000000110 (transition $0 \to 1$) and 400121330034 (transition $1 \to 1$). We see that dominating positions for transition $0 \to 1$ are 1 and 4, and for transition $1 \to 1$, the dominating position is 3. When entropy exceeds 2.5 and yet remains below 3.75, a resting cell excites more likely when it has 2 excited neighbours and remains excited if it has 1 excited neighbour. For the highest values of entropy, $3.75 < H \leq 5$, we observe dominating number of neighbours 1 for transition $0 \to 1$ and 0 for $1 \to 1$.

Proposition 2.1. *Shannon entropy of space–time configurations generated by actin automata is proportional to sensitivity of actin chain nodes: the higher the sensitivity, the larger the values of entropy the configurations have.*

With increase of entropy, the dominating number of neighbours necessary to excite a resting node or to keep an excited node excited decreases as 4 to 1, 4 to 2 to 1 and 4 to 3 to 1 to 0. This is a reflection of increased sensitivity of actin nodes. Decrease of local excitation necessary to keep a node excited ($4 \to 3 \to 1 \to 0$) can be also interpreted as increase in sustainability of excitation, or even as a transition from lateral excitation to lateral inhibition. Namely, for entropy in the interval $H \in [0, 1.25]$, an excited node stays excited if it has four excited neighbours, and $H \in [1.25, 2.5]$ if it has three excited neighbours. That is, excited nodes in actin automata stimulate each other and thus stay excited longer. The situation is changed from lateral stimulation to rather lateral inhibition when entropy exceeds 2.5: an excited node remains excited if one neighbour, $H \in [2.5, 3.75]$, or no neighbours, $H \in [3.75, 5]$, are excited; this can be interpreted as that an excessive amount of excited neighbours inhibits excitation of a node.

We could expect that morphological richness R, ratios of different 3×3 patterns in space–time configurations, will be proportional to entropy H. As we can see in Fig. 2.7(a), this is indeed the case. Richness R grows exponentially with increase of the entropy. Activity A is not a good indicator of morphological complexity of the automata dynamics, as we can see in Fig. 2.7(b), activity remains rather stable for the entropy below 3.5 and only starts to substantially drop down when H exceeds 4. This is because A depends on a ratio of excited nodes, and truly complex structures have rather low

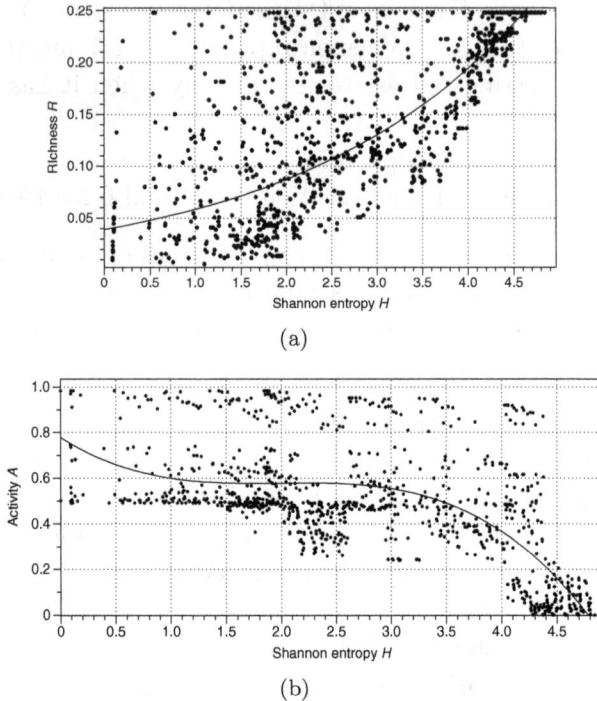

Fig. 2.7. Shannon entropy H versus morphological richness R and activity A. (a) H vs. R, line is an exponential fit, (b) H vs. A, line is a cubic fit.

levels of excitations yet elaborately arranged interacting patterns of excitation.

There is a poor correlation between Shannon entropy H and incoherence I (Fig. 2.8(a)). The incoherence indicates how strongly two chains of actin polymers are desynchronised and might reflect that there are different types of patterns propagating almost independently on the parallel chains of actin units. With regards to compressibility Z, the relation could be approximated by quadratic polynomial, where rules with H between 2 and 2 have lowest indicators of compressibility (Fig. 2.8(b)).

2.4 Localisations

Distribution of rules on a number of seeds supporting localisations (Fig. 2.9) is quite uniform, there are rules which develop 50 seeds into

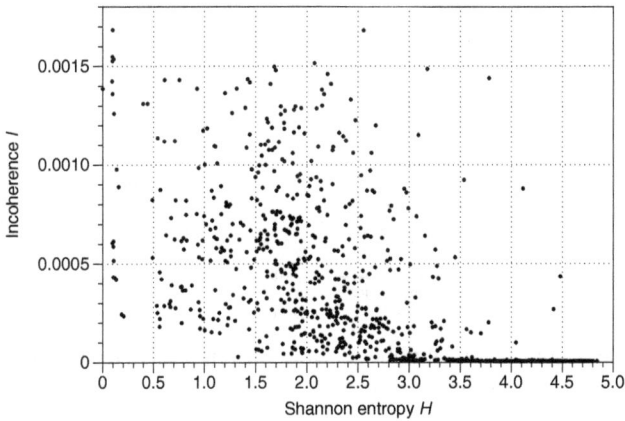

(a)

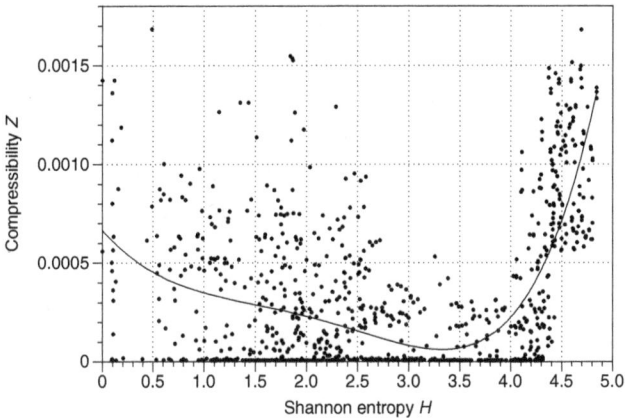

(b)

Fig. 2.8. The entropy versus incoherence and compressibility. (a) H vs. $I_,$. (b) H vs. Z.

localisations and there are rules where almost every seed leads to a stationary or propagating localisation. About 705 of 1024 rules, e.g. almost 69%, do not support any localisations, neither stationary nor travelling. By arranging rules in the descending order on a number of seeds, they develop into into localisations, we can consider three important cases: rules that support travelling localisations, rules that support stationary localisations, rules that support both stationary and mobile localisations.

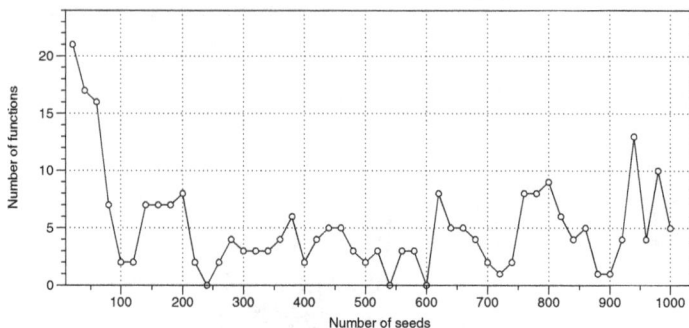

Fig. 2.9. Distribution of a number of rules supporting localisations on a number of seeds, which leads to the localisations.

2.4.1 Travelling localisations

Rules supporting travelling localisations and the integral character-istics of the rules are shown in Table 2.1. A plot of entropy H versus T (Fig. 2.10) shows that rules with lowest values of the localisation richness T show high dispersion in H, they exhibit oscillations in entropy, with amplitude circa 1.5. The rules with T exceeding 40, i.e. those that develop 40 seeds into the localisations, show entropy values around 4.

Proposition 2.2. *Higher values of Shannon entropy are typical for rules supporting large number of travelling localisations.*

Top 14 rules, with values of T exceeding 100, are shown in Table 2.2.

Exemplar space–time configurations generated by some rules from Table 2.2 are shown in Figs. 2.11–2.13. There actin automata were excited by a random configuration (100 nodes) of state '1' and '0' assigned to nodes with probability 0.5. In most cases we see that original pattern of perturbation spreads on the actin chain. The pattern front can propagate with a speed of light (one per node per iteration) as, e.g. in rule (7, 20) in Fig. 2.13(a), or slightly slower, e.g. rule (4, 26) in Fig. 2.11(a). In majority of the rules in Table 2.2, initial perturbations, perturbation produces a great variety of stationary and travelling, which collide with each other, form new localisations, etc. and thus the space-time configurations are well

Table 2.1. Rules supporting travelling localisations.

Rule	T	H	D	R	P	A	I	Z
$(7, 20)$	271	4.31	0.98	0.22	0.84	0.32	0.28	6.67E-006
$(5, 26)$	240	4.70	0.99	0.25	0.92	0.39	0.39	6.40E-006
$(4, 26)$	210	4.54	0.99	0.25	0.86	0.31	0.34	6.92E-006
$(5, 25)$	197	3.83	0.95	0.25	0.73	0.14	0.19	1.37E-005
$(6, 20)$	197	4.37	0.98	0.25	0.86	0.37	0.28	6.33E-006
$(8, 0)$	191	3.50	0.96	0.08	0.83	0.23	0.26	6.82E-006
$(7, 21)$	179	4.31	0.98	0.22	0.97	0.56	0.39	5.97E-006
$(8, 1)$	170	3.50	0.96	0.09	0.84	0.23	0.26	6.81E-006
$(8, 24)$	146	3.71	0.97	0.19	0.80	0.25	0.18	7.39E-006
$(7, 4)$	145	2.75	0.92	0.20	0.54	0.06	0.07	1.82E-004
$(8, 16)$	145	3.54	0.96	0.13	0.77	0.24	0.17	7.61E-006
$(7, 5)$	116	4.31	0.98	0.21	0.93	0.29	0.24	8.71E-006
$(6, 21)$	110	4.40	0.98	0.25	0.70	0.14	0.14	1.28E-005
$(9, 0)$	101	3.69	0.97	0.09	0.92	0.30	0.36	6.17E-006
$(9, 1)$	85	3.70	0.97	0.09	0.91	0.30	0.36	6.15E-006
$(9, 16)$	85	3.63	0.96	0.14	0.81	0.26	0.18	7.18E-006
$(10, 16)$	78	3.96	0.98	0.16	0.90	0.32	0.29	6.19E-006
$(10, 24)$	77	4.20	0.98	0.23	0.92	0.34	0.29	6.14E-006
$(10, 0)$	72	3.69	0.97	0.09	0.93	0.30	0.39	6.15E-006
$(8, 2)$	64	4.02	0.98	0.18	0.90	0.29	0.32	6.19E-006
$(8, 3)$	63	4.03	0.98	0.18	0.90	0.29	0.32	6.18E-006
$(10, 1)$	58	3.69	0.97	0.09	0.92	0.30	0.40	6.14E-006
$(9, 24)$	57	3.79	0.97	0.21	0.82	0.27	0.20	7.21E-006
$(5, 6)$	52	4.20	0.98	0.25	0.88	0.46	0.20	6.43E-006
$(6, 26)$	44	4.67	0.99	0.25	0.96	0.49	0.45	6.19E-006
$(5, 24)$	38	2.57	0.83	0.22	0.54	0.08	0.14	5.61E-005
$(8, 17)$	38	4.25	0.98	0.18	0.91	0.33	0.26	6.17E-006
$(4, 28)$	37	2.83	0.86	0.19	0.94	0.50	0.39	2.35E-005
$(5, 28)$	36	4.24	0.98	0.23	0.91	0.41	0.34	6.11E-006
$(8, 8)$	36	4.26	0.98	0.20	0.93	0.32	0.39	6.26E-006
$(5, 21)$	32	3.31	0.92	0.25	0.36	0.06	0.08	1.08E-005
$(8, 9)$	32	4.31	0.98	0.21	0.93	0.33	0.40	6.17E-006
$(9, 2)$	32	4.11	0.98	0.19	0.94	0.31	0.36	6.05E-006
$(9, 3)$	32	4.13	0.98	0.19	0.94	0.32	0.36	6.02E-006
$(10, 17)$	31	4.38	0.99	0.21	0.95	0.37	0.36	5.91E-006
$(10, 25)$	29	4.57	0.99	0.25	0.94	0.39	0.35	5.90E-006
$(6, 10)$	28	4.62	0.99	0.25	0.90	0.46	0.45	6.36E-006
$(8, 25)$	27	4.28	0.98	0.22	0.91	0.33	0.28	6.25E-006
$(10, 2)$	27	4.19	0.98	0.23	0.95	0.34	0.42	5.92E-006
$(10, 3)$	27	4.28	0.98	0.22	0.95	0.36	0.41	5.90E-006

(Continued)

Table 2.1. (*Continued*)

Rule	T	H	D	R	P	A	I	Z
$(6, 24)$	25	3.35	0.92	0.21	0.84	0.16	0.25	2.00E-005
$(4, 27)$	22	4.20	0.97	0.25	0.88	0.27	0.23	1.02E-005
$(6, 11)$	22	1.07	0.39	0.24	0.98	0.84	0.04	1.54E-004
$(7, 24)$	22	3.96	0.97	0.19	0.93	0.43	0.71	1.27E-005
$(4, 6)$	21	2.18	0.85	0.23	0.97	0.60	0.58	1.21E-004
$(6, 5)$	20	3.60	0.95	0.25	0.47	0.07	0.08	1.51E-004
$(10, 8)$	20	4.48	0.99	0.23	0.95	0.37	0.39	5.90E-006
$(11, 16)$	19	4.03	0.98	0.16	0.95	0.36	0.34	5.98E-006
$(10, 9)$	18	4.61	0.99	0.25	0.96	0.39	0.41	5.84E-006
$(11, 24)$	17	4.27	0.98	0.23	0.96	0.40	0.35	6.00E-006
$(6, 4)$	16	2.55	0.90	0.23	0.44	0.05	0.04	4.65E-004
$(6, 25)$	15	3.25	0.94	0.23	0.98	0.61	0.33	1.03E-005
$(8, 21)$	15	4.53	0.99	0.23	0.95	0.39	0.41	5.87E-006
$(4, 21)$	13	3.05	0.87	0.25	0.39	0.03	0.06	1.83E-004
$(5, 20)$	13	2.74	0.85	0.25	0.39	0.03	0.06	1.87E-004
$(5, 11)$	11	1.81	0.55	0.25	0.45	0.19	0.04	3.46E-004
$(7, 10)$	11	4.35	0.98	0.22	0.94	0.50	0.46	6.47E-006
$(4, 25)$	10	3.34	0.91	0.23	0.60	0.10	0.14	1.98E-005
$(9, 21)$	10	4.57	0.99	0.25	0.96	0.40	0.42	5.83E-006
$(4, 20)$	8	2.57	0.84	0.25	0.38	0.03	0.06	2.71E-006
$(7, 8)$	8	1.84	0.69	0.17	0.92	0.42	0.83	1.71E-006
$(5, 10)$	6	4.41	0.98	0.25	0.09	0.00	0.00	2.68E-004
$(8, 10)$	6	4.48	0.98	0.23	0.96	0.37	0.45	5.92E-006
$(8, 11)$	6	4.43	0.98	0.23	0.96	0.39	0.44	5.97E-006
$(5, 27)$	5	0.54	0.13	0.25	0.98	0.32	0.04	2.51E-004
$(4, 22)$	4	2.62	0.89	0.24	0.93	0.41	0.39	1.74E-004
$(5, 4)$	4	2.95	0.92	0.23	0.25	0.02	0.02	8.79E-004
$(5, 5)$	4	3.45	0.95	0.25	0.24	0.04	0.04	5.32E-004
$(5, 14)$	3	2.80	0.88	0.12	0.96	0.37	0.25	7.01E-005
$(4, 10)$	2	4.48	0.98	0.25	0.05	0.00	0.00	4.36E-004
$(7, 9)$	2	4.19	0.98	0.22	0.95	0.48	0.58	8.30E-006
$(7, 11)$	2	0.78	0.25	0.15	0.98	0.85	0.01	1.99E-006
$(7, 19)$	1	2.93	0.90	0.15	0.93	0.48	0.17	1.40E-004
$(8, 20)$	1	4.42	0.98	0.23	0.95	0.37	0.37	5.99E-006
$(12, 0)$	1	2.85	0.92	0.08	0.84	0.24	0.10	1.45E-005
$(12, 1)$	1	2.85	0.92	0.08	0.84	0.24	0.27	1.25E-005
$(12, 16)$	1	2.85	0.92	0.09	0.84	0.24	0.13	1.45E-005

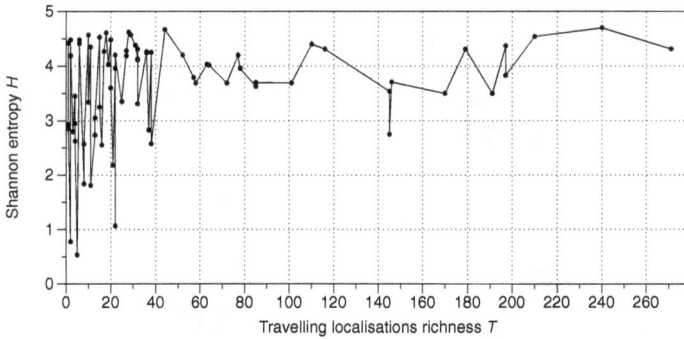

Fig. 2.10. Travelling localisations' richness T vs. Shannon entropy H.

Table 2.2. Top 14 rules supporting travelling localisations with $T > 100$.

Rule	F^0	F^1	T
$(7, 20)$	00111	10100	271
$(5, 26)$	00101	11010	240
$(4, 26)$	00100	11010	210
$(5, 25)$	00101	11001	197
$(6, 20)$	00110	10100	197
$(8, 0)$	01000	00000	191
$(7, 21)$	00111	10101	179
$(8, 1)$	01000	00001	170
$(8, 24)$	01000	11000	146
$(7, 4)$	00111	01110	145
$(8, 16)$	01000	10000	145
$(7, 5)$	00111	00101	116
$(6, 21)$	00110	10101	110
$(9, 0)$	01001	00000	101

filled with excitation activities. In some cases, e.g. rule $(5, 25)$ in Fig. 2.11, the original perturbation does not propagate: colliding travelling localisations produce stationary, still, patterns, which in turn limit further propagation of new travelling localisations. In such rules, travelling localisations bounce between still localisations.

Let us calculate frequency vectors V^0 and V^1 from F^0 and F^1. There are 14 rules in Table 2.2, V_i^0 is a ratio of '1' in F_i^0 to 7, the same for V_1. Thus, we have $V^0 = \left(0 \frac{5}{14} \frac{9}{14} \frac{6}{4} \frac{7}{14}\right)$ and $V^1 = \left(\frac{9}{14} \frac{5}{14} \frac{6}{14} \frac{3}{14} \frac{5}{14}\right)$.

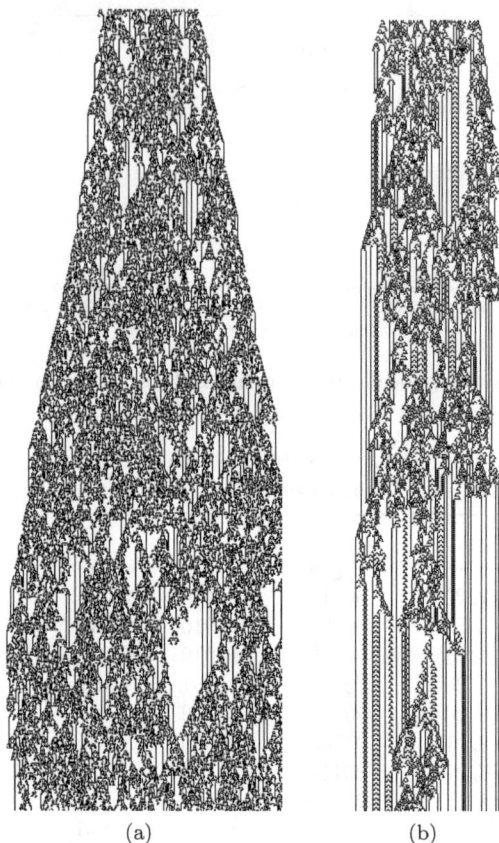

(a)   (b)

Fig. 2.11. Exemplar space–time configurations generated by (a) rule (4, 26) and (b) rule (5, 25).

Probabilities of excitation can be represented via a polynomial approximation of vectors V^0 and V^1 shown in Fig. 2.14(a), compared with the probability of excitation of the rules not supporting any localisations in Fig. 2.14(b).

Assuming 0.5 is a cut-off frequency, we can transform the vector to a simplified form: $^*V^0 = (00101)$ and $^*V^1 = (10000)$.

Proposition 2.3. *A rule more likely supports large number of travelling localisations if it has the following node state transitions. A resting node becomes excited if it has two or four excited neighbours. An excited node remains excited if it has no excited neighbours.*

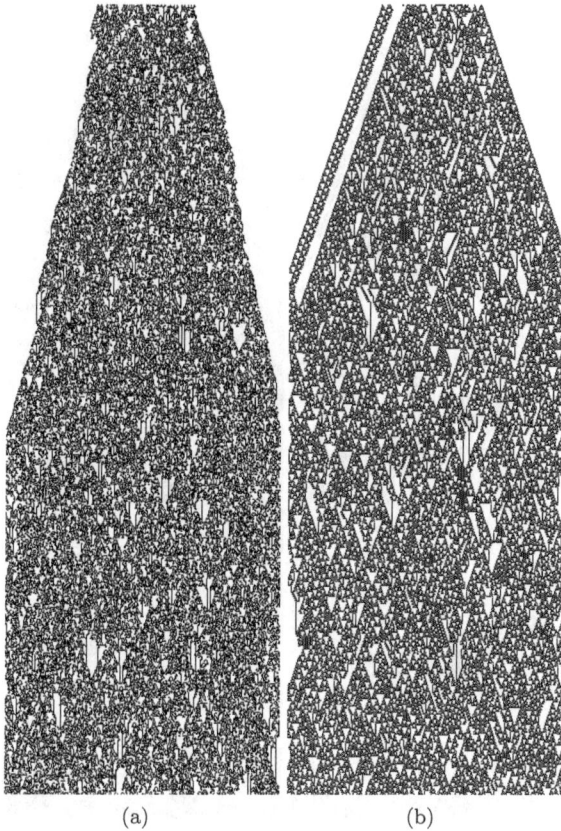

(a)   (b)

Fig. 2.12. Exemplar space–time configurations generated by (a) rule (5, 26) and (b) rule (6, 20).

2.4.2 Stationary localisations

Let us calculate frequency vectors V^0 and V^1 from F^0 and F^1 for rules supporting stationary localisations. There are 39 rules that exhibit stationary localisations for over 900 (of 1024) seeds. V_i^0 is a ratio of '1' in F_i^0 to 39, the same for V_1. Thus, we have $V^0 = \left(00\frac{15}{39}\frac{20}{39}\frac{19}{39}\right)$ and $V^1 = \left(\frac{27}{39}\frac{27}{39}\frac{30}{39}\frac{21}{39}\frac{19}{39}\right)$. Probabilities of excitation can be represented via a polynomial approximation of vectors V^0 and V^1 shown in Fig. 2.15(a), compared with the probability of excitation of the rules that support travelling localisations in Fig. 2.14(b) and rules that do not support any localisations in Fig. 2.14(b). Assuming

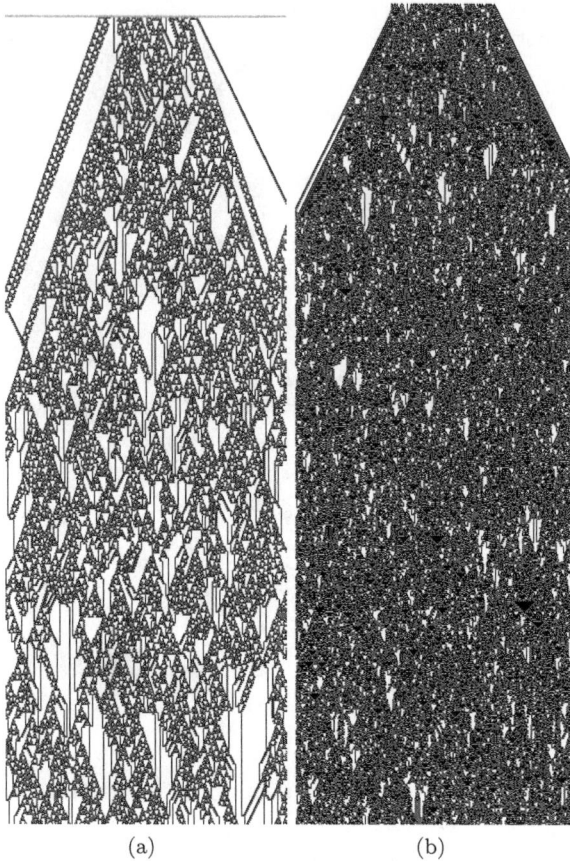

(a)　　　　　　　　　　　　　　　　(b)

Fig. 2.13. Exemplar space–time configurations generated by (a) rule (7, 20) and (b) rule (7, 21).

0.5 is a cut-off frequency, we can transform the vectors to a simplified form: $^*V^0 = (00010)$ and $^*V^1 = (11110)$.

Proposition 2.4. *A rule more likely supports large number of stationary localisations if it has the following node state transitions. A resting node becomes excited only if three of its four neighbours are excited. An excited node remains excited if it has less than four excited neighbours.*

(a)

(b)

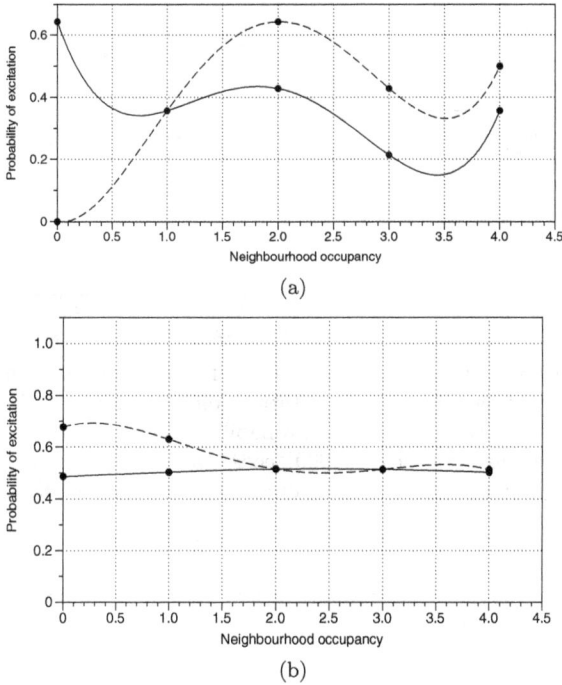

Fig. 2.14. Dependence of a probability of excitation of a node in actin automaton (a) supporting travelling localisations and (b) not supporting any localisations (neither stationary nor travelling) on a number of excited neighbours of the node. The polynomial of the probability (a) is calculated on 14 rules that exhibit travelling localisations for largest number of seeds, and the polynomial of the probability (b) is calculated on 705 rules not supporting any localisations. Dashed line shows probability of excitation of a resting node, solid line, of an excited node, i.e. of an excited node to remain excited.

2.4.3 Rules supporting both stationary and travelling localisations

Scatter plot of a number T of seeds supporting travelling localisations in each rule R versus number T of seeds supporting stationary localisations in the same rule R is shown in Fig. 2.16. The plot shows that typically the more travelling localisations a rule exhibits, the less stationary the localisations the rule shows. There are however rules that show high number of seeds leading to travelling localisations,

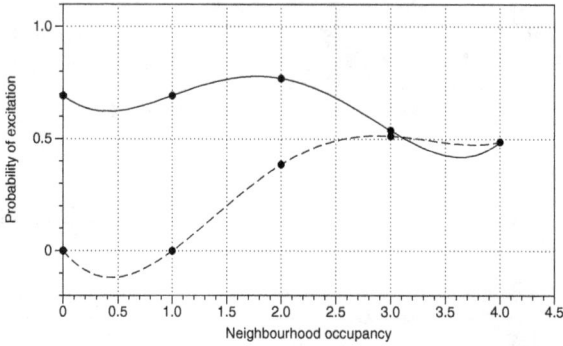

Fig. 2.15. Dependence of a probability of excitation of a node in actin automaton supporting stationary localisations on a number of excited neighbours of the node. The polynomial of the probability is calculated on 39 rules that exhibit stationary localisations for 900 seeds. Dashed line shows probability of excitation of a resting node, solid line, of an excited node, i.e. of an excited node to remain excited.

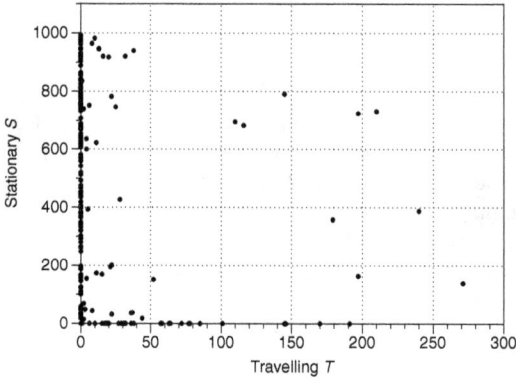

Fig. 2.16. Scatter of a number T of seeds supporting travelling localisations in each rule R vs. number T of seeds supporting stationary localisations in the same rule R.

and seeds leading to stationary localisations. They are shown in Table 2.3.

Configurations generated by rules in Table 2.3 from selected seeds are shown in Fig. 2.17. Rules (4, 26) and (5, 25) show rich dynamics of travelling localisations (Fig. 2.17(a–g)). For most seeds, the developments lead to the formation of several propagating

Table 2.3. Rules developing a maximum number of seeds into travelling and stationary localisations.

Rule	F^0	F^1	T	S	H	D	R	P	A	I	Z
(4, 26)	00100	11010	210	731	4.54	0.98	0.25	0.86	0.31	0.34	6.9E-6
(5, 25)	00101	11001	197	724	3.83	0.95	0.25	0.73	0.14	0.19	1.3E-5
(7, 4)	00111	00100	145	791	2.75	0.92	0.19	0.54	0.06	0.07	1.8E-4
(5, 26)	00101	11010	240	388	4.70	0.99	0.25	0.92	0.39	0.39	6.4E-6
(6, 21)	00110	10101	110	694	4.40	0.98	0.25	0.70	0.14	0.14	1.3E-5
(7, 5)	00111	00101	116	682	4.31	0.98	0.21	0.93	0.29	0.24	8.7E-6
(7, 21)	00111	10101	179	358	4.31	0.98	0.22	0.97	0.56	0.39	6E-6

localisations, some of them could be classed as glider guns, and often travelling localisations produce stationary localisations during their interactions with each other, e.g. Fig. 2.17(b, e). In some cases, e.g. Fig. 2.17(c, f, g) we can observe very elaborate trajectories of travelling localisations.

In rule (7, 4), a seed typically leads to formation of a single glider (Fig. 2.17(h)) or a pair of gliders moving in opposite directions (Fig. 2.17(i)); some seeds can produce a glider and a stationary localisation (Fig. 2.17(j)). Rule (7, 5) exhibits combinations of travelling localisations and breathers: periodically oscillating stationary localisations (Fig. 2.17(k, l)).

A good variety of travelling and stationary localisations are given in space–time dynamics of rule (7, 21). These include large gliders propagating with a speed less than the speed of light ('speed of light' glider translates one node per iteration) (Fig. 2.17(m)), very slowly propagating localisations, or complex clusters of gliders, emitting stationary and travelling localisations (Fig. 2.17(n)), combinations of large still lifes and gliders (Fig. 2.17(p)) and generation of two gliders of different sizes from a single seed (Fig. 2.17(q)).

What is important for a rule to be supportive for both travelling and stationary localisations?

Let us calculate frequency vectors V^0 and V^1 from F^0 and F^1. There are seven rules in Table 2.3, V_i^0 is a ratio of '1' in F_i^0 to 7, the same for V_1. Thus, we have $V^0 = \left(001\frac{4}{7}\frac{5}{7}\right)$ and $V^1 = \left(\frac{5}{7}\frac{3}{7}\frac{4}{7}\frac{2}{7}\frac{4}{7}\right)$. Probabilities of excitation can be represented via a polynomial

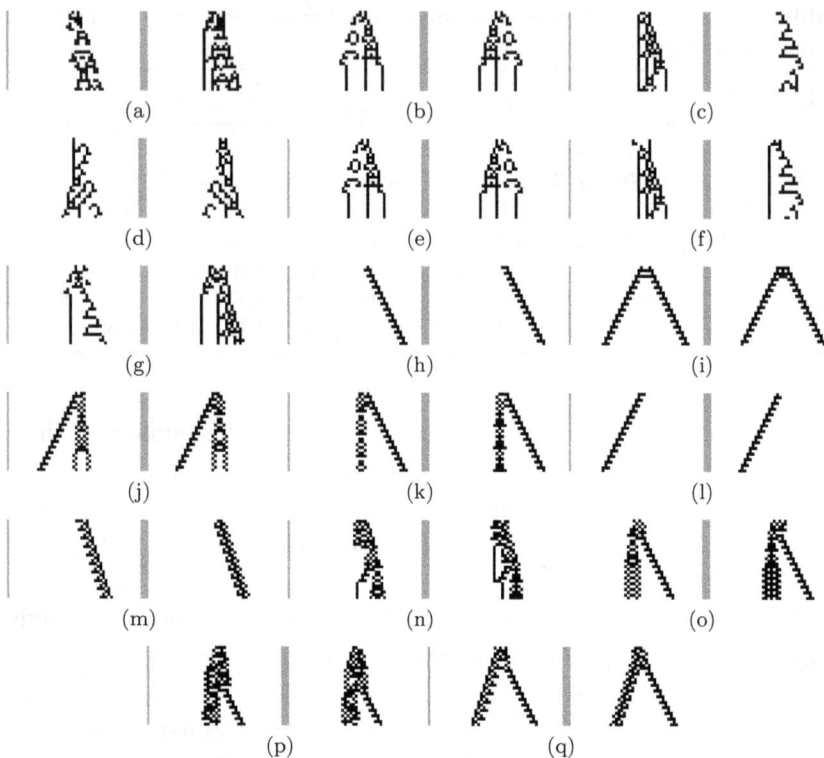

Fig. 2.17. Exemplar configuration of rules from Table 2.3. In each subfigure, space–time states of chain x, left, and y, right, are shown. (a) rule (4, 26), seeds 01100 and 10001, (b) rule (5, 25), seeds 00100 and 00011, (c) rule (5, 25), seeds 00101 and 00011, (d) rule (5, 25), seeds 00101 and 01101, (e) rule (5, 25), seeds 00110 and 00001, (f) rule (5, 25), seeds 10001 and 10111, (g) rule (5, 25), seeds 11010 and 10011, (h) rule (7, 4), seeds 00001 and 00111, (i) rule (7, 4), seeds 00001 and 01111, (j) rule (7, 4), seeds 00001 and 11100, (k) rule (7, 5), seeds 00101 and 00111, (l) rule (7, 5), seeds 00111 and 00100, (m) rule (7, 21), seeds 00011 and 10110, (n) rule (7, 21), seeds 00011 and 11111, (o) rule (7, 21), seeds 00100 and 10001, (p) rule (7, 21), seeds 01001 and 01000, (q) rule (7, 21), seeds 10110 and 11100.

approximation of vectors V^0 and V^1 shown in Fig. 2.18. Assuming 0.5 is a cut-off frequency, we can transform the vector to a simplified form: $^*V^0 = (00111)$ and $^*V^1 = (10101)$.

Proposition 2.5. *An actin automaton supports highest number of travelling and stationary localisations if its local activity is governed*

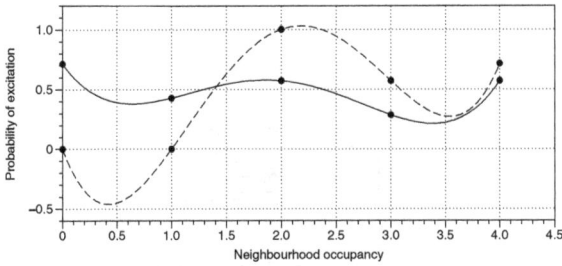

Fig. 2.18. Dependence of a probability of excitation of a node in actin automaton supporting stationary and travelling localisations in a number of excited neighbours of the node. The polynomial of the probability is calculated on seven rules that exhibit stationary and travelling localisations for largest number of seeds. Dashed line shows probability of excitation of a resting node, solid line, of an excited node, i.e. of an excited node to remain excited.

as follows. A resting node excites if it necessarily has two excited neighbours or likely three or more, likely five, excited neighbours. An excited node remains excited if it has no excited neighbours or two or four excited neighbours.

Thus, rule $(7, 4)$ is most representative in terms of F_0 and rule $(6, 12)$ is most representative in terms of F_1.

What integral characteristics are typical for the above localisation supporting rules?

The Shannon entropy H is over 4 for most rules in Table 2.3, except rule $(7, 4)$. Simpson diversity index D is over 0.9. Thus, the rules that support both stationary and travelling localisations show very rich space–time dynamics of excitations, R, when perturbed with a random initial pattern. They also show a moderate degree of space filling D, where, in case of random stimulation, the dynamical excitation fills over 50% of the automaton chains. Rule $(7, 21)$ is the highest space filler, with almost 97% of nodes being excited and Rule $(7, 4)$ is the lowest space filler with just a half of nodes typically occupied by excitation during the automaton development.

The space–time configurations are not overly rich morphologically with only circa 25% of possible local configurations presented. Activity levels vary substantially between the rules in Table 2.3. Rule $(7, 4)$ shows lowest level of overall excitation activity followed by

Table 2.4. Statistical characterisation of main groups of rules. For each characteristics, the table shows average and standard deviation.

	$\overline{H}, \sigma$	$\overline{D}, \sigma$	$\overline{R}, \sigma$	$\overline{P}, \sigma$	$\overline{A}, \sigma$	$\overline{I}, \sigma$
All rules	2.8, 1.2	0.8, 0.2	0.1, 0.1	0.9, 0.2	0.5, 0.2	0.2, 0.2
Rules supporting travelling localisations	3.6, 0.9	0.9, 0.1	0.2, 0.0	0.8, 02	0.3, 0.2	0.3, 0.2
Rules supporting stationary localisations	2.3, 0.6	0.8, 0.1	0.1, 0.1	0.7, 0.2	0.1, 0.1	0.1, 0.0
Rules supporting travelling and stationary localisations	4.1, 0.7	1.0, 0.0	0.2, 0.0	0.8, 0.2	0.3, 0.2	0.2, 0.1
Rules supporting no localisations	2.9, 1.3	0.8, 0.2	0.2, 0.1	0.9, 0.1	0.6, 0.2	0.3, 0.2

rules (5, 25) and (6, 21). Highest level of activities are observed in space–time configurations of rule (7, 21). rules (7, 4), (5, 21), and (6, 21) show highest compressibility Z, which is due to lower richness R and, partly, space filling P.

Proposition 2.6. *Rules supporting both travelling and stationary localisations are characterised by high Shannon entropy and Simpson diversity, low levels of activity, and high degree of compressibility.*

Can we detect rules supporting localisations from integral characteristics of the space–time configurations generated by the rules? Let us look at Table 2.4. We see that space filling P is not a reliable discriminator, but Shannon entropy H, activity level A, and incoherence I are good discriminators on absence/presence and mobility of localisations.

Proposition 2.7. *Actin automata rules supporting highest number of travelling localisations show high Shannon entropy H, medium activity A, and high incoherence I. The rules supporting highest number of stationary localisations show low entropy H, low activity A, and low incoherence I.*

2.5 Discussion

Through a comprehensive computational analysis of automaton models that simulate hypothetical actin filament communication events, we have conducted an in-depth investigation to identify the rules that govern the emergence of stationary and travelling localisations based on the global characteristics of the generated space–time configurations. Our findings indicate that approximately one-third of the node-state transition rules provide support for either stationary or travelling localisations. More importantly, the majority of rules that support localisations demonstrate specialisation, focusing on either stationary or travelling phenomena. Additionally, we have discovered that approximately 3% of the rules exhibit dynamic behaviours characterised by both travelling and stationary localisations, which represent rare occurrences in the space of actin automaton node-state transitions.

Among the identified rules, those that support a significant number of travelling localisations tend to generate space–time configurations with higher Shannon entropy values, indicating a greater potential for information content. Conversely, rules that support both travelling and stationary localisations result in configurations with high Shannon entropy and Simpson diversity index, while exhibiting low levels of excitation activity and a high degree of compressibility. In contrast, rules exclusively supporting stationary localisations demonstrate low levels of excitation activity, limited incoherence, and low Shannon entropy. These findings provide valuable insights into the dynamics and characteristics of actin automaton models, highlighting the diverse behaviours exhibited by different sets of node-state transition rules.

Our research has identified an alternative and more efficient approach for determining whether a rule supports the occurrence of localisations in biopolymer chains such as actin. Instead of relying on exhaustive computational analysis, we propose a structural analysis of the rule itself to make this determination. Specifically, we have discovered that a rule supports travelling localisations when a resting node becomes excited if it has either two or four excited neighbours (50% or 100% neighbourhood excitation), and an excited node remains excited even in the absence of any excited neighbours. This suggests that additional modes of excitability may be necessary to facilitate the propagation of localisations along the chain. Similarly, a rule supports stationary localisations when a resting node becomes excited if it has exactly three excited neighbours (approximately 70% neighbourhood excitation), and an excited node remains excited if it has fewer than four excited neighbours.

By analysing the structural characteristics of the rules, we can efficiently determine their potential to support different types of localisations, streamlining the process of identifying relevant rules for specific scenarios involving biopolymer chains. This approach offers a more practical and targeted method for investigating localisations in such systems.

Chapter 3

Patterns and Dynamics of Rule 22 Cellular Automaton

Genaro J. Martínez[*,‖], Andrew Adamatzky[†,**], Rolf Hoffmann[‡,††],
Dominique Désérable[§,‡‡], and Ivan Zelinka[¶,§§]

*Laboratorio de Ciencias de la Computación, Escuela Superior de
Cómputo, Instituto Politécnico Nacional, México
†Unconventional Computing Lab, UWE Bristol, UK
‡Technische Universität Darmstadt, Darmstadt, Hessen, Deutschland
§Institut National des Sciences Appliquées, Rennes, France
¶Fakulta Elektrotechniky a Informatiky, Technická Univerzita Ostrava,
Czechia
‖gjuarezm@ipn.mx
**andrew.adamatzky@uwe.ac.uk
††hoffmann@informatik.tu-darmstadt.de
‡‡domidese@gmail.com
§§ivan.zelinka@vsb.cz

Abstract

Rule 22 of the elementary cellular automaton (ECA) exhibits non-trivial properties and dynamics. This rule operates on a binary cell state system with a 3-cell neighbourhood. In this rule, a cell takes the State '1' if there is exactly one neighbouring cell, including itself, in State '1'. The transition of cell states can be described as an XOR function of the three cell states involved. From a physico-chemical perspective, Rule 22 can be interpreted as the propagation of self-inhibiting quantities or species. The space–time dynamics of this rule exhibit non-trivial patterns and demonstrate quasi-chaotic behaviour. To understand and characterise these phenomena, various analytical tools and techniques are employed. Mean field theory, attractors, de Bruijn diagrams, subset diagrams, filters, fractals, and memory are used to analyse and describe the observed patterns and behaviours in Rule 22. These tools provide insights into the system's dynamics, the emergence of attractors, the

structure of de Bruijn diagrams, the presence of subsets, the effects
of filters, the presence of fractal patterns, and the system's memory
properties.

3.1 Introduction

3.1.1 Rule 22: History

Elementary cellular automata (ECA) [273] are one-dimensional
arrays of finite state machines, or cells, which take States '0' or '1' and
update their state depending on their own current state and on the
state of their two immediate neighbours. Rule 22 ECA has a simple
cell-state transition function: a cell takes State '1' if exactly one of
its neighbours, including the cell itself, is in State '1'; otherwise, the
cell takes State '0'. When perturbed at a single site, the automaton
exhibits something similar to recurrent wave–fronts in excitable
media, which develop into fractal structures of Sierpiński gasket [210,
275]. Due to countless generations and annihilations of wave–fronts,
a dynamics of Rule 22 is sometimes characterised as chaotic [274],
which rather reflects its unpredictability than any relation to noise
or bifurcations. Most results of studies about Rule 22 were in
algebraic properties or statistical approximations of the automaton
dynamics. Thus, Zabolitzky [280] reported results of an extended
probabilistic analysis estimating non-trivial behaviour on very large
arrays perturbed by configurations with low densities of State '1'.
He discovered critical properties that cannot be reproduced when
the automaton is perturbed by a random configuration. McIntosh
provided a systematic analysis of small configurations emerging in
Rule 22 [280]; he proposed similarities with configurations observable
in Conway's Game of Life. A topological analysis linked to chaotic
behaviour of the rule can be found in Ref. [139].

3.1.2 Rule 22: Definition

A one-dimensional cellular automaton $CA(k, r)$ is an array of cells
x_i where $i \in \mathbb{Z}$. Each cell takes on a value from an alphabet $S = \{0, 1, ..., k-1\}$ with k symbols. A chain of cells $\{x_i\}$ of finite length
n represents a string or global configuration c on Σ. The set of finite

configurations is represented as Σ^n. An evolution is a sequence of configurations $\{c_i\}$ given by the mapping $\Phi : \Sigma^n \to \Sigma^n$ and their global relation is provided by $\Phi(c^t) \to c^{t+1}$ where t is a discrete time and every global state of c is a sequence of cell states. Cells of each configuration c^t are updated to the next configuration c^{t+1} simultaneously by a local transition function $\varphi : S^{2r+1} \to S$ as

$$\varphi(x^t_{i-r}, \ldots, x^t_i, \ldots, x^t_{i+r}) \to x^{t+1}_i$$

acting on a neighbourhood of x_i of length $2r + 1$. For (elementary) ECA$(2, 1)$, $\varphi : S^3 \to S$ becomes

$$\varphi(x^t_{i-1}, x^t_i, x^t_{i+1}) \to x^{t+1}_i \tag{3.1}$$

and for Rule 22, its local cell-state transition is given by:

$$\varphi_{R22} = \begin{cases} 1 & \text{if } 100, 010, 001 \\ 0 & \text{if } 111, 110, 101, 011, 000. \end{cases} \tag{3.2}$$

Rule 22 displays a typical chaotic global behaviour from random initial conditions. Figure 3.1(a) shows the evolution with an initial condition starting with a single cell in State '1'. A pattern growing is a fractal, similar to a Sierpiński gasket. Figure 3.1(b) shows a development from a random initial configuration with a density 0.5 of cells in State '1'.

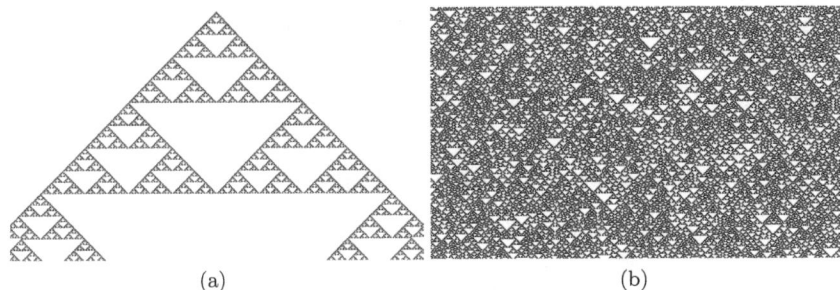

(a)   (b)

Fig. 3.1. Exemplar dynamics in ECA Rule 22. (a) Development from a single cell in State '1'. (b) Development from a random configuration with density of 1-cells 0.5. Both space–time diagrams evolve on a ring of 600 cells for 350 generations. Time evolves from top to bottom.

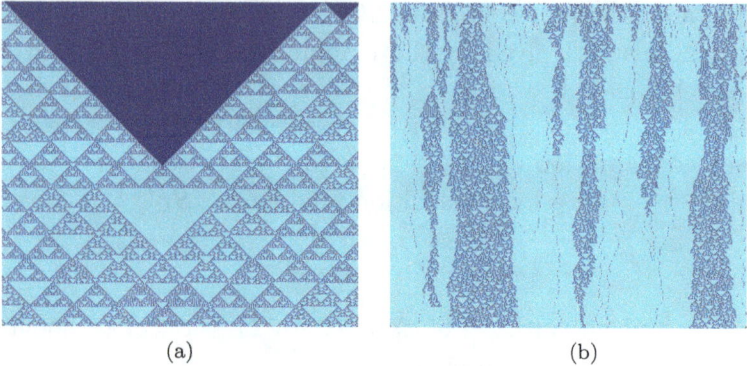

(a)   (b)

Fig. 3.2. Actin cellular automaton rule (14, 9) for (a) $x = y$, and (b) $x\&y$. Both evolutions have a ring of 500 cells to 450 generations.

3.1.3 Rule 22 as an actin automaton

Actin cellular automaton is a variant of conventional cellular automaton. In Ref. [27], a number of actin rules were enumerated and researched. These rules display well-known global behaviour related to complex, chaotic, periodic, and uniform dynamics. Particularly, actin cellular automaton rule (14, 9) is a function with chaotic behaviour able to yield fractals, multi-fractals, and at least two different densities on its evolution. This double density induces non-trivial behaviour in some conditions; also a phase transition is irreversible only if the small density changes to high density. In this chapter, we discuss the actin elementary cellular automaton dynamics as elementary cellular automaton Rule 22. This way, the composition of rules $F_0 = (1110)$ and $F_1 = (1001)$ relates to a negation of Rule 22 when filaments are $x = y$ (see Fig. 3.2).

3.1.4 Patterns and dynamics in Rule 22

Rule 22 is considered chaotic because

(1) future configuration of the automaton is completely determined from its initial state owing to the deterministic rule and synchronous updating;
(2) development of the automaton is sensitive to initial conditions (tiny perturbations might lead to dramatic events);

(3) global transition graph has dense periodic orbits (attractors);
(4) configurations evolved can be characterised as random.

We undertake an extensive and systematic analysis of Rule 22 using different approximations aiming at discovering an emergence of novel non-trivial patterns, periodic patterns, and Garden of Eden configurations. These configurations are discovered with the help of encoding initial conditions into regular expressions, de Bruijn diagrams, subset diagrams, cycle diagrams, fractals, and jump-graphs. We also show an effect of memory upon dynamics of Rule 22.

3.2 Mean field theory

Mean field theory allows us to describe statistical properties of CA without analysing evolution spaces of individual rules [108,176]. This approximation assumes that elements of a set of states Σ are *independent* and not correlated with each other in the rule's evolution space. One can study probabilities of states in the neighbourhood in terms of probability of a single state (the state in which the neighbourhood evolves), thus, a probability of the neighbourhood-state is a product of the probabilities of each cell-state in the neighbourhood. A polynomial on the probabilities is derived and its curve can be used to classify the rules, as proposed by McIntosh in [177].

3.2.1 Mean field in the Game of Life

Using this approach we can construct a mean field polynomial for a two-dimensional CA with a semi-totalistic evolution rule:

$$p_{t+1} = \sum_{v=S_{\min}}^{S_{\max}} \binom{n-1}{v} p_t^{v+1} q_t^{n-v-1} + \sum_{v=B_{\min}}^{B_{\max}} \binom{n-1}{v} p_t^{v} q_t^{n-v}, \quad (3.3)$$

where n represents the number of cells in Moore's neighbourhood, v (resp. $n-v$) the number of occurrences of State '1' (resp. '0'), p_t (resp. q_t) the probability of a cell being in State '1' (resp. '0') and with $q_t = 1 - p_t$. B and S are minimum and maximum of an interval for born and survival conditions in Conway's Game of Life (GoL),

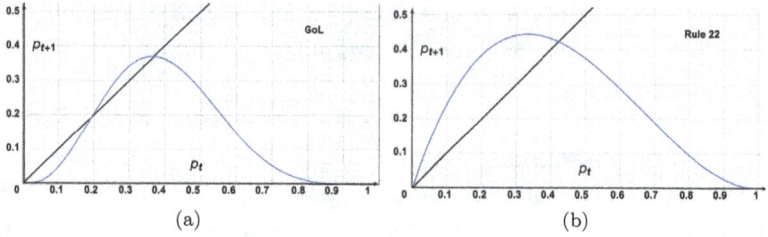

Fig. 3.3. Mean field curves for (a) Game of Life and (b) Rule 22.

respectively. The GoL's polynomial is the following:

$$p_{t+1} = 84\, p_t^3\, q_t^6 + 56\, p_t^4\, q_t^5. \tag{3.4}$$

The mean field curve $\mathcal{F}$ of Eq. (3.4) displayed in Fig. 3.3(a) shows three fixed points $p_{t+1} = p_t$ when crossing the identity. The first stable fixed point at the origin guarantees its stable state, the second unstable point $\mathcal{F} = 0.1986$ relates to areas of densities where the space–time dynamic is unknown. The last stable point in $\mathcal{F} = 0.37$ indicates that GoL will converge almost surely to configurations with small densities of '1'.

3.2.2 Mean field in ECA Rule 22

For one dimension, we adjust Eq. (3.3) to a full local rule and not only a semi-totalistic one. All states of the neighbourhood must be considered, thus p, q, n, and v have the same representation as above. But now the product will be with the value of each neighbourhood, whence, for a 1d CA(k, r) the mean field polynomial

$$p_{t+1} = \sum_{j=0}^{k^{2r+1}-1} \varphi_j(X)\, p_t^v\, q_t^{n-v},$$

which gives

$$p_{t+1} = \sum_{j=0}^{7} \varphi_j(X)\, p_t^v\, q_t^{3-v} \tag{3.5}$$

for ECA(2, 1) and where $\varphi_j(X)$ denotes the jth transition of S^3 in Eq. (3.1). Finally, the mean field polynomial for Rule 22

$$p_{t+1} = 3\, p_t\, q_t^2 = 3\, p_t\, (1 - p_t)^2 \tag{3.6}$$

is deduced from (3.2).

In Rule 22, State '1' appears with probability $\frac{3}{8} = 0.375$ (which is close to the fixed stable point 0.37 of GoL). The mean field curve f of Eq. (3.6) displayed in Fig. 3.3(b) shows a slope $f'(0) = 3$ at the origin. Density is maximal at $f(1/3) = \frac{4}{9} \approx 0.444$ before reaching the stable fixed point $p_{t+1} = p_t$ when crossing the identity at $p_t = 1 - \sqrt{3}/3 \approx 0.423$. It then crosses the inflection point $f(2/3) = \frac{1}{2} \cdot f(1/3)$ with tangential slope $f'(2/3) = -1$ and decreases until $f(1) = f'(1) = 0$. Based on the mean field curves classification, Rule 22 is a chaotic ECA (Fig. 3.4).

3.2.3 Mean field behaviour of Rule 22

Various scenarios of evolution are displayed in Fig. 3.5:

- (a_1) From initial density $d_0 = 1/3$ reaching spontaneously the maximum before evolving rapidly towards the fixed point with density $d_F = 0.423$. Transition from d_0 to d_F is not perceptible. (a_2) From initial density $d_0 = d_F$ reaching spontaneously the fixed point.
- (b_1) From initial density $d_0 = 0.8$ evolving towards the fixed point after an early phase transition. (b_2) From initial density $d_0 = 0.95$ evolving towards the fixed point after a later phase transition delimited by a polygonal broken line.[1] There exists an interval between two thresholds $d_0' \approx 0.01$ and $d_0'' \approx 0.92$ such that any (pseudo–)random initial distribution with density $d_0' < d_0 < d_0''$ converges almost surely towards fixed point d_F.
- (c_1) From initial density $d_0 = 0.97$ evolving towards a *dense* pattern with observable "backbones"[2] after crossing the phase

---

[1] Phase transitions and critical exponents for Rule 22 leading to non-trivial long-range effects are reported in Refs. [103, 236]. Asymptotic properties where described in Ref. [281].

[2] Also observable in the *"Exactly 1"* ECA [104] Fig. 1(d).

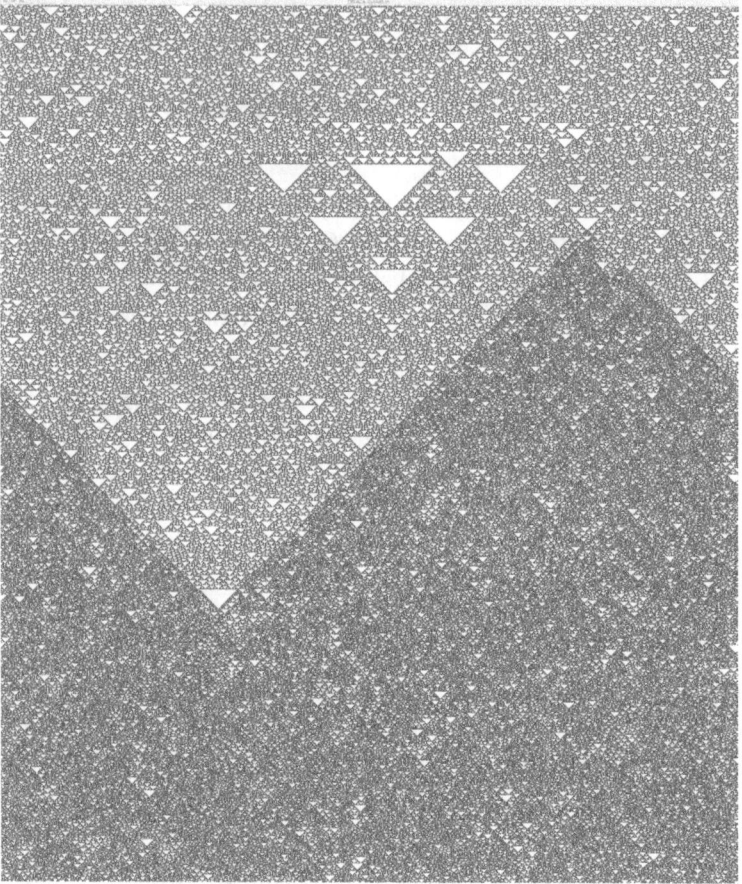

Fig. 3.4. Irreversible phase transition in ECA Rule 22 from a specific initial condition (a regular expression) that after 20,000 generations steps up increases significantly the density of '1', related to the fixed points calculated by mean field theory in Eq. (3.6). This phase transition merges possible complex dynamics to chaos in ECA Rule 22.

transition polygon; two backbones arise from polygon vertices and their patterns are symmetric from either side. (c_2) From the same density $d_0 = 0.97$ and another initial distribution yielding a *sparse* pattern with backbones wherein no phase transition line appears.

Outside interval $]d'_0, d''_0[$ sensitivity to initial conditions is high, with a positive Lyapunov exponent and a chaotic behaviour.

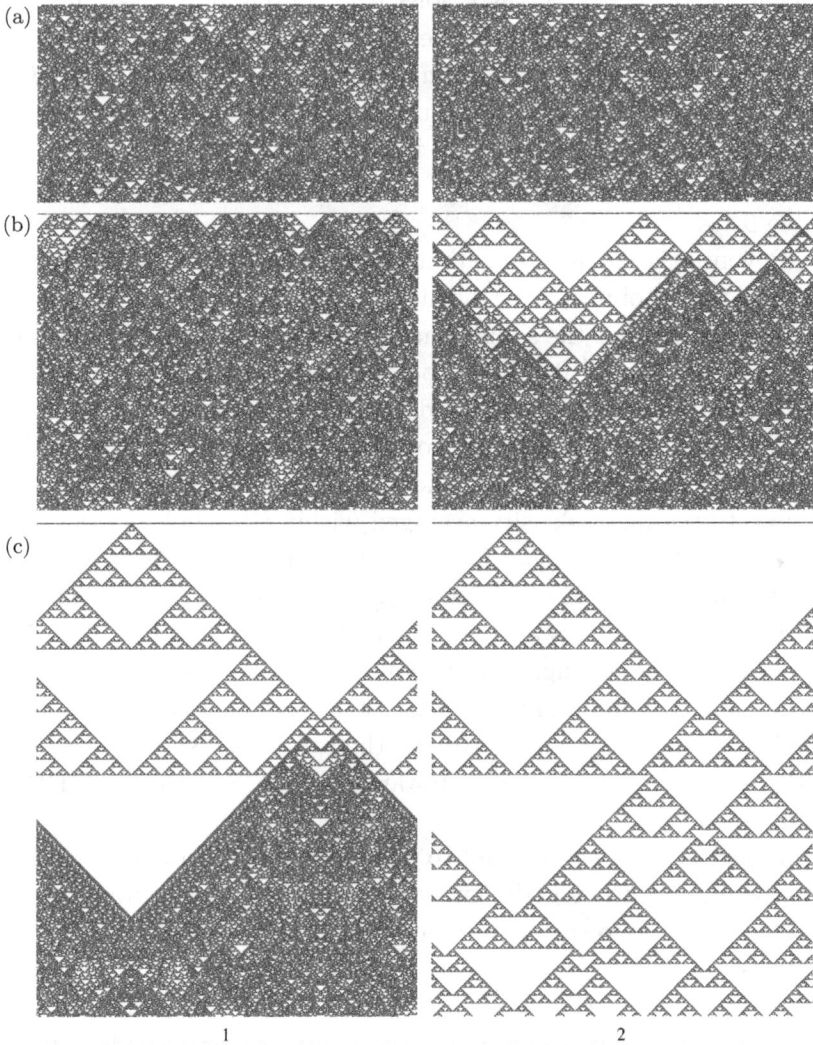

Fig. 3.5. Evolution in a ring of 800 cells: (a) 400 generations (a_1) from initial density $d_0 = 1/3$ (a_2) from density $d_0 = d_F$ reaching spontaneously the fixed point (b) 600 generations (b_1) from density $d_0 = 0.8$ evolving towards d_F after an early phase transition (b_2) from density $d_0 = 0.95$ evolving towards d_F after a later phase transition (c) 1000 generations from density $d_0 = 0.97$ evolving towards (c_1) a dense pattern with thin backbones (c_2) a sparse pattern with large backbones.

Depending on small perturbations from initial configuration, four other evolutions are also possible: (i) disordered sparse fractals (ii) convergence towards fixed point d_F but after a long period (Fig. 3.6) (iii) initial configuration vanishing at first step (iv) rare events of periodic patterns (Fig. 3.7). That is, six types of evolution altogether. Their estimations of occurrence in interval $[d_0'', 1]$ are displayed in Table 3.1.

Beyond the phase transition polygon in case of convergence towards fixed point d_F, the evolution becomes ergodic, in the sense that the system has the same behaviour either averaged over time or averaged over space [109]. The process is stationary and homogeneous at mesoscopic scale. In other words, their exists a smallest macro-cell $\mathcal{C}$ of size $\xi \times \xi$, where ξ is the correlation length, as representative (or statistical) volume element such that density $d_{\mathcal{C}}$ in the macro-cell is close to the mean density averaged within the whole system [79]. Thus, $d_{\mathcal{C}} \approx \frac{3}{8} = 0.375$, that is, the exact ratio of '1' filling φ_{R22}.

It should be observed that a disordered sparse fractal pattern (DSF) may evolve towards ergodicity (ERG) as in Fig. 3.6, but sometimes after a long, unpredictable time. We denote as (DSF) such evolution remaining in this state at least within a time window of arbitrary length 10^3. In the same way, sparse backbone (SBB) patterns may evolve towards a dense backbone (DBB) landscape.

A simple way to check (not to prove) whether unstable evolutions become eventually ergodic or to get a more global overview of evolution is to skip some timesteps with skip time–lengths Δt. This transformation yields a projective view upon the (x, t)–landscape with angle $\arctan 1/\Delta t$. Various skipped scenarios of evolution from initial critical density $d_0 = 0.97$ in a ring of 800 cells and within a time window of length $10^3 \cdot \Delta t$ are displayed in Fig. 3.8:

- (a) Phase transitions (a_1) DSF $\rightarrow$ ERG with $\Delta t = 32$ from disordered sparse fractal to ergodicity — up to 32000 generations, transition occurs after about 8000 timesteps (a_2) SBB $\rightarrow$ DBB with $\Delta t = 31$ from sparse to dense backbones — up to 31000 generations; transition occurs after about 4000 timesteps. Note that apparent discrepancies between densities in (a_1) and (a_2)

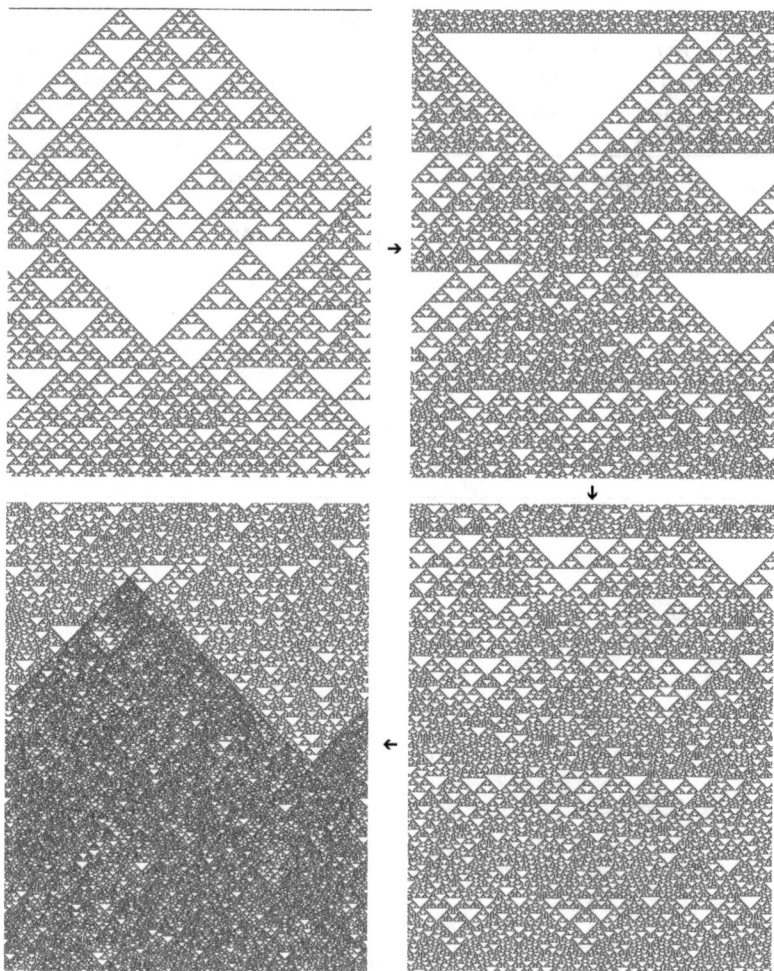

Fig. 3.6. Disordered sparse fractal (DSF) evolving from initial density $d_0 = 0.93$ towards fixed point d_F after a long period: $(0 - 1000) \rightarrow (2000 - 3000) \downarrow (3000 - 4000) \leftarrow (27000 - 28000)$. DSF pattern appears as a long transient state before ergodicity. Compare with landscape in Fig. 3.4 derived from a regular expression.

before phase transition are no more than a side effect resulting from even or odd skip length parity.

- (b) Stratified landscapes SBB with $\Delta t = 64$ (b_1) up to 64,000 generations with observable backbones... and sub-backbones evolving like a Cantor dust (b_2) up to $4 \cdot 10^6$ generations with perpetual phase transitions.

Table 3.1. Statistical estimations of evolutions (%) in interval $[d_0'', 1]$ from samples of 100 initial configurations for each density: ergodicity (ERG), disordered sparse fractals (DSF), dense backbones (DBB), sparse backbones (SBB), vanishing (VAN), rare periodic patterns (RPP). Items in bold font reflect an irreversible state.

Density	ERG	DSF	DBB	SBB	VAN	RPP
0.920	100	0	0	0	0	0
0.930	94	5	1	0	0	0
0.940	93	1	3	1	1	1
0.950	85	2	7	4	2	0
0.960	47	4	13	24	12	0
0.970	21	2	26	26	25	0
0.980	7	0	18	21	54	0
0.990	1	0	5	8	86	0
0.995	0	0	0	0	100	0

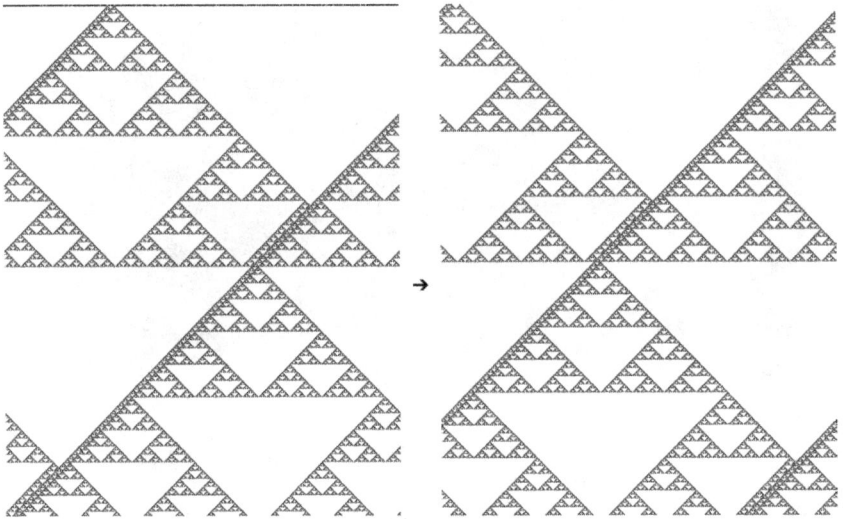

Fig. 3.7. Rare periodic event occurring from initial density $d_0 = 0.94$ with a complete pattern moving leftward like a glider: $(0 - 1000) \rightarrow (200000 - 201000)$. Compare with the patchwork of periodic patterns in Fig. 3.16. The probability of occurrence of such a pattern from a random initial distribution is about 10^{-3}.

Failing to prove the existence or not of ergodic evolution, this skipping approach nevertheless emphasises several chaotic behaviours with long-range correlations. As well as in statistical physics,

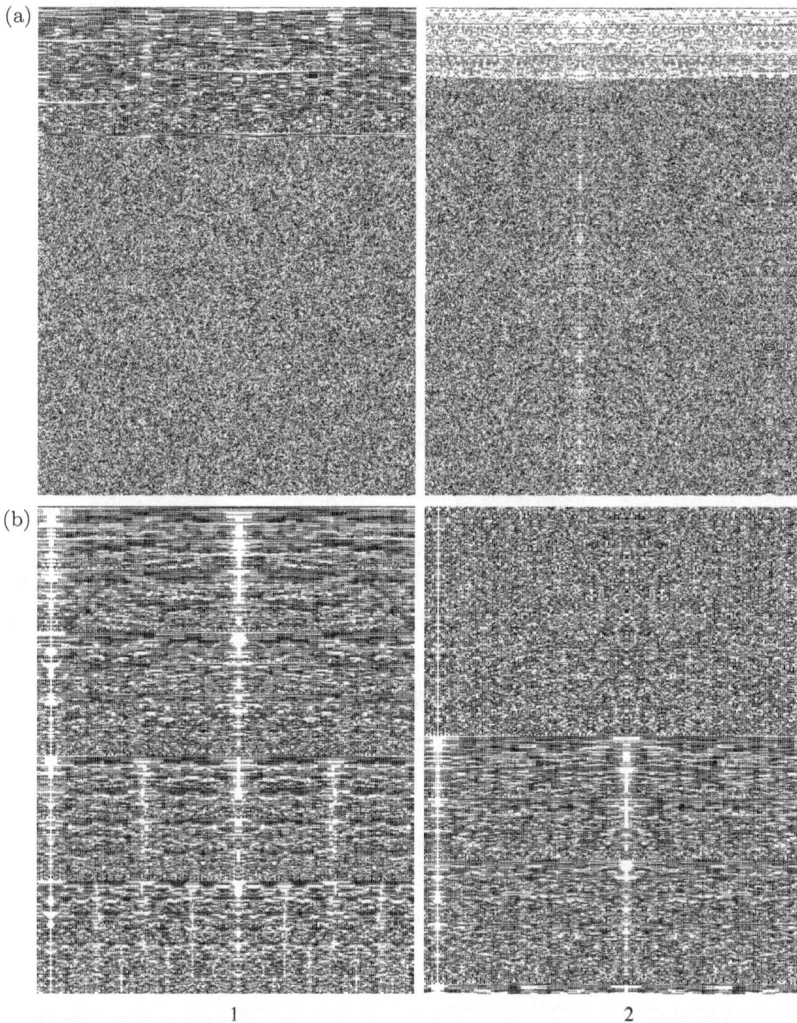

Fig. 3.8. Skipped scenarios in a ring of 800 cells with $d_0 = 0.97$ within a time window of length $10^3 \cdot \Delta t$. (a) Phase transitions (a_1) with $\Delta t = 32$, DSF → ERG from disordered sparse fractal to ergodicity (a_2) with $\Delta t = 31$, SBB → DBB from sparse to dense backbones. (b) Stratified sparse backbone (SBB) landscapes with $\Delta t = 64$ (b_1) up to 64000 generations with observable backbones... and sub-backbones evolving like a Cantor dust (b_2) up to $4 \cdot 10^6$ generations with perpetual phase transitions: a transition line separates a dense backbone (DBB) regime from an SBB regime; evolution remains still unstable.

renormalisation methods overcome the weaknesses of mean field approximations that may fail or, at least, produce insufficient information. Mean field theory is a rough approximation that assumes *independence* between neighbouring sites. On the contrary, other deterministic approaches like de Bruijn diagrams assume *dependence*. They will be discussed thereafter.

3.3 Chaos or determinism?

Despite the fact that Rule 22 is based on simple and determined interactions, the behaviour generated by such a system is visibly complex and seems to be non-deterministic. To test the kind of the rule's behaviour, chaos test 0–1 [100] (classifier returning value near 1 if series is chaotic and near to 0 if deterministic) was used to determine whether a behaviour is chaotic/random or deterministic. It is applied directly to the time series data and does not require phase space reconstruction. These two kinds of behaviours are significantly different. Physically, randomness has a stochastic nature, while deterministic chaos is generated by even a simple system that does not contain any source of randomness, as commonly understood by the public. If executed on a personal computer (PC), then algorithms simulating such behaviour generate only pseudo-random/chaotic behaviour, and series generated in such a way are essential deterministic and periodic, but with very long periods. The period is usually long enough to simulate randomness/chaos.

The term *chaos* covers a rather broad class of phenomena whose behaviour may seem erratic, chaotic at first glance. Till now chaos was observed in many systems (including evolutionary ones) and, in the last few years, it has been also used to replace pseudo-random number generators (PRGNs) in evolutionary algorithms (EAs). Let us mention, for example, research papers like Ref. [280], one of the first use of chaos inside EAs [58, 197] discussing the use of deterministic chaos inside particle swarm algorithm instead of PRGNs, [156] investigating relations between chaos and randomness or the latest ones [119, 248], using chaos with EAs in applications, among others.

Another research joining deterministic chaos and pseudo-random number generator has been done, for example, in Ref. [156]. The possibility of generation of random or pseudo-random numbers by use of the ultra weak multidimensional coupling of one-dimensional (1D) dynamical systems is discussed there. Another paper [193] deeply investigates a logistic map as a possible pseudo-random number generator and it is compared with contemporary pseudo-random number generators. A comparison of logistic map results is made with conventional methods of generating pseudo-random numbers. The approach is used to determine the number, delay, and period of the orbits of the logistic map at varying degrees of precision (3–23 bits). The logistic map we are using here was also used in Ref. [82], like chaos-based true random number generator embedded in re-configurable switched-capacitor hardware.

For a long time, various PRNGs were used inside evolutionary algorithms. During the last few years, deterministic chaos systems (DCHS) have been used instead of PRNGs. Very often the performance of EAs using DCHS is better or fully comparable with EAs using PRNGs. See, for example, Ref. [197].

The chaos test 0–1 [100] was used on different series in order to verify and test the nature of Rule 22. Figure 3.9 visualises results of our experiments. For evaluation 2000, Rule 22 behaviour strings of length 2000 have been used. The same was repeated for the Mersenne twister random number generator [172] (MTPRNG) (Fig. 3.9(c)), chaos generated by logistic equation (LE) (Fig. 3.9(b)), and periodic series (PS): the sinus function generated from the randomly selected position (Fig. 3.9(d)). As clearly visible, test 0–1 has clearly classified Rule 22, MTPRNG, and LE as a non-deterministic series, while series based on periodic pattern are classified as deterministic. The random series were not distinguished from chaotic ones. This was probably caused by insufficient length (2000 is likely not enough) of series; however, this was not a matter for this experiment.

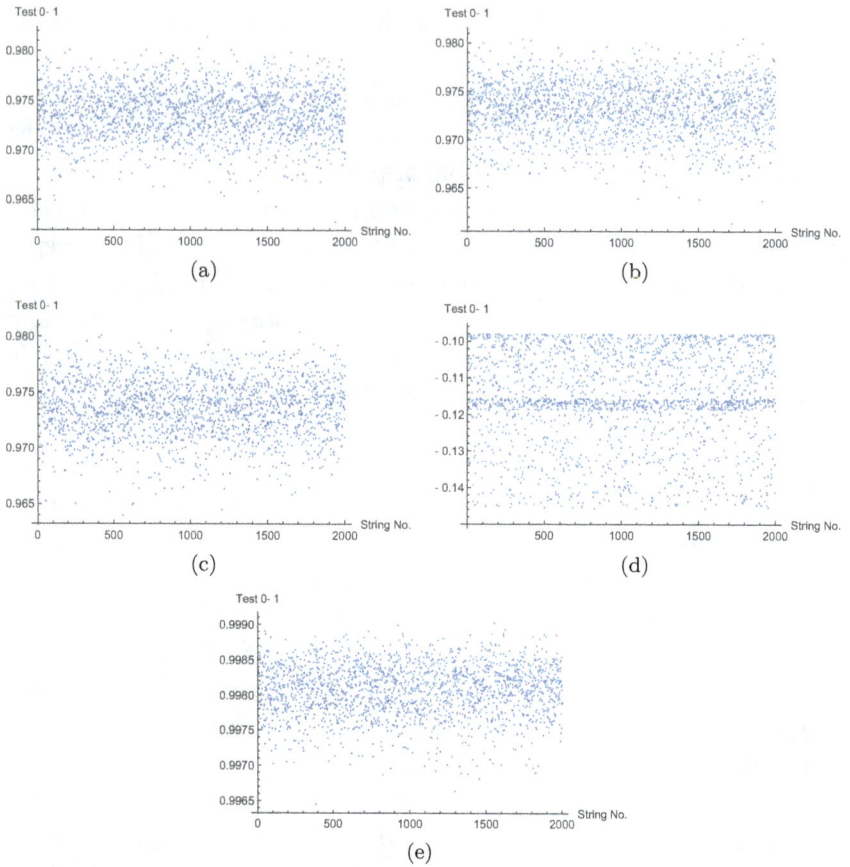

Fig. 3.9. Chaos test 0–1 of (a) logistic equation, (b) Mersenne Twister generator, (c) standard PRNG, (d) deterministic series (returned values are all around 0, i.e. evaluated process is deterministic), (e) Rule 22. All results are concentrated around 1. Chaos based on chaos test 0–1 was proved.

3.4 Attractors

3.4.1 Basins of attraction

Basins of attraction have been studied by Andy Wuensche in the framework of ECA and random Boolean networks [277]. A string of cell-states x_i^t is a configuration c. An evolution is represented by a sequence of configurations $\{c_0, c_1, c_2, \ldots, c_{m-1}\}$, such that $\Phi : \Sigma^n \to \Sigma^n$, and the global transition can be represented as $\Phi(c^t) \to c^{t+1}$. A number of all global states of c is determined by the length of a

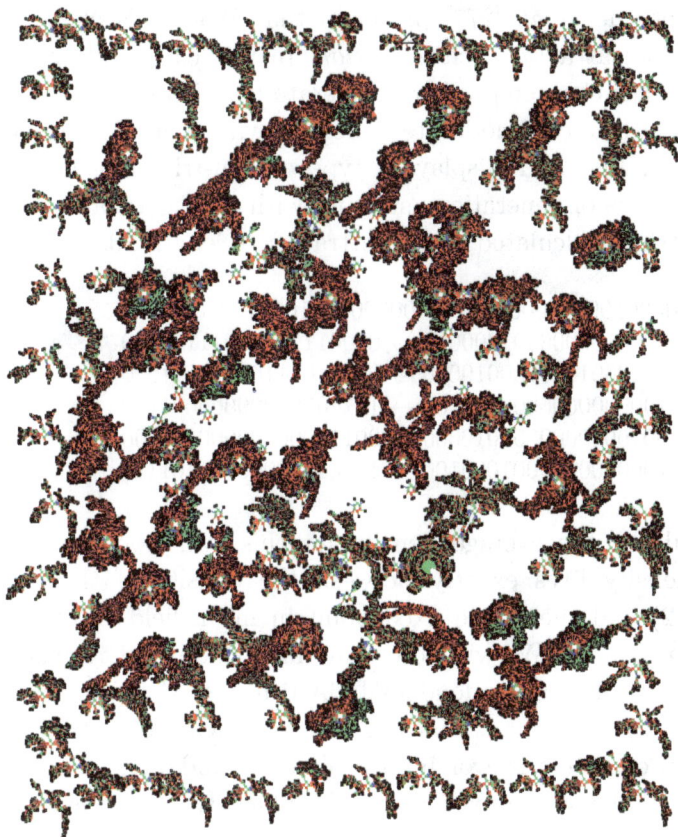

Fig. 3.10. Basin of attractors in ECA Rule 22 for rings of size 20. The number of attractors are 108 with 12 non-equivalent types. Based on attractors' characterisation, Rule 22 displays chaotic behaviour with highly dense, not long transients, and several symmetric trees.

string m^n (where n is the length and m, the number of symbols). The structure of an attractor (Fig. 3.10) is given in three parts. Leaves represent Garden of Eden, i.e. unreachable in the evolution but only as initial global states (these states have no ancestors). Branches are configurations that have at least one ancestor and just one successor. Height of branches determines a number of generations to reach the attractor. An attractor is the final state of a string of length n. Numbers labelling vertices represent the decimal values of the strings.

Wuensche *et al.* [277] proposed that Wolfram's classes can be represented as a *basin classification*. In this classification, complex behaviour is characterised by moderate transients, moderate-length periodic attractors, moderate in-degree, with small density of leaves. This way, Fig. 3.11 displays a type of non-trivial behaviour later of thousands of generations starting with a concatenation of one of these strings calculated by one attractor of length 20:

00000001000100000000 → 00000011101110000000
→ 00000100000001000000 → 00001110000011100000
→ 00010001000100010000 → 00111011101110111000
→ 01000000000000000100 → 11100000000000001110
→ 00010000000000010000 → 00111000000000111000
→ 01000100000001000100 → 11101110000011101110.

Particularly, the average density for this evolution space is 4/15. This density lives exactly between the density of the rule itself (Eq. 3.2) and the stable fixed point in mean field theory (Eq. 3.6 and Fig. 3.3(b)). We will note that this value is not reachable from the statistical analysis done for ECA Rule 22. Also, it is the average where we report non-trivial behaviour in ECA Rule 22. Of course, when an evolution is evolving to this value and later switches to the density of the stable fixed point, the phase transition is irreversible (see Fig. 3.4).

By calculating large attractors, we can discover landscapes of complexity in basins featured with non-symmetric, high and dense ramifications: these kinds of ramifications are indicators of 'unpredictable' behaviour on most large configurations. No rarely chaotic rules tend to have symmetric basins. Basins of attraction can be connected into a meta-network, called the 'jump-graph'. Jump-graphs determine the next level of CA complexity by showing a probability to jump from one attractor to another attractor given a mutation on the same domain of strings [164].

Let us consider a one-bit value mutation $\Psi(\Phi(c_i)) \rightarrow \Phi(c_j)$. A configuration c_i is expressed as a string $w_i = a_0 a_1 \ldots a_{n-1}$, such than it can jump into an other configuration c_j expressed as a string $w_j = b_0 b_1 \ldots b_{n-1}$. Hence, a_i can mutate to one b_i,

Fig. 3.11. Discovering non-trivial patterns emerging in ECA Rule 22 displaying a family of tilings of different sizes from a string of a basin of length 20 (Fig. 3.10). A lot of these patterns can be reached with concatenation of the string 00000001000100000000.

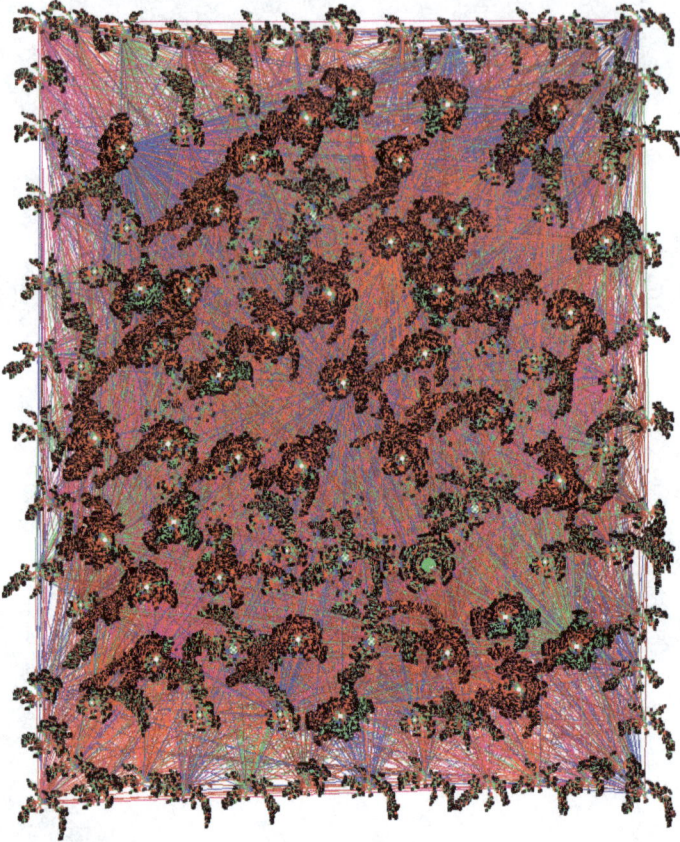

Fig. 3.12. Jump-graph in ECA Rule 22 constructed with a base of attractors of length 20. The connection is determined by mutation of one bit in the strings. The chaotic behaviour from jump-graphs is characterised to the high connectivity between all attractors (for details, see Ref. [164]).

where each configuration c belongs at the same field of attractors Ψ. Also, the mutation represents a loop in the same basin if $a_i = b_i$. Figure 3.12 shows a jump-graph from the basins of attraction of length 20 (Fig. 3.10). This way, a chaotic system presents a high density of connectivity with all the attractors in the jump-graph.

3.4.2 Longest paths and representative cycles

In this section, we will use the term *state* as an alias for global state c_i, configuration, or $1d$ string. CA(n) denotes an ECA Rule 22 with

n cells and a cyclic boundary. The aim of this section is to study the following:

- What is the length of the longest path until the zero-state (00...0) is reached, and how does a related initial Garden of Eden state look?
- How many cycles exist, how long are they, and what is a representative state for each cycle? Cycles that belong to the same class (cycles' states are equivalent under shift and mirroring) shall be listed only once as a *representative cycle*, the kind of cycles we are interested in here.
- Are there similar states (cyclically shifted) that appear periodically within a cycle?
- What is the length of the longest path until a certain cycle is reached, and how does a related initial state look?

The following terms and functions are used here:

- *path(A, C)* is a sequence of states (from State A to State C).
- *length(path)* gives the number of states of a path or cycle.
- *prefix(C)* is a path(A, B) where B is a direct predecessor of C.
- *maxprefix(C)* is a prefix(C) of maximum length.
- $\alpha = $ *length(maxprefix(0))* where (0) is the zero-state (00...0), the longest prefix(0).
- *cycle* denotes a periodic attractor, a cyclic path.
- *k-cycle* is a cycle of length k.
- $\omega(cycle)$ gives the length of a cycle.
- *similar(S)* is a state that can be derived from state S by cyclic shift and optional mirroring.
- ϵ is called *intra-cycle period*.
- In some cycles, similar states appear again after ϵ time steps.
- *k/e-cycle* is a k-cycle where $e = \epsilon(k$-cycle). We may call k/e-cycles *strong* if $k = e$, and *weak* if $k > e$.
- *cycle-prefix* is a prefix(D) where D belongs to a cycle.
- $\lambda(cycle)$ gives the length of the longest cycle-prefix.

First method and results. The CA(n), $n = 3, \ldots, 20$, were simulated for all possible initial states and then analysed by special

Table 3.2.   Longest path to the zero state.

n	α: Length max prefix(0)	Normalised initial state
3	2	001
4	2	0001
5	6	01011
6	5	010111
7	6	0001111
8	8	00101011
9	2	000100101
10	7	0010111101
11	6	00101010101
12	23	001001100111
13	20	0001001100111
14	24	00010101100111
15	32	001001010100111
16	41	0000000101010011
17	53	00101010101110011
18	8	000001011000010011
19	17	0000101010111010101
20	18	00001000101010111101

programs and manual inspection. To avoid unnecessary simulations, similar initial states were excluded. Similar states are states are equivalent under cyclic shift and mirroring. For instance, the number of different cases to be simulated for $n = 18$ is only 7685, which is significantly lower than 2^{18}. It should be noted that the cycles for n up to 34 were already computed by McIntosh [177].

Table 3.2 shows α (the length of the *maxprefix*(0)), and the related *normalised* initial state. A normalised state is a representative of all states that are equivalent under cyclic shift and mirroring. It is found by selecting the state with the smallest binary number among all equivalents. For example, *normalise*(100011) = 000111, and *normalise*(110110100) = 001011011 (by mirror and shift). The α values are much smaller than 2^n, and not monotonically increasing with n.

Table 3.3 shows the results obtained by analysing all the simulations. The operator *shlP*(c) means shift c to the left by P positions, and *shlPm*(c) means that first the mirror operator is applied before shifting.

Table 3.3. Representative cycles for $n = 3, \ldots, 20$. ω: cycle length, ϵ: intra-cycle period, λ: length of longest cycle prefix. Not all leading zeroes are displayed.

n	ω	Representative cycle state (normalised)	ϵ	Repetition within cycle	λ	Initial state of longest cycle prefix
3, 5, 6		no cycle $\omega > 1$				
4	2	(0011)*	1	$c^{t+1}=\mathrm{shl}2(c^t)$	0	no prefix
7	7	(0001011)*	1	$c^{t+1}=\mathrm{shl}4(c^t)$	0	no prefix
8	2	(0011)*	1	$c^{t+1}=\mathrm{shl}2(c^t)$	0	no prefix
	4	00000101	2	$c^{t+2}=\mathrm{shl}4(c^t)$	2	00100111
	6	00000011	3	$c^{t+3}=\mathrm{shl}4(c^t)$	2	00101101
9	4	000000101			9	000100111
10	4	0000001001	2	$c^{t+2}=\mathrm{shl}5(c^t)$	7	0011010101
	4	0000000101			5	0010110011
	6	0000010011	3	$c^{t+3}=\mathrm{shl}3m(c^t)$	1	0000010011
11	4	00000001001			7	00000100111
	5	00000001111			7	00101010011
	11	00001001111	1	$c^{t+1}=\mathrm{shl}7(c^t)$	9	00010110011
12	2	(0011)*	1	$c^{t+1}=\mathrm{shl}2(c^t)$	0	no prefix
$= 3 \times 4$	4	000000010001	2	$c^{t+2}=\mathrm{shl}6(c^t)$	6	000100110011
	5	000000001111			9	001010101101
13	5	0000000001111			13	0000101001101
14	7	(0001011)*	1	$c^{t+1}=\mathrm{shl}3(c^t)$	0	no prefix
$= 2 \times 7$	12	00000010001111	6	$c^{t+6}=\mathrm{shl}9m(c^t)$	4	00000101010011
15	20	000000010000011	4	$c^{t+4}=\mathrm{shl}4(c^t)$	20	001010100101111
16	2	(0011)*	1	$c^{t+1}=\mathrm{shl}2(c^t)$	0	no prefix
$= 2 \times 8$	4	(00000101)*	2	$c^{t+2}=\mathrm{shl}4(c^t)$	2	(00100111)*
$= 4 \times 4$	6	(00000011)*	3	$c^{t+3}=\mathrm{shl}4(c^t)$	2	(00101101)*
	7	0000000000000011			11	0000000010010101
	12	0000000000000101	6	$c^{t+6}=\mathrm{shl}8(c^t)$	29	0000001001010111
	12	0000000000100001			10	0010101010101011
17	4	000010100000101			2	00100101100100111
	12	000000000000101			21	00101010101101011
	26	000000000010011	13	$c^{t+13}=\mathrm{shl}15m(c^t)$	26	00101010010110011
18	4	000100100000101	2	$c^{t+2}=\mathrm{shl}4(c^t)$	2	00101010100101101
$= 2 \times 9$	4	(000000101)*			45	00111001010101111
	4	000010100000101			5	00101010011001111
	12	000000000001001	6	$c^{t+6}=\mathrm{shl}9(c^t)$	35	00000001000011101
	12	000000000000101			55	00010010100110011
	18	000001101001011	9	$c^{t+9}=\mathrm{shl}9(c^t)$	7	01111001100110011
19	4	000100100000101			21	00001011100010111
	4	000101000000101			6	10100101010101011
	4	001001000000101			1	00110111000110111
	12	000000000001001			78	00001100010010111
	28	000001000011101	14	$c^{t+14}=\mathrm{shl}4m(c^t)$	9	01001100110011101

(*Continued*)

Table 3.3. (*Continued*)

n	ω	Representative cycle state (normalised)	ϵ	Repetition within cycle	λ	Initial state of longest cycle prefix
20	2	(0011)*	1	$c^{t+1}=$shl2(c^t)	0	no prefix
$= 2 \times 10$	4	(0000001001)*	2	$c^{t+2}=$shl5(c^t)	44	00101100011010011
$= 4 \times 5$	4	001000100000101	2	$c^{t+2}=$shl4(c^t)	5	00110110011001000
	4	(0000000101)*			12	11010001110101111
	4	001001000001001			62	01111001010101111
	4	001001000000101			10	00010000100001111
	6	100110000101111	3	$c^{t+3}=$shl7m(c^t)	1	01011010100101111
	6	(0000010011)*	3	$c^{t+3}=$shl7m(c^t)	8	00001010100010111
	8	000000001000001	4	$c^{t+4}=$shl10(c^t)	42	00000010001110101
	12	000000000010001	6	$c^{t+6}=$shl10(c^t)	24	00100101100100011
	24	000000000100001			98	01010101110101101

Note: *means iteration.

We find always the trivial $(00..0) \rightarrow (00..0)$ cycle of length 1. For even n, there always exists the fixed point $(01)^* \rightarrow (01)^*$, a lonely 1-cycle with no prefix. We will not mention further or pay special attention to these basic 1-cycles.

For $n = 4$, and multiples of 4, we get the 2/1-cycle $(0011)^* \leftrightarrow (1100)^*$. The two strings are similar under the shifting of two positions, so the inherent pattern is the same.

For $n = 7$, and multiples of 7, we get a 7-cycle. To characterise this cycle, one representative is chosen, it is the one with the smallest normalised value, i.e. $(0001011)^*$.

For $n = 8$, we get three cycles with lengths $\omega/\epsilon = 2/1, 4/2, 6/3$.

For $n = 10$, there exists a 6/3-cycle. After every three time steps, the same string appears in mirrored form and shifted three positions to the left.

For $n = 12$, the cycles of CA(4) form a subset (to be included if not detected), which is the 2/1-cycle (0011 0011 0011) $\leftrightarrow$ (110 1100 1100).

For $n = 14$, the cycles of CA(7) form a subset, which is shown as 7/1-cycle.

Figure 3.10 shows all possible cycles where many of them are similar (equivalent under shift and mirroring). From that figure we may anticipate a very complex attractor structure, but we should realise that the number of representative cycles in CA(20) is only 11.

In general, because of the cyclic boundary, if k is a factor of n, then the CA(k) cycles are a subset of the CA(n) cycles. For example, the cycles of CA$(k = 4, 5, 10)$ form a subset of the CA$(n = 20)$ cycles. However, there is a difference: the strings of CA(n) are cyclic repetitions of the CA(k) strings, and the original intra-cycle period ϵ may not appear in CA(n).

Second method and results. For larger n, the first method cannot further be applied due to extensive computational costs. Therefore, now only a relatively small random subset of all possible 2^n initial states is used to find a subset of all cycles and paths that are not necessarily the longest ones.

About 100000 random initial states were generated for $n = 25, 30, 35, ..., 60$. For 5000 of the states, the probability 0.125 was used for each cell to generate a cell-state 1. For the other 5000 initial states, at first a probability p between 0 and 1 (in steps of $1/1000$) was randomly selected. Then p was used for each cell to generate a cell-state 1, otherwise 0. This technique of randomising gave better results in experiments for Rule 22 compared to the usage of a fixed probability of 0.5. CAs were simulated and cycles and path lengths were computed and processed in a semi-automatic mode. In addition, a genetic algorithm was used to find near-optimal α-values. The results are presented in Table 3.4. Note that because of the statistical approach, the listed cycles are not complete and the true maximum path lengths could be longer, e.g. for CA(60), all cycles already found for the factors CA$(4, 5, 10, 12, 15, 20, 30)$ have to be included.

We can summarise that for $n \leq 60$, the longest paths are much smaller than $2^n - 1$ (which is achievable with other ECA rules [2]). Further work remains to find a general formula or at least boundaries for the longest paths and the cycle distribution, depending on a number of cells.

3.5 de Bruijn diagrams

de Bruijn diagrams [73] were originally proposed in shift–register theory [99]. For a $1d$ CA(k, r), the de Bruijn diagram is defined as a

Table 3.4. Representative cycles for $n = 25, \ldots, 60$. ω: cycle length, λ: length of longest cycle prefix, $\omega + \lambda$: length of longest path detected. The values were obtained by simulation of 10000 random initial states.

n	ω Length of cycles detected	$\omega + \lambda$ Longest path ending in a cycle	α Longest prefix(0)
25	$4, 5, 26, 28, 50, 55, 150$	$150 + 57 = 207$	152
30	$1, 4, 6, 14, 20, 40, 70, 86, 120, 240, 1070$	$1070 + 153 = 1223$	419
35	$4, 5, 12, 28, 64, 1015$	$1015 + 302 = 1317$	179
40	$1, 4, 8, 12, 16, 24, 52, 80, 124, 206, 320$	$124 + 2551 = 2675$	303
45	$4, 8, 16, 19, 2295$	$4 + 4815 = 4819$	540
50	$1, 12, 28, 31, 55, 56, 100, 117, 150,$ $252, 700, 3150$	$252 + 13956 = 14208$	750
55	$12, 28, 30, 56, 60, 330, 440, 660,$ $990, 4620, 36190, 148225$	$148225 + 7124 = 155349$	13904
60	$1, 12, 60, 120, 138, 395, 476,$ $480, 22740, 40980$	$40980 + 1004 = 41984$	35579

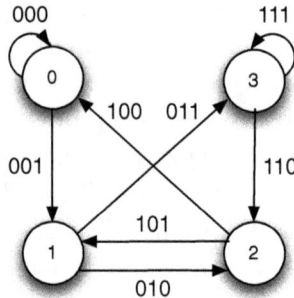

Fig. 3.13. Generic de Bruijn diagram for ECA (2,1).

directed graph with k^{2r} vertices and k^{2r+1} edges. Vertices are labelled with the elements of symbols of length $2r$. An edge is directed from vertex i to vertex j, if and only if, the $2r - 1$ final symbols of i are the same as the $2r - 1$ initial ones in j forming a neighbourhood of $2r + 1$ states represented by $i \diamond j$. In this case, the edge connecting i to j is labelled with $\varphi(i \diamond j)$ (the value of the neighbourhood defined

$$M_{R22} = \begin{bmatrix} 0 & 1 & . & . \\ . & . & 1 & 0 \\ 1 & 0 & . & . \\ . & . & 0 & 0 \end{bmatrix}$$

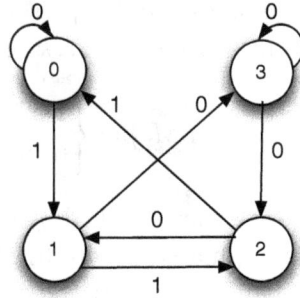

Fig. 3.14. Connection matrix and de Bruijn diagram for ECA Rule 22.

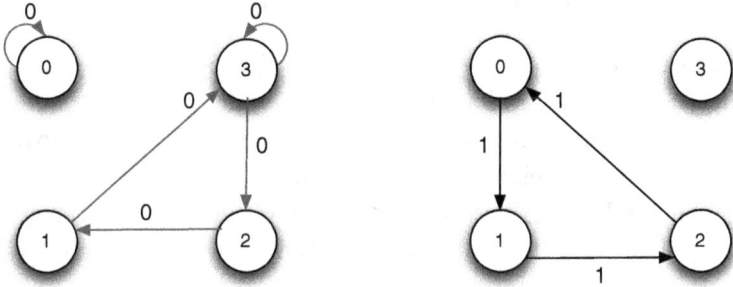

Fig. 3.15. de Bruijn subdiagrams showing unreachable states.

by the local function) [265], as shown in Fig. 3.13 for ECA(2,1) and for Rule 22 in Figs. 3.14 and 3.15.

3.5.1 Basic de Bruijn diagram

The connection matrix M corresponding to the de Bruijn diagram [178] is as follows:

$$M_{i,j} = \begin{cases} 1 & \text{if } j \in \{k\,i, k\,i+1, \ldots, k\,i+k-1 \ (\text{mod } k^{2 \cdot r})\}, \\ 0 & \text{otherwise} \end{cases} \tag{3.7}$$

wherein module $k^{2 \cdot r}$ represents the number of vertices and j takes on values in $\{k\,i, k\,i+1, \ldots, k\,i+k-1 \pmod{k^{2 \cdot r}}\}$. Hence, for ECA(2,1)

$$M_{i,j} = \begin{cases} 1 & \text{if } j \in \{2\,i, 2\,i+1 \pmod 4\} \\ 0 & \text{otherwise} \end{cases} \tag{3.8}$$

and vertices are labelled by fractions of the overlapping of neighbourhoods originated by 00, 01, 10, and 11, and the overlaps of the full neighbourhood determine each connection:

$$
\begin{aligned}
(0,\mathbf{0}) \diamond (\mathbf{0},0) &\to 000 & (1,\mathbf{0}) \diamond (\mathbf{0},0) &\to 100 \\
(0,\mathbf{0}) \diamond (\mathbf{0},1) &\to 001 & (1,\mathbf{0}) \diamond (\mathbf{0},1) &\to 101 \\
(0,\mathbf{1}) \diamond (\mathbf{1},0) &\to 010 & (1,\mathbf{1}) \diamond (\mathbf{1},0) &\to 110 \\
(0,\mathbf{1}) \diamond (\mathbf{1},1) &\to 011 & (1,\mathbf{1}) \diamond (\mathbf{1},1) &\to 111.
\end{aligned}
$$

They are the edges of the generic de Bruijn diagram in Fig. 3.13. The de Bruijn diagram has four vertices, which can be renamed as $\{0, 1, 2, 3\}$ corresponding to four partial neighbourhoods of two cells $\{00, 01, 10, 11\}$, and eight edges representing neighbourhoods of size $2\,r+1$. The de Bruijn diagram for Rule 22 is derived from the generic one (Fig. 3.13) and it is calculated in Fig. 3.14, where the edges are labelled by the next state.

Paths in the de Bruijn diagram may represent chains, configurations, or classes of configurations in the evolution space. Vertices are sequences of symbols in the set of states and the strings are sequences of vertices in the diagram. The edges represent overlapping of the sequences. Different intersection degrees evoke different de Bruijn diagrams (Fig. 3.15). Thus, the connection takes place between an initial symbol, the overlapping symbols, and a terminal one. For practical reasons, we can use colours, thus the colour of an edge represents the next state to which each neighbourhood, as shown in Fig. 3.13, evolves.

3.5.2 Extended de Bruijn diagram

An extended de Bruijn diagram takes into account wide overlapping of neighbourhoods. We represent $M_{R22}^{(2)}$ by indexes $i = j = 2\,r\,n$, where $n \in \mathbb{Z}^{+}$, $M_{R22}^{(3)}$ and $i = j = 3\,r\,n$, $M_{R22}^{(4)}$ and $i = j = 4\,r\,n$, and

Table 3.5. Regular expressions derived in ECA Rule 22. The set of equations is calculated using the recursive function $R_{i,j}^k$ (Eq. 3.9) to recognise k paths between nodes i to j in the de Bruijn diagram (Fig. 3.14).

Eq.	Expression	Evolution
1	$(0+1)^*$	Stable state
2	$(01)^*$	Stable periodic
3	$(001)^*$	Stable state
4	$11(01)^*00$	Still life & symmetric complex behaviour
5	$(01)^*(0+1)$	Stable periodic & symmetric complex behaviour
6	$(000111)^*$	Stable state
7	$(0+0(01)^*00)^*$	Stable periodic, chaos & big gaps
8	$(((01)^* + 0)00^*1)^*$	Chaos, complex behaviour
9	$(0+1) + 11(01)^*(0+1)$	Complex behaviour
10	$((01)^*00)(0+0(01)^*00)^*$	Chaos, stable periodic, chaos & big gaps
11	$(11(01)^*00)(0+0(01)^*00)^*$	Chaos, still life & symmetric complex behaviour
12	$(0+0(01)^*00)^*0(01)^*(0+1)$	Stable state, stable periodic, chaos & big gaps
13	$(0+1(01)^*00)(0+0(01)^*00)^*$	Chaos, stable periodic, chaos & big gaps
14	$((0+1) + 11(01)^*1) + (11(01)^*00)$ $(0+0(01)^*00)^*(0(01)^*(0+1))$	—
15	$(0^*10^*)(10^*10 + (10^* + 10^*10)0^*)^*$	—

so up to $M_{R22}^{(m)}$ with $i = j = m\,r\,n$; consequently, the basic de Bruijn diagram is obtained when $m = 1$. The regular expressions derived from the de Bruijn diagram for Rule 22 (Table 3.5) can be linked to space–time dynamics phenomena exhibited by the rules. These include symmetric complex behaviour, chaos, and stable periodic behaviour. Figure 3.16 shows in detail every periodic pattern yielded from extended de Bruijn diagrams. To read the diagram, we use notation (i, j), where i is a displacement (left or right) and j is a number of generations. Thus, the pattern in position $(0, 0)$ (upper centre) displays a periodic pattern without both displacement and period, the expression reproducing this pattern is $(01)^*$ (de Bruijn subdiagram in Fig. 3.16 and Eq. (3.2) in Table 3.5). In this way, a

G.J. *Martínez* et al.

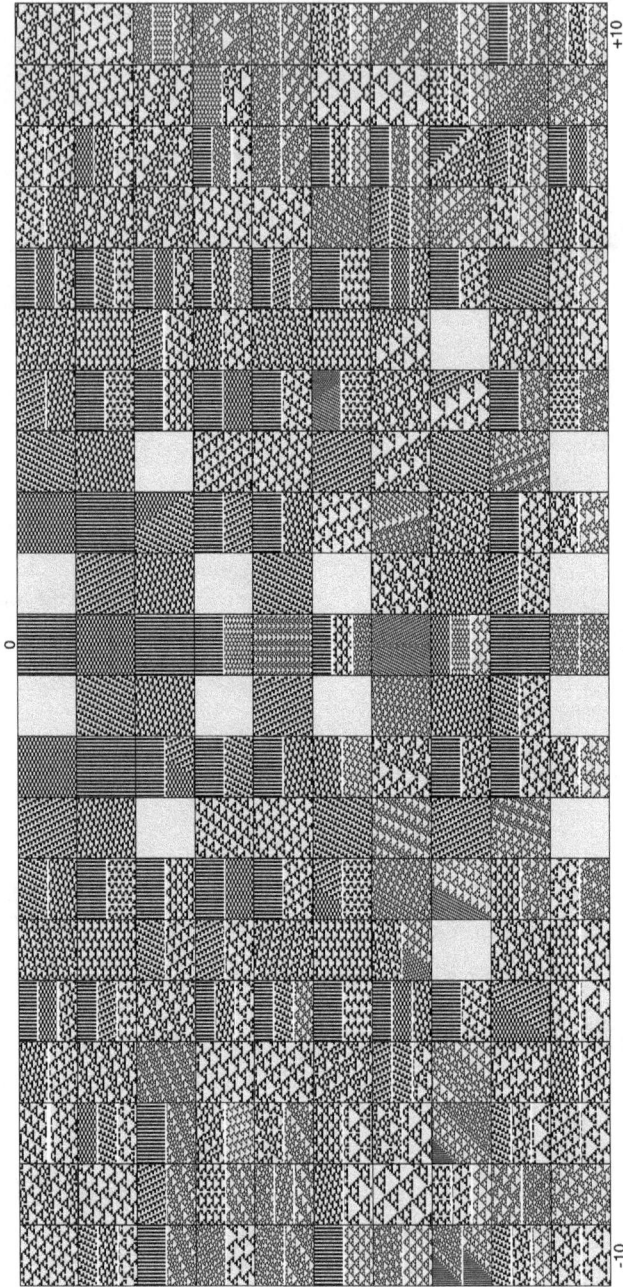

Fig. 3.16. The whole set of periodic patterns yielded from extended de Bruijn diagrams to 10 generations with positive and negative shifts to 10 cells.

lot of non-trivial patterns can be extracted. Let us consider few examples.

- $(0, 7)$. The graph is characterised by two cycles, the small cycle yielding simple still life patterns, and the large cycle representing configurations emitting travelling localisations, or particles. These configurations cannot evolve naturally, i.e. from an initial random condition, because they are destroyed when a certain limit of their size is reached (Fig. 3.17). The configurations are expressed for the following cycle–ways:

  $4515 \rightarrow 9080 \rightarrow 167 \rightarrow 3354 \rightarrow 6708 \rightarrow 13416 \rightarrow 10449 \rightarrow$
  $4515 \rightarrow 9031 \rightarrow 1679 \rightarrow 3358 \rightarrow 6716 \rightarrow 13432 \rightarrow 10481 \rightarrow$
  $4578 \rightarrow 9157 \rightarrow 1931 \rightarrow 3862 \rightarrow 7724 \rightarrow 15448 \rightarrow 14513 \rightarrow$
  $12642 \rightarrow 8901 \rightarrow 1419 \rightarrow 2838 \rightarrow 5676 \rightarrow 11352 \rightarrow 6321 \rightarrow 12642$

  The regular expression to reproduce the same pattern is calculated as Eq. (3.8) in Table 3.5.
- $(4, 8)$. The graph has several paths between different cycles. Some of these cycles calculate trivial patterns. Other cycles represent configurations developed from a 'fusion' of two periodic regions competing for the space (Fig. 3.18). Other fused configurations can be found in coordinates $(6, 9)$, $(-4, 6)$, $(-4, 8)$, $(-5, 7)$, $(-6, 9)$, $(-8, 7)$, and $(-10, 8)$.
- $(10, 2)$. Here we observe composed triangular polygons. The polygons sustain in a periodic mobile background. The background features small particles crossing the space (Fig. 3.19). This complex behaviour emerges with probability between $1/7$ and $3/7$. This complex behaviour was discovered with the help of expressions derived from the basins of attraction (see Fig. 3.11). One more example of interaction of large triangular domains evolving in the periodic background is shown in Fig. 3.20.
- $(0, 4)$. The periodic background is stationary. Several types of triangular domains and several families of small tiles emerge (Fig. 3.21). Other coordinates leading to similar configurations are $(0, 8)$, $(-6, 2)$, $(-6, 6)$, $(6, 2)$, and $(6, 6)$.

During the analysis of configurations derived with the de Bruijn diagrams and basin of attractors, we have referred to some specific

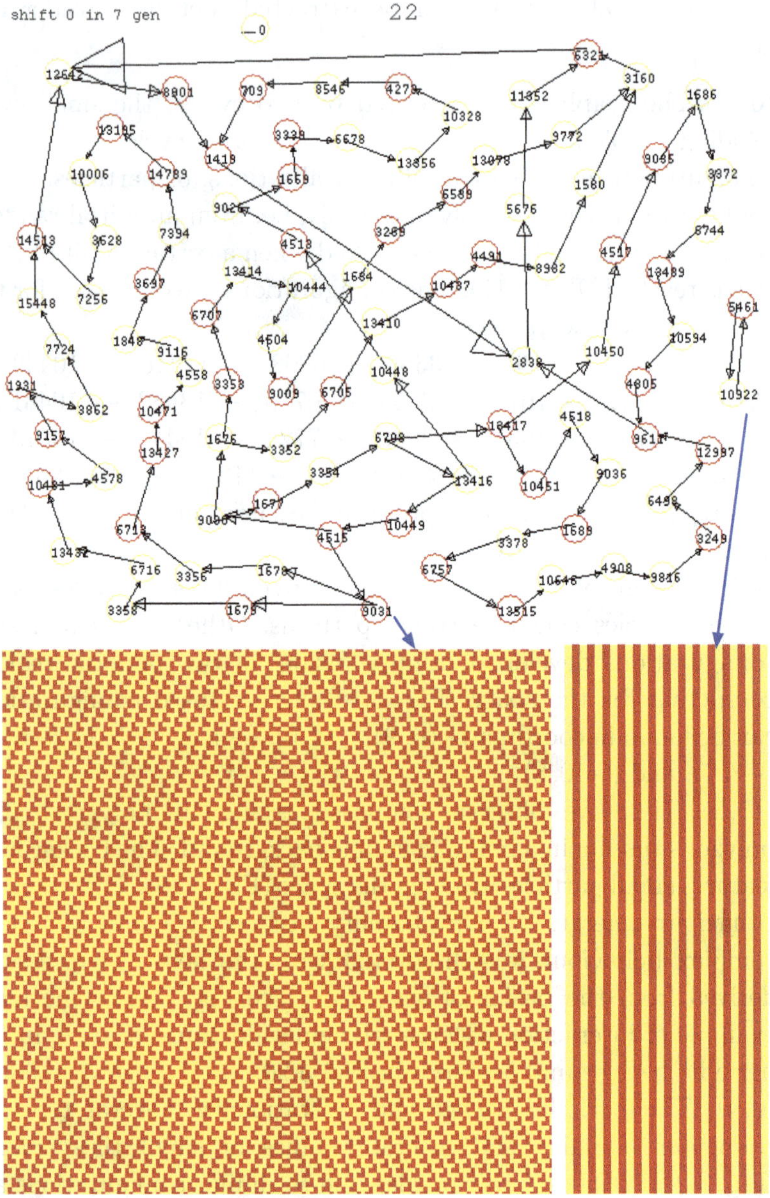

Fig. 3.17. The extended de Bruijn diagram (0,7) calculating a pattern emitting mobile self-localisations. To reproduce this pattern, we concatenate the expressions $(1010001)^n$-111000101100010-$(1100010)^n$, where $n > 0$ is the number of copies. The small cycle represents still life patterns.

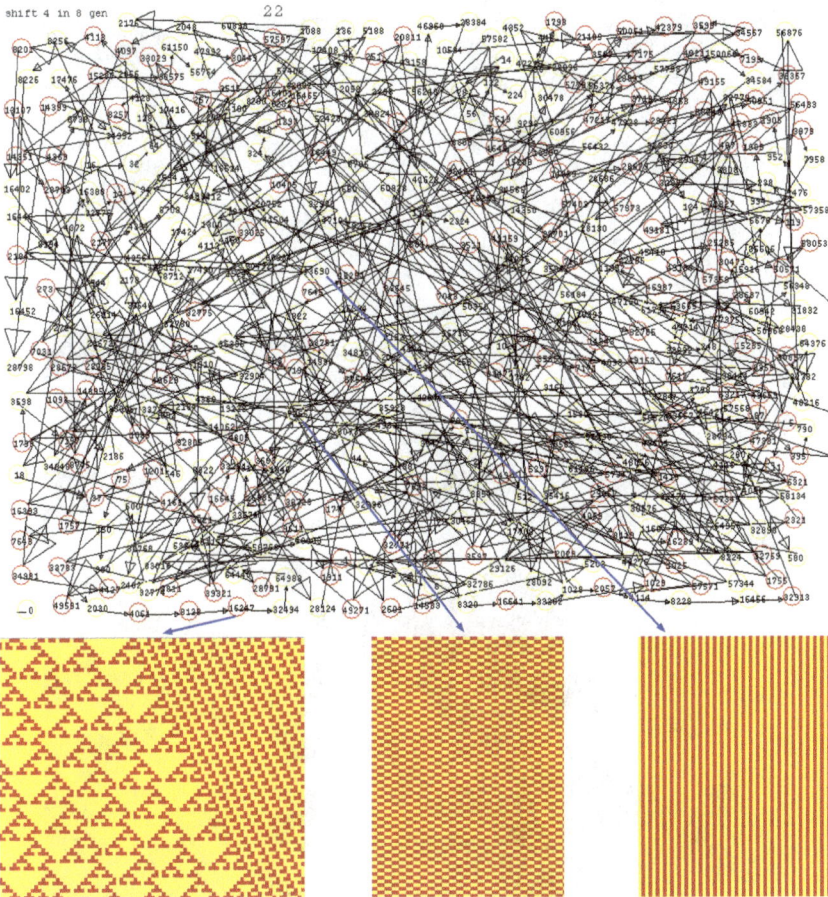

Fig. 3.18. The extended de Bruijn diagram (4,8) calculating mobile self-localisations, small tilings, and meshes. The first large cycle calculates a configuration known as fuse [178] because two periodic patterns with different densities can evolve together without perturbing each other's boundaries.

regular expressions. Regular expressions can be calculated recursively by following paths on a graph with [120]

$$R_{i,j}^k = R_{i,j}^{k-1} + R_{i,k}^{k-1}(R_{k,k}^{k-1})^* R_{k,j}^{k-1}, \qquad (3.9)$$

where i is the initial state and j, the final state. Base case when $k = 0$ is the direct path to every node. This way, by using the basic de Bruijn diagram in Fig. 3.14, we have calculated the whole set of

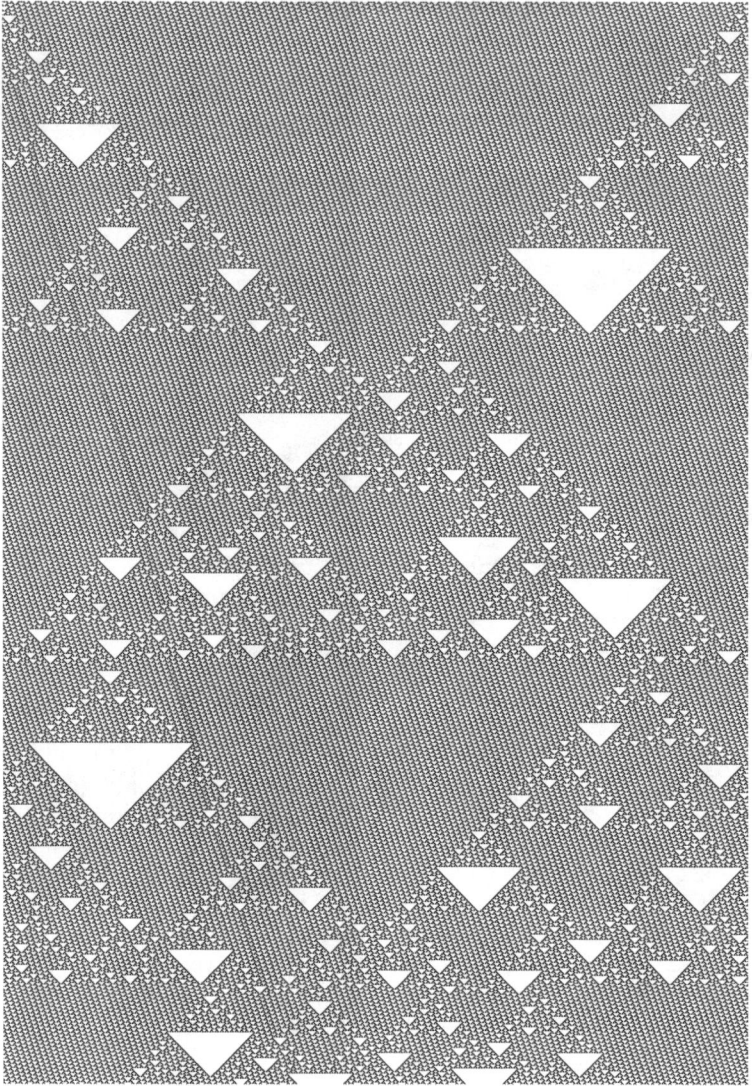

Fig. 3.19. Non-trivial dynamics emerging in ECA Rule 22 on a ring of 1,198 cells during 1,679 generations. Mobile localisations emerge as triangular polygons travelling in a mobile periodic background. The localisations conserve their shape when collide with each other. This dynamics was discovered with the extended de Bruijn diagram order (10,2), see Fig. 3.14. These configurations cannot be reached from a random initial condition.

Fig. 3.20. Non-trivial dynamics emerging in ECA Rule 22 with the mobile periodic background (10, 2) evolving with a high density of small tilings. Different large triangular polygons can be constructed from the interactions of other polygons.

regular expressions, summarised in Table 3.5. The first column shows an equation number, the second column, the regular expression, and the third column, the kind of behaviour that emerges when we codify configurations by these regular expressions. This way, we could evaluate these equations and explore an unlimited number of configurations.

3.6 Garden of Eden

The question "does a complex CA contain a universal constructor?" is a classic problem appearing in the CA literature since

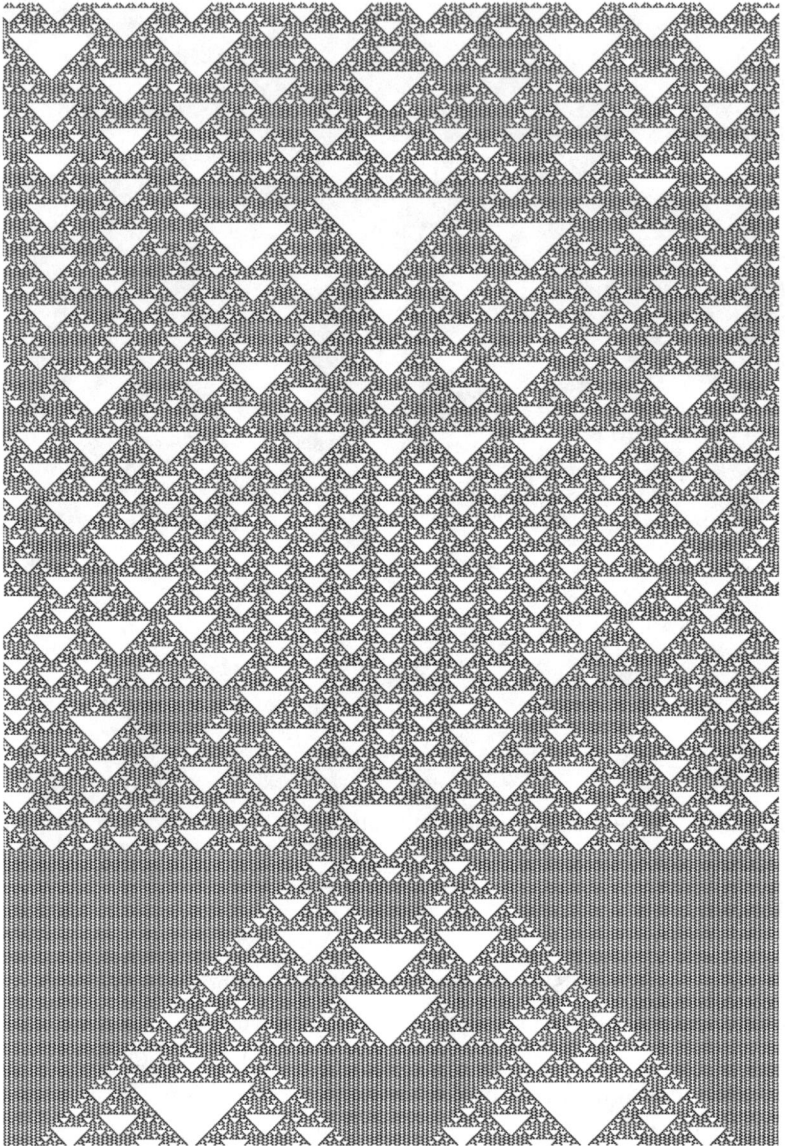

Fig. 3.21. Non-trivial behaviour emerging in a fixed periodic background, i.e. this background does not move. In this periodic background, complex large triangular polygons can emerge as well, including several types of small tiles. This fixed periodic background can be calculated from the extended de Bruijn diagram (0,4), see Fig. 3.14.

von Neumann's works [265]. A configuration of a universal constructor in the Game of Life CA was proposed by Goucher in 2010 [102]. In this context, our aim is to know if ECA Rule 22 is able to construct any string. Previously, this problem was studied by McIntosh [178], who found that Rule 22 has a global injective relation and, therefore, configurations without ancestors exist.

We use a subset diagram to calculate *Garden of Eden* configurations [178], i.e. the configurations without ancestors [36]. A subset diagram has $2^{k^{2r}}$ vertices with k states and r neighbours. If all the configurations of a certain length have ancestors, then all the configurations with extensions both to the left and to the right with the same equivalence must have ancestors. But if this is not the case, then the vertices represent Garden of Eden configurations.

We can define the subset diagram as the power set of $2^{k^{2r}}$. Such that each subset $S \in U_S$ (where U_S is a power set) and one symbol $a \in \Sigma$:

$$\alpha(S, a) = \bigcup_{q_i \in S} \varphi(q_i, a). \tag{3.10}$$

Vertices of the subset diagram are formed by the combination of each subset formed from the states of the de Bruijn diagram. Symbolic de Bruijn matrices $M_{k,s}$ or M_s are characterised by k states and s number of states in the partial neighbourhood. Thus, for Rule 22 we can obtain symbolic matrices, derived from the de Bruijn subdiagrams shown in Fig. 3.15. For any ECA, we have four sequences of states in the de Bruijn diagram enumerated as 0, 1, 2, and 3 (see Fig. 3.13).

Union between subsets is represented by the state in which each sequence evolves and is assigned to the states (subsets that form it) as governed by Eq. (3.10). Relations between subsets for Rule 22 are constructed in Table 3.6. Figure 3.22 shows the full scalar subset diagram. Each class of edges defines a function on Σ_0 or Σ_1. The subset diagram describes the union $\Sigma_0 \cup \Sigma_1$ that by itself is not functional [177].

We must distinguish four types of subsets, where it is possible to make a transition between its four unit classes. Also, we should

Table 3.6. Relations between states of
the subset diagram in Rule 22.

S	Label	0	1
ϕ	0	0	0
$\{0\}$	1	1	2
$\{1\}$	2	0	12
$\{2\}$	4	1	2
$\{3\}$	8	8	4
$\{0,1\}$	3	1	14
$\{0,2\}$	5	1	2
$\{0,3\}$	9	9	6
$\{1,2\}$	6	1	14
$\{1,3\}$	10	8	12
$\{2,3\}$	12	9	6
$\{0,1,2\}$	7	1	14
$\{0,1,3\}$	11	9	6
$\{0,2,3\}$	13	9	6
$\{1,2,3\}$	14	9	14
$\{0,1,2,3\}$	15	9	14

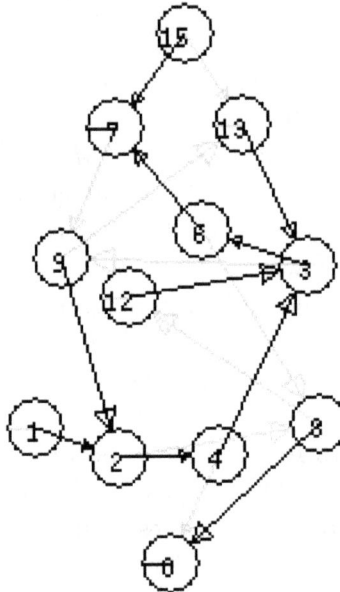

Fig. 3.22. The simplified subset diagram for ECA Rule 22.

observe that a residual of the de Bruijn diagram can be found in the subset diagram. This is because a unit class is precisely defined by the nodes of the original diagram. At first instance, we can see some relations are more frequent than others. Also there are nodes without inputs, or nodes with most connections including loops. Most important are cycles of different lengths. They are used to infer words, or sequences, that a CA could recognise. Thus, the subset diagram can be used as a general machine to recognise the universe of words in which a CA could evolve.

By analysing the full diagram, we can derive a small subset diagram that is deduced from the original diagram (de Bruijn diagram). This diagram includes only vertexes with cycles, the universal and empty sets and the subset with one element, yielding a new diagram that will be more practical for us. The reduction gives a yet smaller diagram to read quickly the strings belonging to the Garden of Eden configurations. The reduction is also useful to calculate the degree of Welch indices for reversible CAs [225]. The expressions that determine Garden of Eden configurations in ECA Rule 22 are listed as follows:

- 10110
- 01111*01101*
- 11*(011111)*1*0110

3.7 Fractals

3.7.1 Iterated functions in Rule 22

A fractal is constructed recursively from a self-replication of a pattern [84]. Chaotic systems often bear properties of fractals.

ECA Rule 22 produces a fractal pattern, known as Sierpiński triangle, starting from a single cell in State '1' (Fig. 3.1). Figure 3.23 shows a triangle constructed with three small tiles derived in Rule 22, this triangle grows in power of two with respect to the number of cells. The main triangle (Fig. 3.23(a)) has three replicas in the next iteration (Fig. 3.23(b)), and the following iteration produces nine base replicas (Fig. 3.23(c)). The fractal dimension D can be

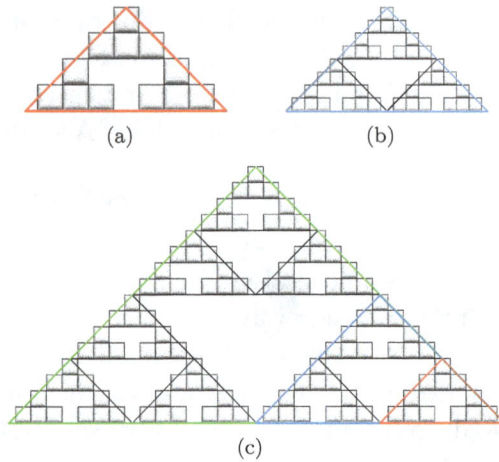

Fig. 3.23. Iterated function determines a fractal defining a Sierpiński triangle in ECA Rule 22 from a composition of three tiles starting with a 1.

calculated given the number of replicates N and the scaling factor m [60]. The fractal dimension of the patterns generated by Rule 22 is the following:

$$D = \frac{\log(N)}{\log(m)} = \frac{\log(3)}{\log(2)} = 1.5849625. \tag{3.11}$$

Also, ECA Rule 22 displays non-trivial behaviour via fractals where they emerge in different stages during the evolution. These fractals are combined with other fractals constructed from Rule 22 over thousands of generations.

Using regular expressions, we found two different periodic backgrounds emerging in ECA Rule 22, as discussed in Sections 5, 6, and 7. Let us illustrate two fractals growing in intervals of other fractals with different compositions of tiles. Figure 3.24(a) shows the initial state of fractals growing in a periodic background without displacement conserving the same fractal dimension. Figure 3.24(b) shows the same iterated function over thousands of generations. The same behaviour is tested on a periodic background with displacement in Figs. 3.25(a) and 3.25(b). Composed fractals emerging in periodic backgrounds with or without displacement are disjoint.

(a)

(b)

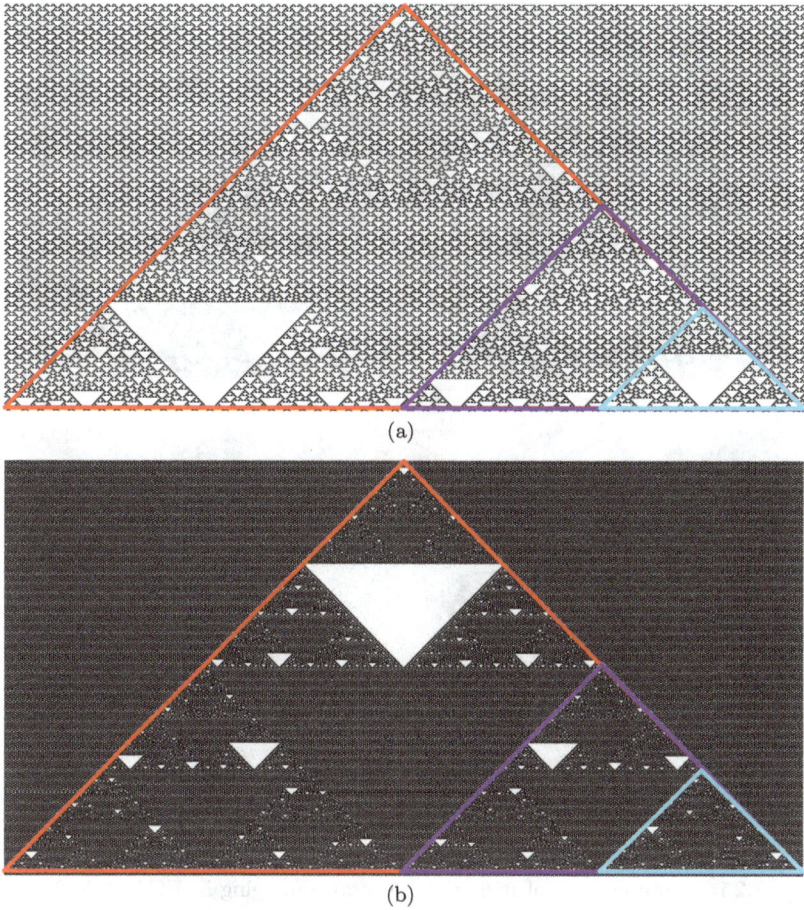

Fig. 3.24. Composition of non-trivial fractals emerging in ECA Rule 22 after thousands of generations. The iterated function preserves its fractal dimension. These fractals evolve on a periodic background without displacement.

3.7.2 Rule 18, mutations, gaskets, and seashells

Rules 18 and 22 are complex rules widely reported in the literature. Let us redefine them as follows: a cell takes State '1' if exactly one

- (R18): of its neighbours is in State '1': $(100, 001) \rightarrow 1$
- (R22): in its neighbourhood is in State '1': $(100, 010, 001) \rightarrow 1$

(a)

(b)

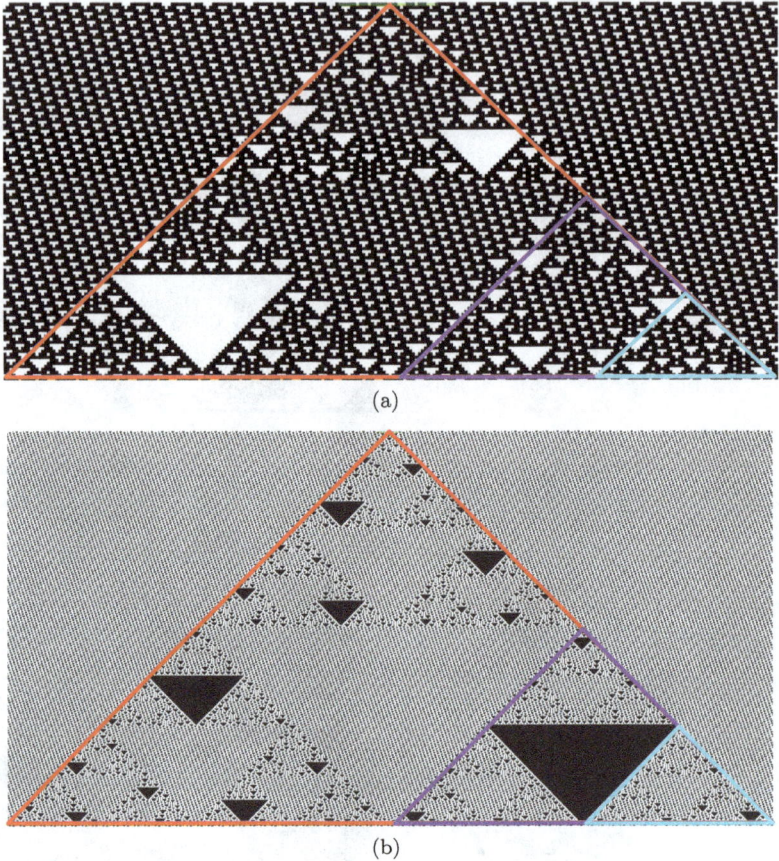

Fig. 3.25. Composition of non-trivial fractals emerging in ECA Rule 22 after thousands of generations. The iterated function preserves its fractal dimension. These fractals evolve on a periodic background without displacement.

and consider the subset of the 256 ECA

$$\Psi_{\widehat{R18}} = \begin{cases} 1 & \text{if } 100,001 \\ 0 & \text{if } 111,000 \end{cases} \tag{3.12}$$

that defines the 16 rules displayed in Table 3.7. Referring to the genotype paradigm in with a rule defined by the sequence $(b_7 b_6 ... b_1 b_0)$, a rule R *mutates* into Rule R' through bit (or '*gene*') b_i, or $(R|b_i \rightsquigarrow R')$ with exactly a 1-bit change, that yields the mutation (or inheritance) tree in Fig. 3.26.

Table 3.7. Mutation table from Rule 18 in the ECA subset $(b_7 b_6 b_5 b_4 b_3 b_2 b_1 b_0) = (0\ b_6\ b_5\ 1\ b_3\ b_2\ 1\ 0)$. A rule R *mutates* into Rule R' through bit b_i $(R|b_i \rightsquigarrow R')$ with exactly a 1-bit change.

Rule	111	110	101	100	011	010	001	000	Mutation
18	0	0	0	1	0	0	1	0	
22	0	0	0	1	0	1	1	0	$18\|b_2 \rightsquigarrow 22$
26	0	0	0	1	1	0	1	0	$18\|b_3 \rightsquigarrow 26$
30	0	0	0	1	1	1	1	0	$22\|b_3 \rightsquigarrow 30$
50	0	0	1	1	0	0	1	0	$18\|b_5 \rightsquigarrow 50$
54	0	0	1	1	0	1	1	0	$22\|b_5 \rightsquigarrow 54$
58	0	0	1	1	1	0	1	0	$26\|b_5 \rightsquigarrow 58$
62	0	0	1	1	1	1	1	0	$30\|b_5 \rightsquigarrow 62$
82	0	1	0	1	0	0	1	0	$18\|b_6 \rightsquigarrow 82$
86	0	1	0	1	0	1	1	0	$22\|b_6 \rightsquigarrow 86$
90	0	1	0	1	1	0	1	0	$26\|b_6 \rightsquigarrow 90$
94	0	1	0	1	1	1	1	0	$30\|b_6 \rightsquigarrow 94$
114	0	1	1	1	0	0	1	0	$50\|b_6 \rightsquigarrow 114$
118	0	1	1	1	0	1	1	0	$54\|b_6 \rightsquigarrow 118$
122	0	1	1	1	1	0	1	0	$58\|b_6 \rightsquigarrow 122$
126	0	1	1	1	1	1	1	0	$62\|b_6 \rightsquigarrow 126$

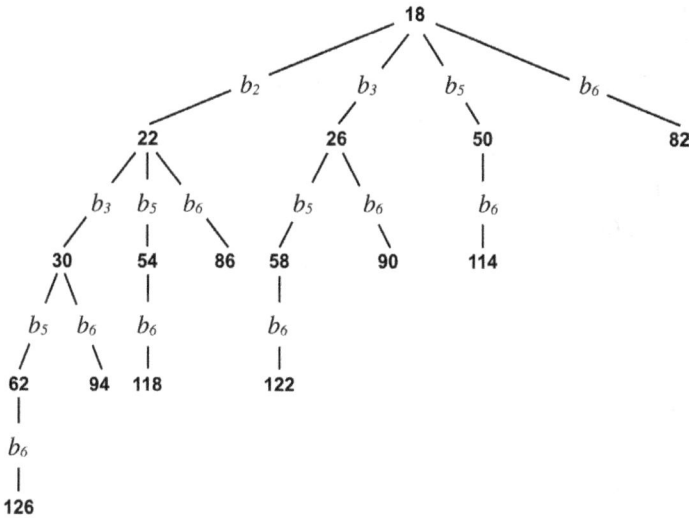

Fig. 3.26. Mutation tree from Rule 18 derived from Table 3.7. A branch from Rule R to Rule R' $(R|b_i \rightsquigarrow R')$ bears exactly the 1-bit genetic change b_i and $d_H(R, R') = 1$, where d_H is the Hamming distance.

This tree contains a set of rules with complex behaviour inherited from Rule 18 and where Sierpiński patterns often appear. Rules 18, 22, 122, and 126 reveal a similar behaviour towards ergodic regime and unstable areas, as does Rule 90, but that displays a multilayered pattern, whereas Rule 30 and, by reflection, Rule 86 exhibit small striped patterns near the polygonal border. Rule 26 and, by reflection, Rule 82 exhibit large striped patterns at equilibrium, that may evolve towards sparse backbones at low or high densities, or periodic patterns similar to those in Fig. 3.7. Rule 94 displays a Sierpiński gasket, but evanescent and quickly entering a uniform polygon with vertical stripes. Finally, it should be emphasised that Rules 50, 54, 58, and by reflection 114, 62, and by further reflection 118, display patterns somewhat far from the sieve, nevertheless, all of them enter the phase transition polygon.

This tree does not contain the subset of rules associated either by conjugation or by conjugation-reflection. Beginning from Rule 129 — conjugate with Rule 126 — this subset would display a large collection of Sierpiński-like patterns, but this time in white on a black background.

Even in the synchronous "1nCA" [46] with a minimal neighbourhood of two cells — the cell itself and either the left or the right adjacent cell alternating at either odd or even time steps — Sierpiński patterns appear in Rules 6Δ and 9Δ, namely, for the only symmetric rules with an equal number of black and white cells, and for the only rules fulfilling the maximal "sensitivity parameter".

Six rules (18, 22, 30, 90, 122, 126) evolving in their ergodic regime are displayed in Fig. 3.27. All patterns have the Hausdorff dimension $\log_2(3)$ of Eq. (3.11). They only differ from their average density $d_\mathcal{C}$ of their mesoscopic minimal macrocell $\mathcal{C}$.

Referring again to the mean field curves in Section 3.2 now displayed for Rules 18 and 22 in Fig. 3.28, we observe that the Rule 18 curve reaches its stable fixed point $p_{t+1} = p_t$ when crossing

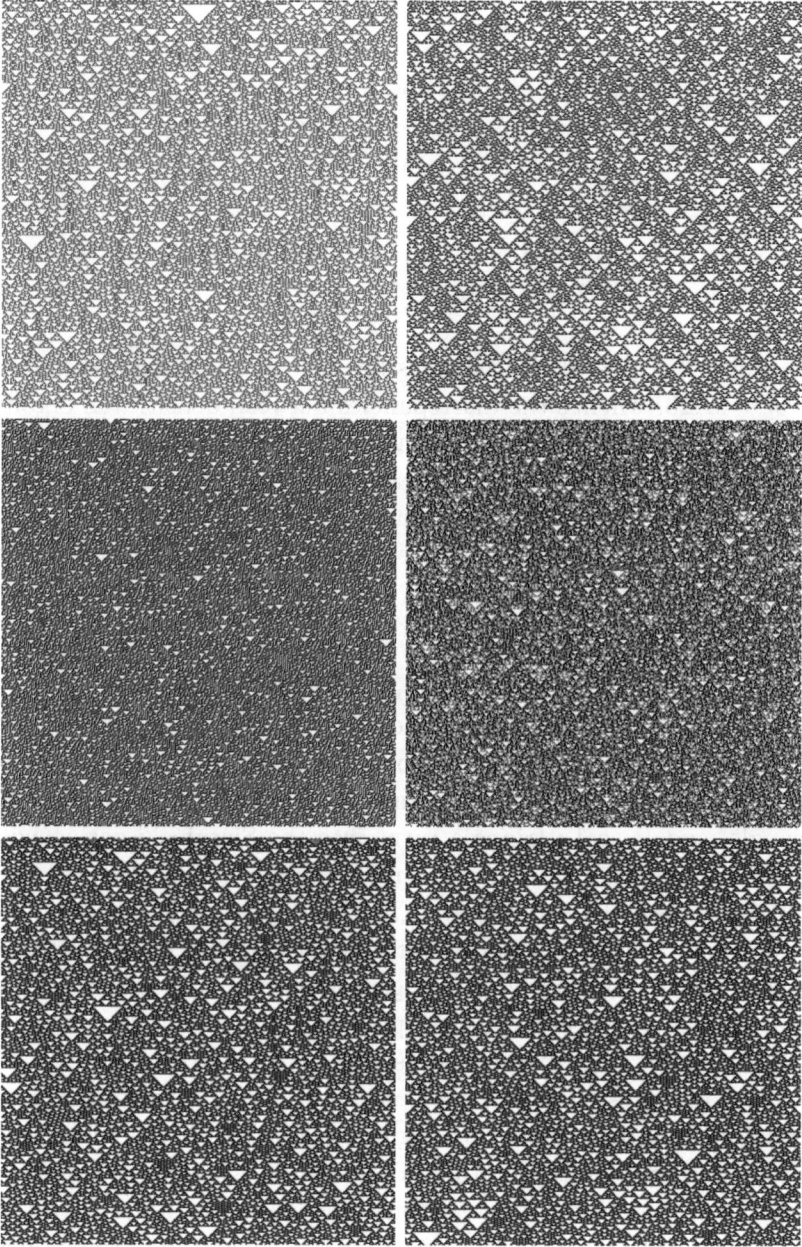

Fig. 3.27. Evolutions in a ring of 400 cells, 400 generations from initial density $d_0 = 0.50$. Left to right, top to bottom: Rules 18, 22, 30, 90, 122, 126 in ergodic regime. All patterns have the Hausdorff dimension $\log_2(3)$ of Eq. (3.11).

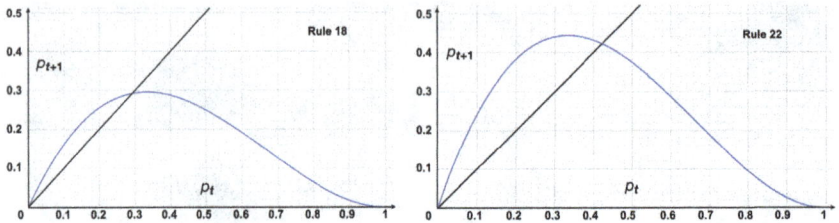

Fig. 3.28. Rule 18 ($p_{t+1} = 2p_t q_t^2$) and Rule 22 ($p_{t+1} = 3p_t q_t^2$) with their mean field curves.

the identity at $p_t \approx 0.293$, whereas for Rule 22 the fixed point is got at $p_t \approx 0.423$, whence the discrepancies between densities in ergodic regime, observable in Fig. 3.27, emerge. Moreover, the Rule 18 curve shows a slope $f'_{R18}(0) = 2$, whereas the Rule 22 one shows a slope $f'_{R22}(0) = 3$ at the origin. That comes from the fact that these rules induce from a single source their following evolution:

- (R18): $0^*10^* \rightarrow 0^*(101)0^*$
- (R22): $0^*10^* \rightarrow 0^*(111)0^*$

and that their density ratio at time steps $2^p - 1$ ($p > 0$) remains 2/3.

The Sierpiński gasket [236] appears in a wide variety of situations [47]. The binomial coefficients can be arranged to form Pascal's triangle, and Pascal's triangle turns into the Sierpiński gasket with coefficients modulo two. It may turn into something like a natural tree by some diffeomorphism. This tree is embeddable into the $2d$ diffusion graphs embedded into the triangulate lattice [78]: its vertex dust forms the Sierpiński gasket patterns. Sierpiński gasket is often known as a Banach fixed point from some contractive affine transformation into three elements.

But the most fascinating is the formation of patterns from random initial distributions of pigmentations on certain varieties of natural pigmentations in nature. This phenomenon can be explained from the Gierer–Meinhardt reaction–diffusion model of the activator–inhibitor type, arising in various situations of pattern formation in morphogenesis [96].

3.8 Memory leads to complexity

In this section, we show that ECA Rule 22 with memory is *strongly chaotic* [163].

Conventional CA are ahistoric (memoryless): i.e. the new state of a cell depends on the neighbourhood configuration solely at the preceding time step of φ. Thus, CA with *memory* (CAM) can be considered as an extension of the standard framework of CA where every cell x_i is allowed to remember some period of its previous evolution. A memory is based on the state and history of the system, thus, we design a memory function ϕ, as follows: $\phi(x_i^{t-\tau}, \ldots, x_i^{t-1}, x_i^t) \to s_i$, such that $\tau < t$ determines the backwards degree of memory and each cell $s_i \in \Sigma$ is a function of the series of states in cell x_i up to time step $t - \tau$. To execute the evolution, we apply the original rule as follows: $\varphi(\ldots, s_{i-1}^t, s_i^t, s_{i+1}^t, \ldots) \to x_i^{t+1}$.

In CAM, while the mapping φ remains unaltered, a historic memory of past iterations is retained by featuring each cell as a summary of its previous states; therefore, cells *canalise* memory to the map φ. As an example, we can take the memory function ϕ as a *majority memory*: $\phi_{\text{maj}} \to s_i$, where in case of a tie given by $\Sigma_1 = \Sigma_0$ in ϕ, we shall take the last value x_i. So ϕ_{maj} represents the classic majority function for three variables on cells $(x_i^{t-\tau}, \ldots, x_i^{t-1}, x_i^t)$ and defines a temporal ring before calculating the next global configuration c. In case of a tie, it is allowed to break it in favour of zero if $x_{\tau-1} = 0$, or to one if $x_{\tau-1} = 1$.

The representation of an ECAM is given as follows:

$$\phi_{\text{CARm}:\tau}, \qquad (3.13)$$

where CAR represents the decimal notation of a particular ECA rule and m, the kind of memory given with a specific value of τ.

Note that memory is as simple as any CA, and that the global behaviour produced by the local rule is rather unpredictable, it can lead to emergent properties and so can be classed as complex. Memory functions were developed and extensively studied by Alonso-Sanz in Ref. [35]. Memory in ECA have been studied, showing

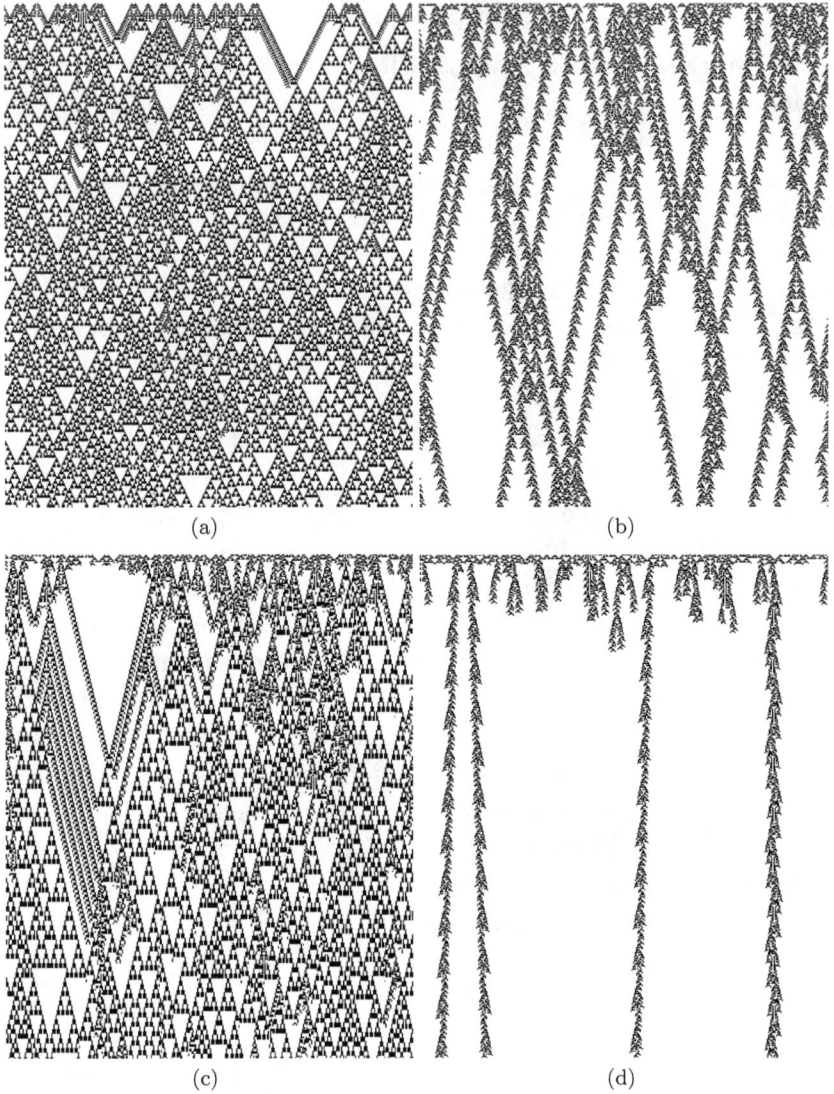

Fig. 3.29. ECA Rule 22 with a memory function reveals complex behaviour. (a) Evolution of the function $\phi_{R22maj:3}$. (b) The function $\phi_{R22maj:4}$ (recently proven to be logically universal by simulating the Fredkin gate. (c) The function $\phi_{R22maj:7}$. (d) The function $\phi_{R22maj:8}$ (particles with long period).

its potential to produce complex behaviour from chaotic systems and beyond in Refs. [59] and [167]. Thus, we can conjecture that a memory function can produce complex behaviour [163] as follows:

$$\phi(\varphi_{\text{chaos}}) \rightarrow \text{complex.} \tag{3.14}$$

Eppstein [314] demonstrates that a CA class IV is a system where mobile self-localizations emerge. We can relate type of classes with memory functions in ECA Rule 22 as:

$$\phi_{R22maj:3} \Rightarrow chaos \rightarrow chaos \tag{3.15}$$

$$\phi_{R22maj:4} \Rightarrow chaos \rightarrow complexity \tag{3.16}$$

$$\phi_{R22maj:7} \Rightarrow chaos \rightarrow complexity \tag{3.17}$$

$$\phi_{R22maj:8} \Rightarrow chaos \rightarrow complexity \tag{3.18}$$

3.8.1 Rareness and unpredictability

The state transition function $\varphi(x_{i-r}^t, \ldots, x_i^t, \ldots, x_{i+r}^t) \rightarrow x_i^{t+1}$ can be re-written as Boolean formula with two XOR operations: $x_{i+r}^t = x_{i-r}^t \oplus x_i^t \oplus x_{i+r}^t$. The XOR gate is the most rare, most hard to find in natural non-linear systems, Boolean gate. Let gates g_1 and g_2 discovered with occurrence frequencies $f(g_1)$ and $f(g_2)$, we say a gate g_1 is easier to develop or evolve than a gate g_2: $g_1 \triangleright g_2$ if $f(g_1) > f(g_2)$. The hierarchies of gates obtained using evolutionary techniques in liquid crystals [118], light-sensitive modification of Belousov–Zhabotinsky system [252], slime mould *Physarum polycephalum* [116] and protein molecules [16, 315]:

- Gates in liquid crystals:
$$\{OR, NOR\} \triangleright AND \triangleright NOT \triangleright NAND \triangleright XOR$$
- Gates in Belousov–Zhabotinsky medium: AND $\triangleright$ NAND $\triangleright$ XOR
- Gates in cellular automata [316]:
$$OR \triangleright NOR \triangleright AND \triangleright NAND \triangleright XOR$$
- Gates in Physarum: AND $\triangleright$ OR $\triangleright$ NAND $\triangleright$ NOR $\triangleright$ XOR $\triangleright$ XNOR
- Gates in protein molecules verotoxin and actin:
$$AND \triangleright OR \triangleright AND\text{-}NOT \triangleright XOR$$

The XOR gate is hard to find and space-time dynamics of Rule 22 automata is hard to predict. A strong link computational difficulty of a problem and its randomness was established by Yao [317]. His famous lemma, rephrased by Impagliazzo and Wigderson [318], can be seen in the framework of predictability of Rule 22 ECA behavior:

> Fix a non-uniform model of computation (with certain closure properties) and a Boolean function $f\colon \{0,1\}^n \to \{0,1\}$. Assume that any algorithm in the model of a certain complexity has a significant probability of failure when predicting f on a randomly chosen instance x. Then any algorithm (of a slightly smaller complexity) that tries to guess the XOR $f(x_1) \oplus f(x_2) \oplus \ldots \oplus f(x_k)$ of k random instances $x_1, \ldots, x_k$ won't do significantly better than a random coin toss.

Potential associations between dynamics in Rule 22 ECA and a role of XOR functions in communication complexity [319, 320] could be explored in future.

Chapter 4

Actin Automata with Memory

Ramón Alonso-Sanz*,‡ and Andy Adamatzky†,§

*Technical University of Madrid, Spain
†Unconventional Computing Lab, UWE Bristol, UK
‡ramon.alonso@upm.es
§andrew.adamatzky@uwe.ac.uk

Abstract
We propose a model that represents actin filaments as a double chain
of finite state machines, referred to as nodes. These nodes can be in
states '0' or '1', which abstractly represent the absence or presence of a
sub-threshold charge on the actin units corresponding to the nodes. The
state of each node is updated in discrete time, with all nodes updating
their states in parallel. Unlike previous models of actin automata that
only considered momentary state transitions of nodes, we enhance the
model by introducing the concept of memory. We assume that the states
of nodes not only depend on the current states of their neighbouring
nodes but also on their past states. This addition allows us to investigate
the impact of memory on the dynamics of actin automata. Through
computational experiments, we demonstrate effects of memory on the
behaviour of actin automata. First, we find that memory slows down
the propagation of perturbations within the system. This means that
perturbations take longer to spread throughout the filament due to the
influence of past states. Second, the introduction of memory decreases
the entropy of the space–time patterns generated by the automata.
Entropy, in this context, refers to the amount of unpredictability or
disorder in the patterns. The inclusion of memory leads to more
ordered and structured patterns. Furthermore, we observe that memory
transforms travelling localisations, which are moving patterns of activity,
into stationary oscillators. This means that the patterns oscillate in
a fixed position rather than propagating through the filament. Lastly,
memory can also transform stationary oscillations into static or still
patterns, where there is no movement or oscillation observed. Overall,
our computational experiments highlight the significant influence of

memory on the dynamics of actin automata. The introduction of memory alters the speed of propagation, decreases entropy, transforms travelling localisations into stationary oscillators, and stationary oscillations into static patterns. These findings deepen our understanding of the complex behaviour exhibited by actin filaments and provide insights into the role of memory in cellular processes.

4.1 Automata with memory

Conventional cellular automata (CA) are memoryless: i.e. the new state of a cell depends on the neighbourhood configuration solely at the preceding time step. The standard framework of CA can be extended by implementing memory capabilities in cells [32–34] as follows:

$$\sigma_i^{(T+1)} = \phi(\{s_j^{(T)}\} \in \mathcal{N}_i),$$

with $s_j^{(T)}$ being a state function of the series of states of the cell j up to T:

$$s_j^{(T)} = s\left(\sigma_j^{(1)}, \ldots, \sigma_j^{(T-1)}, \sigma_j^{(T)}\right). \tag{4.1}$$

Thus, in CA with memory, while the mappings ϕ remain unaltered, historic memory of all past iterations is retained by featuring each cell as a summary of its past states. So to say, cells *canalise* memory to the map ϕ.

Memory (4.1) may be implemented as majority memory, i.e. the most frequent (mode) state:

$$s_i^{(T)} = \text{mode}(\sigma_i^{(1)}, \ldots, \sigma_i^{(T)}) \tag{4.2}$$

with $s_i^{(T)} = \sigma_i^{(T)}$ in case of a tie: card$\{1\}$ = card$\{0\}$.

At variance with the unlimited memory implementation in Ref. (4.2), the length of the trailing memory may be limited to the last τ time steps,

$$s_i^{(T)} = \text{mode}(\sigma_i^{(T-\tau+1)}, \ldots, \sigma_i^{(T-2)}, \sigma_i^{(T-1)}, \sigma_i^{(T)}). \tag{4.3}$$

The shortest operative trailing length is that of $\tau = 3$:

$$s_i^{(T)} = \text{mode}(\sigma_i^{(T-2)}, \sigma_i^{(T-1)}, \sigma_i^{(T)}). \tag{4.4}$$

with initial assignations: $s_i^{(1)} = \sigma_i^{(1)}$, $s_i^{(2)} = \sigma_i^{(2)}$.

Keeping track of the states of the last τ time steps demands τ extra bits per cell to store their corresponding state values. To avoid this drawback, past state values can be weighted in such a way that only the accumulated memory charge needs to be stored. Thus, for example, historic memory can be weighted by applying a geometric discounting process in which the state $\sigma_i^{(T-\tau)}$, obtained τ time steps before the last round, is actualised to $\alpha^\tau \sigma_i^{(T-\tau)}$, α being the *memory factor* lying in the [0,1] interval. This well-known mechanism fully takes into account the last round ($\alpha^0 = 1$), and tends to *forget* the older rounds. Thus,

$$\omega_i^{(T)}\left(\sigma_i^{(1)}, \ldots, \sigma_i^{(T)}\right) = \sigma_i^{(T)} + \sum_{\tau=0}^{T-1} \alpha^\tau \sigma_i^{(T-\tau)} = \sigma_i^{(T)} + \alpha\omega_i^{(T-1)}.$$

(4.5)

Consequently, only one number per cell (ω_i) needs to be stored. This positive property is accompanied by the drawback that it computes with real numbers, which is not in the realm of CA, that claims for integer arithmetic.

Every cell will be featured first by the weighted mean (m) of all its past states, so the memory charge at time step T is

$$m_i^{(T)} = \frac{\omega_i^{(T)}}{\Omega(T)}, \quad \text{with } \Omega(T) = 1 + \sum_{t=1}^{T-1} \alpha^{T-t}.$$

(4.6)

The trait state s is obtained by comparing m to the landmark 0.5 (if $\sigma \in \{0,1\}$), assigning the last state in case of an equality to this value, so that

$$s_i^{(T)} = \mathcal{H}(m_i^{(T)}) = \begin{cases} 1 & \text{if } m_i^{(T)} > 0.5 \quad 2\omega_i^{(T)} > \Omega(T) \\ \sigma_i^{(T)} & \text{if } m_i^{(T)} = 0.5 \equiv 2\omega_i^{(T)} = \Omega(T) \\ 0 & \text{if } m_i^{(T)} < 0.5 \quad 2\omega_i^{(T)} < \Omega(T). \end{cases}$$

(4.7)

The choice of the memory factor α tunes the memory effect: the limit case $\alpha = 1$ corresponds to a memory with equally weighted records (*full* memory, equivalent to unlimited trailing *majority* memory), whereas $\alpha \ll 1$ intensifies the contribution of the most

recent states and diminishes the contribution of the more remote
states (short-term memory). The choice $\alpha = 0$ leads to the ahistoric
model. Due to the rounding (4.7), α-memory is not effective if
$\alpha \leq 0.5$.

4.2 Actin automata with memory

Each globular actin (G-actin) molecule (except those at the ends of
F-actin strands) has four neighbours, as demonstrated in Fig. 4.1.
An actin automaton consists of two chains σ and $[\sigma]$ of semi-
totalistic binary-state automata. Each automaton takes two states,
'0' (resting) and '1' (excited).

Conventional (Markovian) actin CA rules have been proposed to
be implemented in two coupled layers (noted σ and $[\sigma]$) with semi-
totalistic rules [6]:

$$
\sigma_i^{(T+1)} = \begin{cases} \phi(\sigma_{i-1}^{(T)} + \sigma_{i+1}^{(T)} + [\sigma_i^{(T)}] + [\sigma_{i-1}^{(T)}]) & \text{if } \sigma_i^{(T)} = 0 \\ \psi(\sigma_{i-1}^{(T)} + \sigma_{i+1}^{(T)} + [\sigma_i^{(T)}] + [\sigma_{i-1}^{(T)}]) & \text{if } \sigma_i^{(T)} = 1 \end{cases} \tag{4.8a}
$$

$$
[\sigma_i^{(T+1)}] = \begin{cases} \phi([\sigma_{i-1}^{(T)}] + [\sigma_{i+1}^{(T)}] + \sigma_i^{(T)} + \sigma_{i+1}^{(T)}) & \text{if } [\sigma_i^{(T)}] = 0 \\ \psi([\sigma_{i-1}^{(T)}] + [\sigma_{i+1}^{(T)}] + \sigma_i^{(T)} + \sigma_{i+1}^{(T)}) & \text{if } [\sigma_i^{(T)}] = 1. \end{cases} \tag{4.8b}
$$

With the subrules (ϕ, ψ) expressed in binary and decimal forms
as $\phi = (\beta_0\beta_1\beta_2\beta_3\beta_4) \equiv \sum_{i=0}^{4} 2^{4-i}\beta_i$, and $\psi = (\gamma_0\gamma_1\gamma_2\gamma_3\gamma_4) \equiv \sum_{i=0}^{4} 2^{4-i}\gamma_i$.

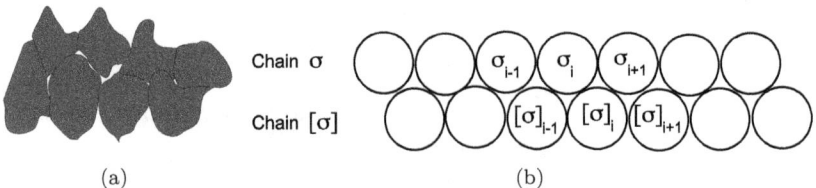

(a)   (b)

Fig. 4.1. Schematic diagram of F-actin strands. (a) Structure of actin detected
by X-ray fibre diffraction. Adapted from Ref. [187]. (b) Actin automata.

Actin CA rules with memory will be implemented from (4.8) as follows:

$$\sigma_i^{(T+1)} = \begin{cases} \phi(s_{i-1}^{(T)} + s_{i+1}^{(T)} + [s_i^{(T)}] + [s_{i-1}^{(T)}]) & \text{if } s_i^{(T)} = 0 \\ \psi(s_{i-1}^{(T)} + s_{i+1}^{(T)} + [s_i^{(T)}] + [s_{i-1}^{(T)}]) & \text{if } s_i^{(T)} = 1 \end{cases} \quad (4.9a)$$

$$[\sigma_i^{(T+1)}] = \begin{cases} \phi([s_{i-1}^{(T)}] + [s_{i+1}^{(T)}] + s_i^{(T)} + s_{i+1}^{(T)}) & \text{if } [s_i^{(T)}] = 0 \\ \psi([s_{i-1}^{(T)}] + [s_{i+1}^{(T)}] + s_i^{(T)} + s_{i+1}^{(T)}) & \text{if } [s_i^{(T)}] = 1. \end{cases} \quad (4.9b)$$

Figure 4.2 shows the effect of memory up to $T = 150$ in two actin rules when starting at random, i.e. the values of sites are initially uncorrelated and chosen at random to be 0 (*blank*) or 1 (*black*) with probability $1/2$ in layers of size $n = 300$. The pictures show also the differences in patterns resulting from reversing the initial centre site value. The *perturbed* region is enhanced with *red* pixels, corresponding

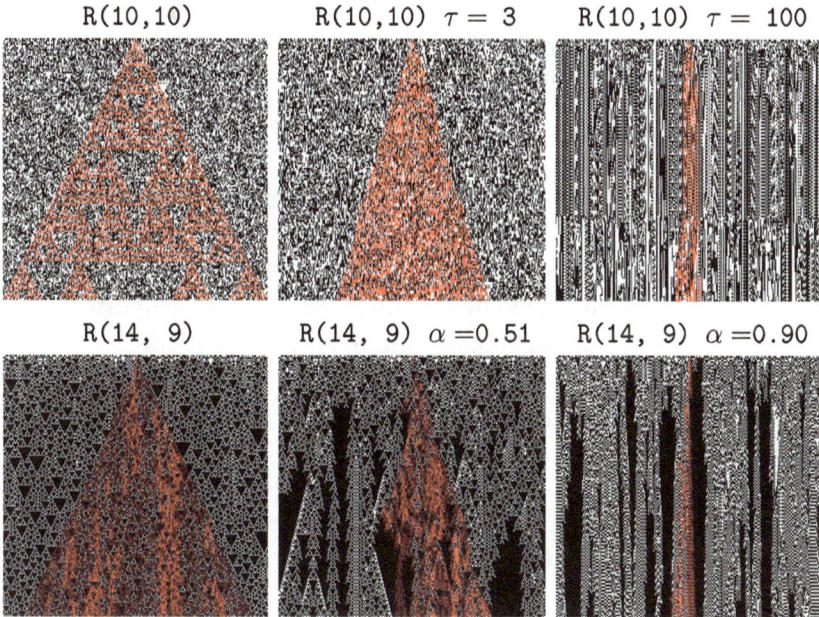

Fig. 4.2. Dynamics in two actin rules up to $T = 150$. Top: rule R(10, 10). Ahistoric (left) and majority memory models. Bottom: rule R(14, 9). Ahistoric (left) and α-memory models.

to the site values that differed among the patterns generated with the
two initial configurations. In the top frame of Fig. 4.2, the actin rule
R(10,10)[1] is shown with ahistoric (left), $\tau = 3$, and $\tau = 100$-majority
memory models. In the bottom frame, the rule R(14, 9) is considered,
with ahistoric (left), $\alpha = 0.51$, and $\alpha = 0.9$ memory models. The
inertial (or conservative) effect that memory exerts tends to slow
down propagation of growing patterns (perturbations, excitations).
This 'slowing down' is manifested by apparent 'shrinking of patterns'
on the space–time configurations. This is so even with low memory
charge, either in the form $\tau = 3$ or as $\alpha = 0.51$, but becomes
fully appreciable with high memory (either $\tau = 100$, or $\alpha = 0.90$).
In correspondence, the perturbation spreading in Fig. 4.2 becomes
highly restrained as the memory charge increases, so that with high
memory charge it remains confined to the proximity of its initial seed.

Figure 4.3 shows the effect of memory up to $T = 1000$ in the
actin rule R(14, 9) when starting at random in layers of size 1500.
In the α memory type (top snapshots), even the very low memory
charge $\alpha = 0.51$ dramatically alters the conventional (ahistoric)
spatio-temporal patterns, with stationary localisations becoming
dominating, whereas when α increases to 0.6, a sophisticated spatio-
temporal pattern emerges where travelling glider guns produce
gliders, and gliders reflect or annihilate in their collisions. With
majority memory (bottom snapshots), the low memory charge $\tau =$
induces a *monotonous* transition between from enormous amount of
gliders generated (so many that space is almost completely filled with
them), to a situation when few stationary glider guns generate gliders
with low frequency, gliders collide and in most cases annihilate. When
τ reaches 8, only stationary (breathing) localisations persist.

The richness of the effect of memory may be envisaged in Fig. 4.4,
where rule R(14, 9) starts from one site active seed. Sophisticated
patterns emerge both in the α and majority scenarios shown in
Fig. 4.4.

---

[1]R(10, 10) is usually referred to as the *parity* rule, where both subrules turn out
to be the sum of their inputs modulo two. The simple in form *parity* rule has
proven to be highly chaotic in CA scenarios.

R(14, 9)   R(14, 9) α =0.51   R(14, 9) α =0.60

R(14, 9)   R(14, 9) $\tau = 4$ R(14, 9) $\tau = 8$

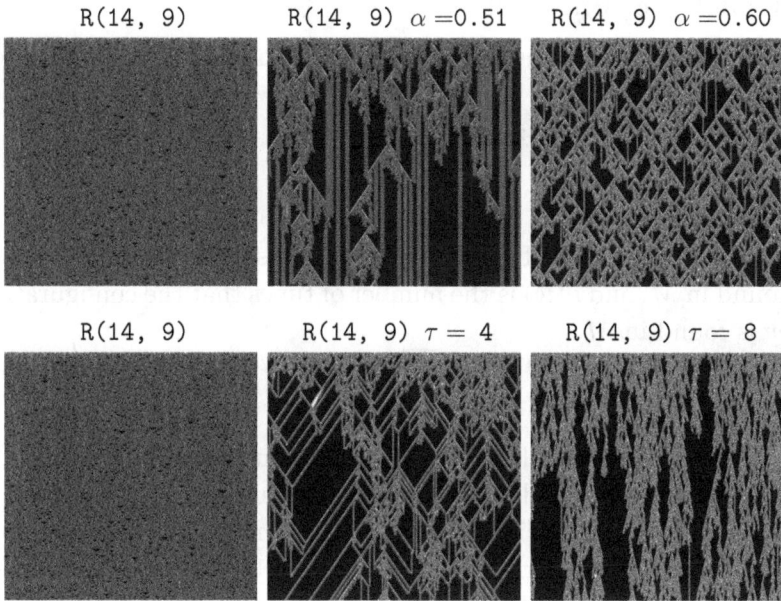

Fig. 4.3. Dynamics in the rule R(14, 9). Top: α-memory, Bottom: majority memory.

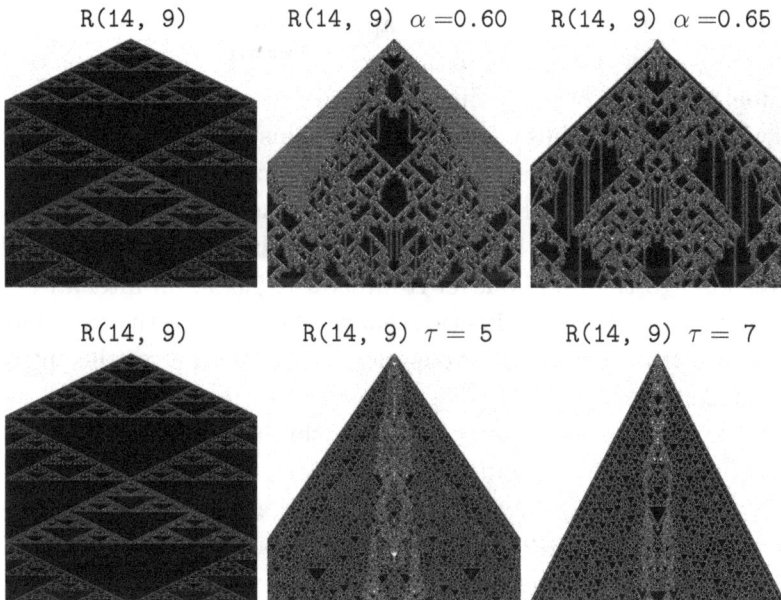

R(14, 9)   R(14, 9) α =0.60   R(14, 9) α =0.65

R(14, 9)   R(14, 9) $\tau = 5$ R(14, 9) $\tau = 7$

Fig. 4.4. Dynamics in the rule R(14, 9) starting from one site active seed. Top: α-memory, Bottom: majority memory.

4.3 Global characteristics

Following Ref. [6], the subsequent integral measures are calculated on the spatio-temporal pattern $M = (m_{ti})$, where m_{ti} is the state of the cell i at time step t:

- Shannon entropy: $H = -\sum_{w \in W} \nu(w)/\eta \ln(\nu(w)/\eta)$,
  where η is the sum of the W possible kinds of 3×3 configurations found in M, and $\nu(w)$ is the number of times that the configuration w is found in M.
- Simpson diversity: $D = 1 - \sum_{w \in W} (\nu(w)/\eta)^2$.

Figure 4.5 shows the entropy (H) versus diversity (D) in the left frames, and the distribution of the number of rules $(\#)$ of given entropy. Ahistoric (black-marked) and τ-majority memory (red-marked) models in runs up to $T = 1000$ are considered in Fig. 4.5: Top: $\tau = 3$, Middle: $\tau = 100$, Bottom: $\tau = 1000$. Memory seems not to dramatically alter the form of the H–D plots, at least at a first glance at the left panels of Fig. 4.5, where the data from the ahistoric and memory simulations appear rather masked. At variance with this, memory appears to notably modify the distribution of the number of rules corresponding to the different levels of entropy in the right panels. Thus, comparing the ahistoric (or even the low level $\tau = 3$ memory charge) to the full memory implementation in the bottom-right panel, become for rules not implementing space filling in full memory implementation, whereas most of the rules correspond to the middle level H-interval [2.5,3.5], an interval fairly low represented in the ahistoric scenario. Let us take the example of rule R(10, 10). In the ahistoric model, R(10,10) achieves the entropy $H = 6.233$, very close to the maximum attainable $H^* = \ln 2^9 = 6.238$. With $\tau = 3$ majority memory, the entropy of R(10, 10) is lowered to $H = 6.005$, and with full memory, to $H = 5.335$. In the scenarios of Fig. 4.5, the actin rule R(14, 9) achieves in the ahistoric model the entropy 3.342, which atypically increases, albeit very little, to 3.835 with $\tau = 3$, then decreases to 3.199 with full memory.

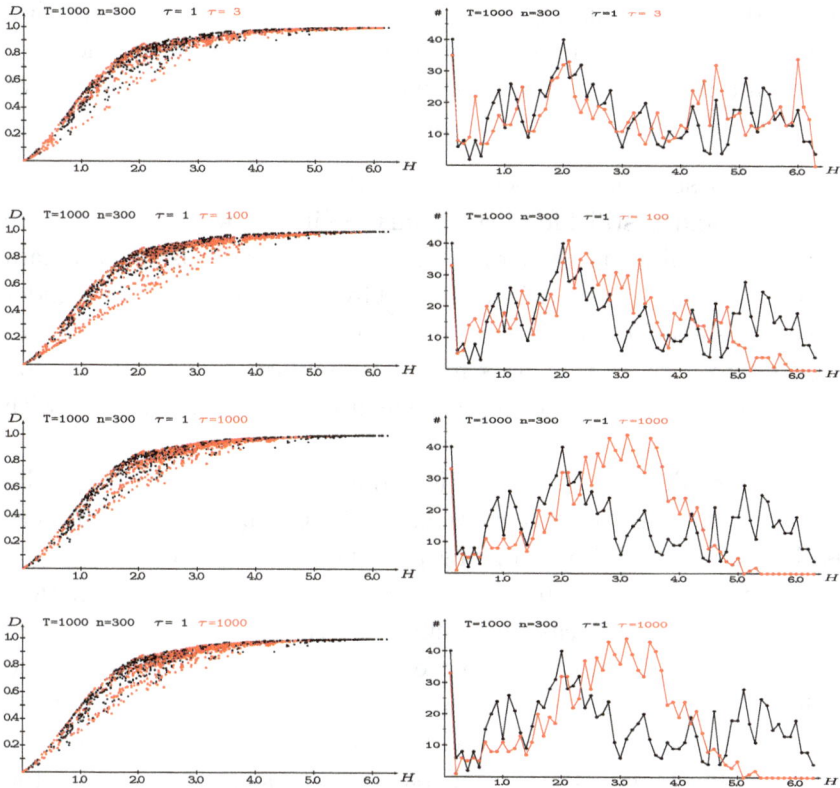

Fig. 4.5. Entropy (H) vs. diversity (D) (left) and entropy (H) vs. number of rules (#) (right) plots. Ahistoric (black) and τ-majority (red) memory models. Top: $\tau = 3$; Middle: $\tau = 100$; Bottom: $\tau = 1000$.

4.4 Localisations

We consider the following three types of localisations:

- **Glider:** travelling localisation, analogues to voltage solitons in cable equation model of actin [254], and discrete travelling breathers [90].
- **Still life:** still stationary patterns, or localised excitation in vibrating granular material [257].

- **Oscillator:** oscillating stationary localisation, a pattern that repeats itself in a finite number of evolution steps; a glider also repeats itself but it is not stationary), stationary breather [90] or immobile voltage soliton [254].

Glider and still life are oscillators as well: a glider is a translating oscillator and a still life is stationary oscillator with period 1. We compare localisations generated by seeds of five cells in automata without memory and with memory. Given a rule $\mathcal{R}$, we say 'glider becomes an oscillator' if an automaton governed by rule $\mathcal{R}$ evolves seed s into a glider that does not have memory and the automaton evolves the seed s into a breather when the rule $\mathcal{R}$ is enriched with memory.

As we discussed before, introduction of memory typically 'shrinks' patterns. In some rules, e.g. R(6, 20), the spreading pattern is shrunk to a still life occupying just one cell (Fig. 4.6).

Oscillating stationary localisation is preserved (Fig. 4.7(a)), sometimes with decrease of oscillation frequency (Fig. 4.7(b)) or change of the oscillations pattern (Fig. 4.7(c)) or annihilated (Fig. 4.7(d)). In some rules, oscillators can emerge, when state-transition rules are enriched with memory, e.g. in (Fig. 4.7(d)), a seed produces an extinguishing pattern in memoryless automaton yet it produces an oscillator when majority memory is introduced. In rule $R(6, 16)$, introduction of memory converts the oscillator into a still life.

Gliders are transformed as follows (Fig. 4.8). A glider can be transformed to an oscillator (Fig. 4.8(a)), or expanding pattern (we

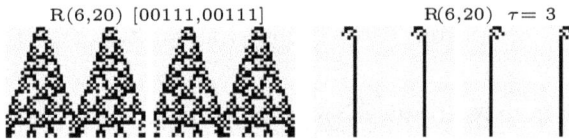

R(6,20) [00111,00111]   R(6,20) $\tau = 3$

Fig. 4.6. Spreading pattern is localised into a still life. Transformation of localisations by majority memory. In each subfigure, we see space–time configurations of both actin chains in automata without memory (left) and with memory (right). Time goes down. Rules are indicated above the space–time configurations. Seeds are indicated in the label of the left (ahistoric) simulation.

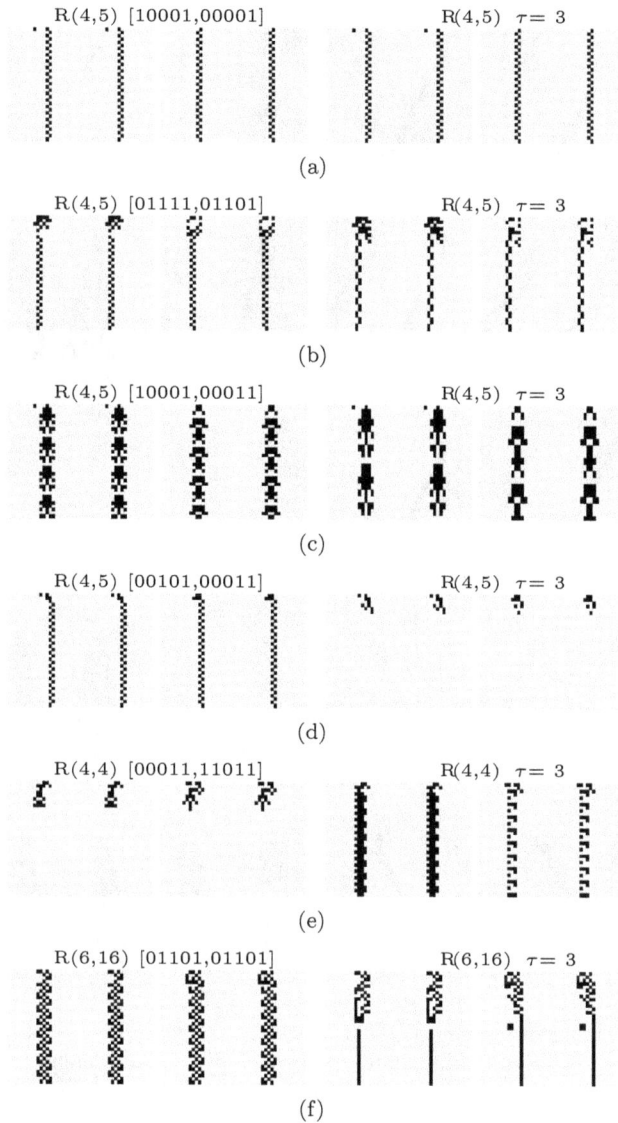

R(4,5) [10001,00001]   R(4,5) τ= 3

(a)

R(4,5) [01111,01101]   R(4,5) τ= 3

(b)

R(4,5) [10001,00011]   R(4,5) τ= 3

(c)

R(4,5) [00101,00011]   R(4,5) τ= 3

(d)

R(4,4) [00011,11011]   R(4,4) τ= 3

(e)

R(6,16) [01101,01101]   R(6,16) τ= 3

(f)

Fig. 4.7. Transformation of oscillators by $\tau = 3$ majority memory. In each subfigure, we see space–time configurations of both actin chains in automata without memory (left) and with memory (right). Time goes down. Rules are indicated above the space–time configurations. (a) Oscillator is preserved. (b) Oscillator remains yet frequency of its oscillation (breathing) decreases. (c) Pattern of oscillation changes. (d) Oscillator is annihilated by memory. (e) Oscillator emerges assisted by memory. (f) Oscillator is transformed into a still life. Seeds are indicated in the label of the left (ahistoric) simulation.

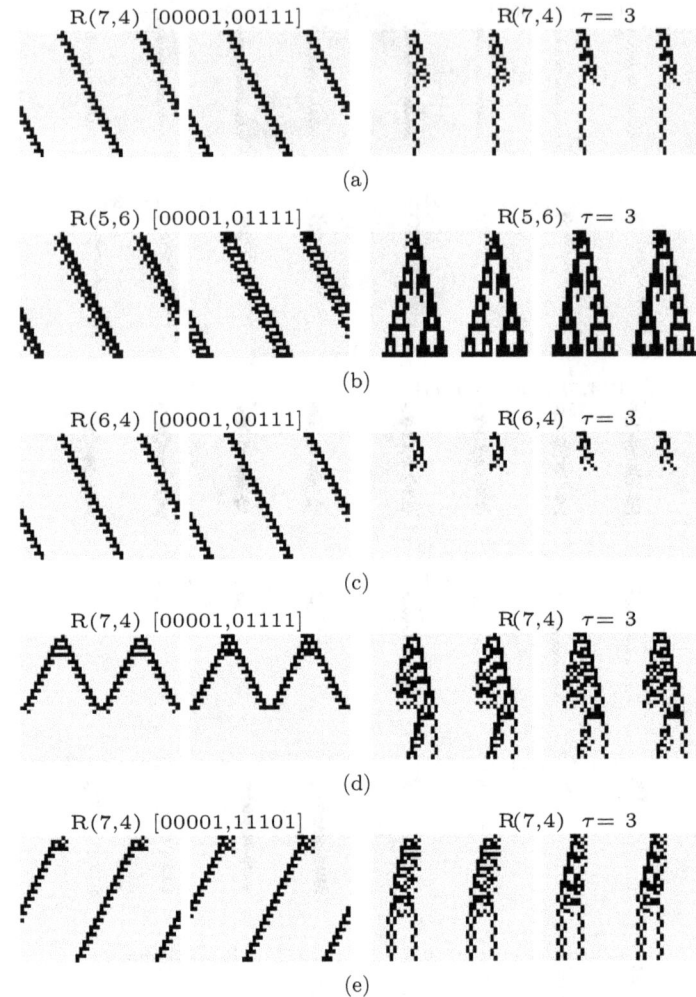

R(7,4) [00001,00111]   R(7,4) τ= 3

(a)

R(5,6) [00001,01111]   R(5,6) τ= 3

(b)

R(6,4) [00001,00111]   R(6,4) τ= 3

(c)

R(7,4) [00001,01111]   R(7,4) τ= 3

(d)

R(7,4) [00001,11101]   R(7,4) τ= 3

(e)

Fig. 4.8. Transformation of gliders by $\tau = 3$ majority memory. In each subfigure, we see space–time configurations of both actin chains in automata without memory (left) and with memory (right). Time goes down. Rules are indicated above the space–time configurations. (a) Glider becomes an oscillator. (b) Glider becomes a propagating and expanding pattern (Sierpinski gasket in this particular example). (c) Glider is annihilated. (d) Glider propagating left is transformed into larger, slow-moving glider and glider propagating right becomes an oscillator. (e) Glider 'splits' into two breathers.

can say that the glider was exploded by memory, Fig. 4.8(b)), or annihilated (Fig. 4.8(c)). In some cases, where a seed generates two gliders — one travels left and another travels right — one of the gliders becomes larger and a slow moving glider while another glider is transformed into an oscillator (Fig. 4.8(d)); or, a single glider is transformed into two oscillators with different patterns of oscillations (Fig. 4.8(e)).

In rare cases, e.g. rule R(14, 24), a spreading pattern is converted to two gliders, propagating to the left and to the right. The gliders annihilate when they collide (Fig. 4.9). We can also observe an entrainment phenomenon (Fig. 4.10) where a localisation 'inhabiting' only one chain in a rule without memory, spreads to the second chain when memory is introduced.

By undertaking exhaustive analysis of the rules, we found that only the following transformations of localisations are possible when

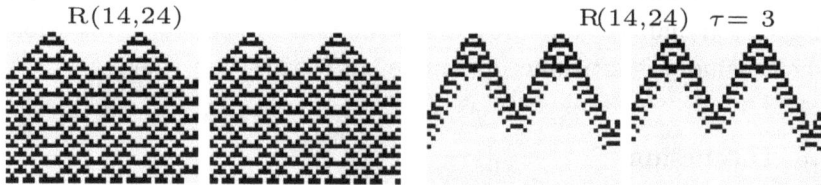

Fig. 4.9. Initial configurations not of the 5-seed type. A spreading pattern is trimmed to gliders. Transformation of gliders by majority memory. In each subfigure, we see space–time configurations of both actin chains in automata without memory (left) and with $\tau = 3$-memory (right). Time goes down. Rules are indicated above the space–time configurations. Seeds: [00110000000000000000000110, 01100000000000000000001100].

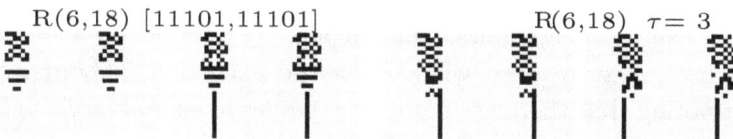

Fig. 4.10. Entrainment. Dynamics of actin chains are asynchronous in rules without memory: still life persists only in one chain. When memory is introduced, both chains host still lifes. In each subfigure we see space–time configurations of both actin chains in automata without memory (left) and with $\tau = 3$-memory (right). Time goes down. Rules are indicated above the space–time configurations.

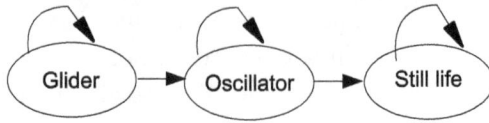

Fig. 4.11. Memory-induced transformations of localisation.

state transition rules are enriched with memory (Fig. 4.11). A glider is transformed to an oscillator. An oscillator is transformed to a still life. In some cases, memory preserves the gliders, oscillators, and still lifes. Transformations of still lifes to oscillators or gliders, and oscillations to gliders have not been observed so far. Notably, elementary cellular automata with memory transitions from glider to oscillation to still life, induced by memory, could be found [162, 167, 169]. We did not find such transitions in actin automata with memory, this could be due to the fact that interaction between two one-dimensional automata forming acting filament leads to 'inhibition' of glider dynamics and prevents formation of mobile localisations when memory is introduced to node state-transition rules.

4.5 Discussion

In our original investigation of actin automata (without memory) [26], we conducted an exhaustive search of localisations throughout the entire rule space. Through this comprehensive analysis, we demonstrated that certain rules exhibit support for travelling localisations when specific conditions are met.

Specifically, we discovered that a rule supports travelling localisations when a resting node becomes excited if it has either two or four excited neighbours, and an already excited node remains in an excited state even when it has no excited neighbours. It is worth noting that there are additional modes of excitability required to facilitate the propagation of gliders or travelling localisations within the actin automata system. Furthermore, we found that a rule supports stationary localisations when a resting node becomes excited if it has three excited neighbours, and an excited node remains excited if it has fewer than four excited neighbours.

By analysing the rules of actin automata with a majority, which we employed to illustrate the transformation of travelling and stationary localisations, we made an important observation. We determined that two excited neighbours are a necessary prerequisite for a resting node to become excited and for an already excited node to remain in an excited state, thus supporting the occurrence of travelling localisations (Fig. 4.12(a)) and stationary localisations (Fig. 4.12(b)). Interestingly, the second most common scenario among these rules involves three excited neighbours.

These findings shed light on the fundamental characteristics and conditions that govern the emergence of travelling and stationary localisations within actin automata, contributing to a deeper understanding of the dynamics and behaviours exhibited by actin-based computational systems.

Through the integration of memory into actin automata, we conducted further investigations and made noteworthy discoveries. We observed that when local transition rules incorporate memory, they give rise to slower propagation or expanding patterns and generate less complex space–time configurations, as quantified by entropy. Furthermore, the introduction of memory leads to a transformation of gliders into oscillatory stationary localisations and eventually into stable stationary localisations.

These findings provide valuable insights that complement our previous findings in the abstract analysis of actin filaments. They are also consistent with parallel advancements, such as the extension of actin automata to quantum actin automata [230] and the utilisation of cable equation models to examine voltage solitons in actin [234]. Gliders, representing propagating signals, and stationary localisations, serving as a form of volatile computer memory, assume distinct roles in these enriched actin automata systems. By adjusting the depth of actin units' memory, we can control whether actin filaments function as signal producers, where computations occur upon collision, or as memory devices.

It is important to note that the fundamental computing properties demonstrated by quantum actin automata, such as Boolean logic gates and binary adders [230], as well as three-valued logic

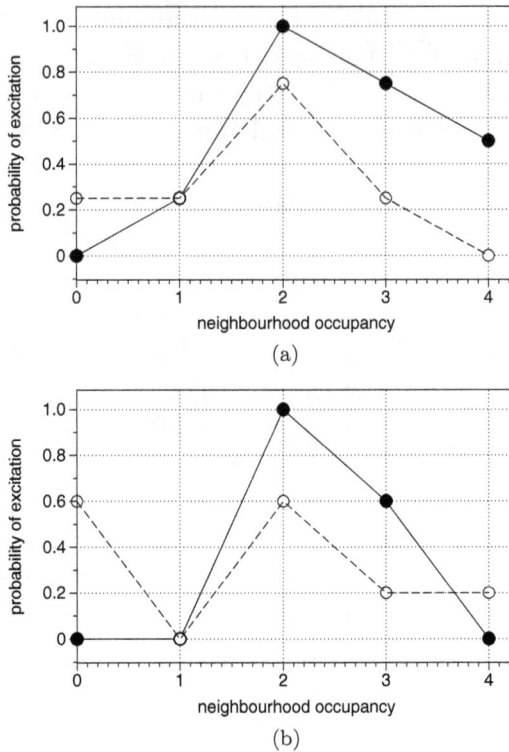

Fig. 4.12. Dependence of a probability of excitation of a node in actin automaton (a) supporting travelling localisations and (b) supporting stationary localisations on a number of excited neighbours of the node. The plots are calculated for rules R(7, 4), R(5, 6), R(6, 4), and R(12, 24), which support travelling localisations shown in Figs. 4.8, 4.9, and 4.13, and rules R(6, 20), R(4, 5), R(4, 4), R(6, 16), and R(6, 18), which support stationary localisations shown in Fig. 4.6, 4.7, and 4.10. Solid discs, connected by solid line, show probability of excitation of a resting node, circles, connected by dashed line, show probability of excitation of an excited node, i.e. of an excited node to remain excited.

operations [4], have not yet been fully replicated in actin automata with memory. These aspects will be the focal point of our future studies as we strive to uncover and comprehend the computational capabilities and limitations of actin-based systems with memory.

Currently, we are able to illustrate interactions that can be interpreted in terms of collision-based computing, where the presence of a localisation signifies logical TRUE and its absence signifies logical

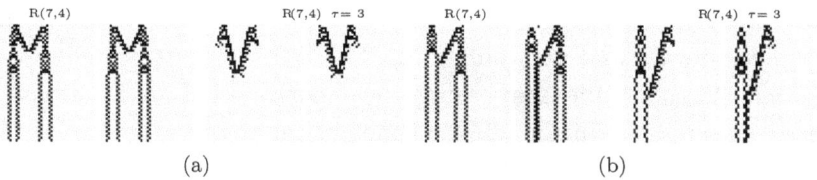

Fig. 4.13. Initial configurations not of the 5-seed type. Interactions of localisations of rule R(7, 4) with (right) and without (left) $\tau = 3$ majority memory. (a) Two gliders collide and annihilate in the result of the collision. (b) Glider is stopped by oscillator. In each subfigure, we see space–time configurations of both actin chains in automata. Seeds: (a) [0011100000000000001, 0100000000000001110] and (b) [11100000000000001,00000010000000111].

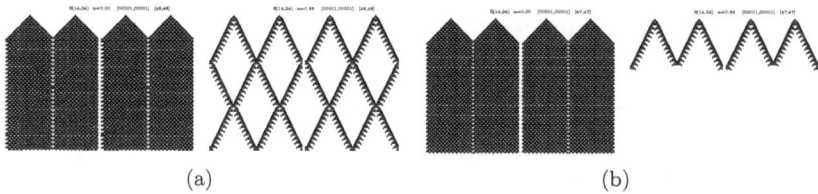

Fig. 4.14. Emergence of localisation induced by memory, rule R(12, 24). Distance between glider seeds is (a) 48 nodes and (b) 51 nodes. In each subfigure, we see space–time configurations of both actin chains in automata without memory (left) and with $\alpha = 0.55$-memory (right). Initially, just a single node is in State '1'. Time goes down. Rules are indicated above the space–time configurations.

FALSE. An example of a 'classical' annihilation gate $\langle x, y \rangle \to \langle \overline{x}y, x\overline{y} \rangle$ is depicted in Fig. 4.13(a), with inputs $x = $ TRUE and $y = $ FALSE. In this gate, when $x = $ TRUE, a glider travels to the right, and when $y = $ TRUE, a glider travels to the left. If only one glider is present in the gate, it continues its movement undisturbed. The trajectory of the undisturbed glider travelling right represents $x\overline{y}$, while the trajectory of the undisturbed glider travelling left represents $\overline{x}y$. These travelling localisations serve as data and represent the computation results. To capture and store the travelling data, we utilise stationary oscillators, as shown in Fig. 4.13(b). In this example, a glider travels left and collides with an oscillator, causing the oscillator to change its state. Thus, the travelling data are effectively converted into stationary data through the collision with the oscillator, allowing for data storage and manipulation.

Another example showcasing the emergence of collision-based gates through the introduction of memory is depicted in Fig. 4.14. In this case, we consider the actin automata rule R(14, 24), where initial seeds of a single node in state '1' give rise to propagating patterns of regular "excitations" in the actin filaments. When memory is introduced, the interior of these propagating patterns is filtered out, and only wave-fronts, referred to as gliders, remain. In the depicted scenario, the gliders exhibit reflective behaviour and undergo phase changes when the distance between the seeds is 48 nodes (Fig. 4.14(a)). Conversely, when the distance between seeds is 51 nodes, the gliders annihilate upon collision (Fig. 4.14(b)). These phenomena demonstrate the ability to manipulate and control glider interactions based on the distance between the initial seeds, highlighting the memory-induced emergence of collision-based gates in actin automata.

Chapter 5

Dynamics of Excitation in F-actin: Automaton Model

Andrew Adamatzky

Unconventional Computing Lab, UWE Bristol, UK

andrew.adamatzky@uwe.ac.uk

Abstract

We propose a model called the F-actin automaton, which represents a filamentous actin molecule as a graph of finite state machines. Each node in the graph can be in one of three states: resting, excited, or refractory. The states of all nodes are updated simultaneously and according to the same rule, in discrete time steps. We consider two different rules in our analysis: the threshold rule and the narrow excitation interval rule. In the threshold rule, a resting node becomes excited if it has at least one excited neighbour. In the narrow excitation interval rule, a resting node is excited only if it has exactly one excited neighbour. To understand the dynamics of the F-actin automaton, we examine the distributions of transient periods and lengths of limit cycles that emerge during its evolution. We investigate the mechanisms underlying the formation of these limit cycles, which represent recurring patterns or oscillations in the system. Additionally, we evaluate the density of information storage in F-actin automata. Information storage refers to the capacity of the system to encode and retain information within its states and transitions. By analysing the characteristics of the F-actin automaton, we gain insights into the amount and efficiency of information that can be stored in this model. Through our analysis, we aim to deepen our understanding of the behaviour and properties of the F-actin automaton. This research contributes to the broader understanding of actin dynamics and provides insights into the potential roles of these automata in cellular processes and information processing.

5.1 Introduction

In Ref. [27], we proposed a model of actin filaments as two chains of one-dimensional binary-state semi-totalistic automaton arrays. We discovered automaton state transition rules that support travelling localisations, which are compact clusters of non-resting states. These travelling localisations are analogous to ionic waves proposed in actin filaments [254]. We speculated that a computation in actin filaments could be implemented when localisations (defects, conformation changes, ionic clouds, solitons), which represent data, collide with each other and change their velocity vectors or states. Parameters of the localisations before a collision are interpreted as values of input variables. Parameters of the localisation after the collision are values of output variables. We implemented a range of computing schemes in several families of actin filament models, from quantum automata to lattice with Morse potential [230–233,235]. These models considered a unit (F-actin) of an actin filament as a single, discrete entity that can take just two or three states, and carriers of information occupied one-two actin units. These were models of rather coarse-grained computation [230–233, 235]. To take the paradigm of computation via interaction travelling localisations at the sub-molecular level, we must understand how information, presented by a perturbation of some part of an F-actin unit from its resting state, propagates in the F-actin unit.

5.2 Model

We utilised a structural model of the F-actin molecule, which was derived from X-ray fibre diffraction intensities obtained from well-oriented sols of rabbit skeletal muscles [187]. The resolution of the calculated structure was 3.3Å in the radial direction and 5.6Å along the axis [187].

To analyse the molecular structure, we converted it into a non-directed graph denoted as $\mathcal{A}$. In this graph, each node represents an atom, and the edges correspond to bonds between the atoms. The graph $\mathcal{A}$ consists of 2961 nodes and 3025 edges. The minimum degree of the nodes is 1, the maximum degree is 4, and the average degree is 2.044 (with a standard deviation of 0.8). The median degree of the nodes is 2. Among the nodes, there are 883 with a degree of 1, 1009 with a degree of 2, 1066 with a degree of 3, and two with a degree of 4. Furthermore, the graph $\mathcal{A}$ exhibits a diameter of 1130 nodes, representing the longest shortest path within the graph. The mean distance, which is the average shortest path between any two nodes, is 376, while the median distance is 338. These structural characteristics provide important insights into the organisation and connectivity of the actin molecule, as represented by the graph $\mathcal{A}$.

We study dynamics of excitation in the actin graph $\mathcal{A}$ using the following models. Each node s of $\mathcal{A}$ takes three states: resting ($\circ$), excited ($\oplus$), and refractory ($\ominus$). Each node s has a neighbourhood $u(s)$, which is a set of nodes connected to the node s by edges in $\mathcal{A}$. The nodes update their states simultaneously in discrete time by the same rule. Each step of simulated discrete time corresponds to one attosecond of real time.

A resting node $s^t = \circ$ excites depending on a number σ_s^t of its excited neighbours in neighbourhood $u(s)$: $\sigma_s^t = \sum_{w \in u(s)} \{w^t = \oplus\}$. We consider two excitation rules. In rule $\mathcal{A}_0$, a resting node excites if it has at least one excited neighbour: $\sigma_s^t > 0$. In rule $\mathcal{A}_1$, a resting node excites if it has exactly one excited neighbour: $\sigma_s^t = 1$ (we do not consider rules where $\sigma_s^t > 1$ because excitation there extincts quickly). Transitions from excited state to refractory state and from refractory state to resting state are unconditional, i.e. these transitions take place independently of neighbourhood states. The rules can be written as follows:

$$
\begin{array}{cc}
\mathcal{A}_0 & \mathcal{A}_1 \\[4pt]
s^{t+1} = \begin{cases} \oplus, & \text{if } \sigma_s^t > 0 \\ \ominus, & \text{if } s^t = \oplus, \\ \circ, & \text{otherwise} \end{cases} &
s^{t+1} = \begin{cases} \oplus, & \text{if } \sigma_s^t = 1 \\ \ominus, & \text{if } s^t = \oplus. \\ \circ, & \text{otherwise} \end{cases}
\end{array}
$$

At the beginning of each computational experiment, the F-actin automaton $\mathcal{A}$ is in a global resting state, every node is assigned state $\circ$. An excitation dynamic is initiated by assigning non-resting states $\oplus$ or $\ominus$ or both to a portion of randomly selected nodes. Three stimulation scenarios are considered:

- **Single node stimulation:** A single node is selected at random and this node is assigned excited state $\oplus$
- **(+)-stimulation:** A specified ratio of nodes is selected at random and the selected nodes are assigned excited state $\oplus$
- **(+−)-stimulation:** A specified ratio of nodes is selected at random and the selected nodes are assigned either excited state $\oplus$ or refractory state $\ominus$ at random.

The automaton $\mathcal{A}$ is deterministic, therefore, from any initial configuration the automaton evolves into a limit cycle in its state space (where its configuration is repeated after a finite number of steps) or an absorbing state (this is limit cycle length one). For the rules selected, there is only one absorbing state — all nodes are in the resting state. A limit cycle comprises of configurations where compact patterns of excitation travel along closed paths. A transient period is an interval of automaton evolution from initial configuration to entering a limit cycle or an absorbing state.

5.3 Dynamics of $\mathcal{A}_0$

5.3.1 Single node stimulation

The excitation propagates as a localised pattern (Fig. 5.1(a–f)). A number of nodes excited at every single step of time varies between one and five (Fig. 5.2(a)). Sometimes an excitation pattern splits into two localisations that travel along their independent pathways. The automaton $\mathcal{A}_0$ always evolves into the absorbing state where all nodes are resting. This is because travelling localisations either cancel each other when they collide or reach cul-de-sacs of their pathways. A distribution of transition periods is shown in Fig. 5.2(b). The mean transition period is 840 time steps, median, 847, minimum, 2 and maximum, 1131. Only 29 nodes, when stimulated, lead to excitation development with a transition period between 2 and 15 steps. Stimulation of all other 2932 trigger excitation dynamics for at least 568 steps. The longest transition period is observed when

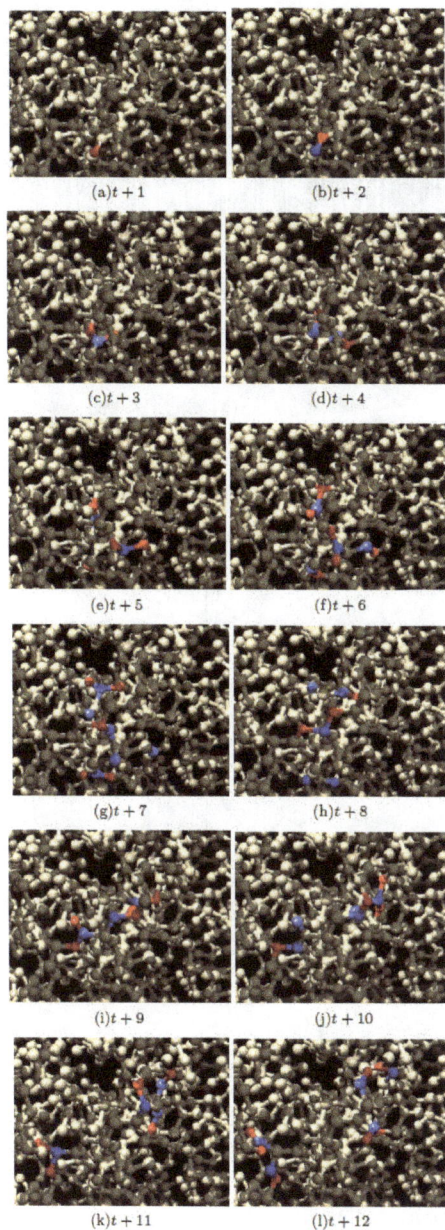

Fig. 5.1. Exemplar excitation dynamics of $\mathcal{A}_0$ in a scenario of a single node stimulation. In the initial configuration, all nodes are resting but one node is excited. (a–f) Snapshots of the simulation. Excited nodes are red, refractory nodes are blue, resting nodes are light-grey.

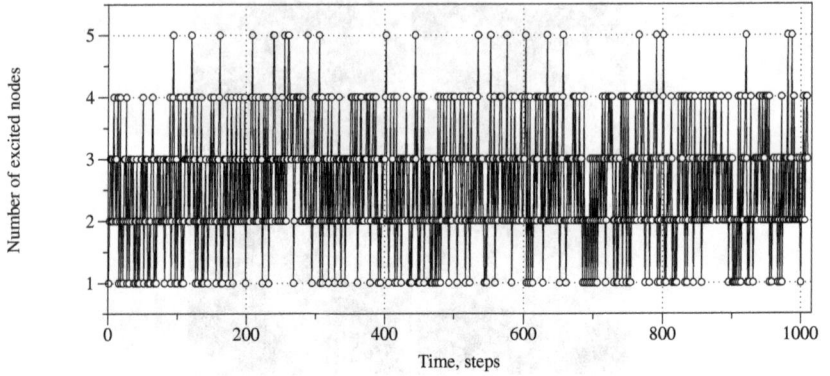

(a)

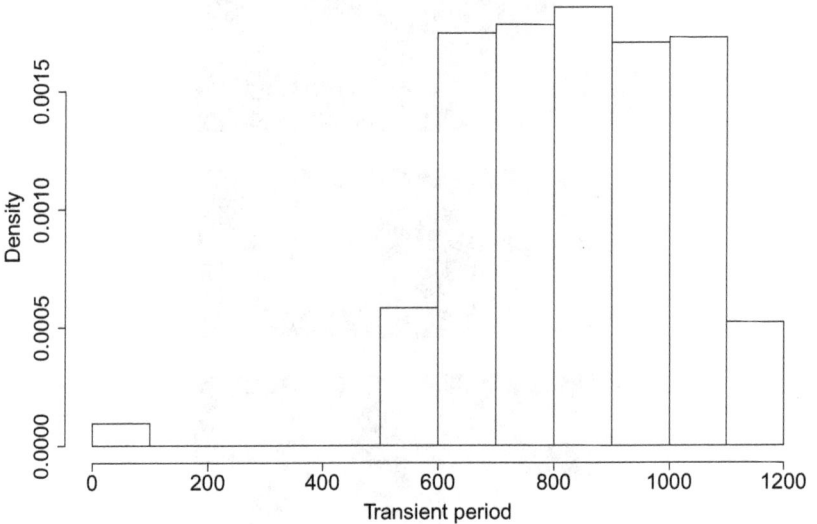

(b)

Fig. 5.2. Integral dynamics of rule $\mathcal{A}_0$ automata in scenarios of a single node stimulation. (a) Number of excited nodes at each step of simulation in a single experiment. (b) Distribution of transient periods.

the localised excitation runs along a longest shortest path where the initially stimulated node is a source. The path of the longest excitation is visualised in Fig. 5.3; the path matches the backbone of the actin unit.

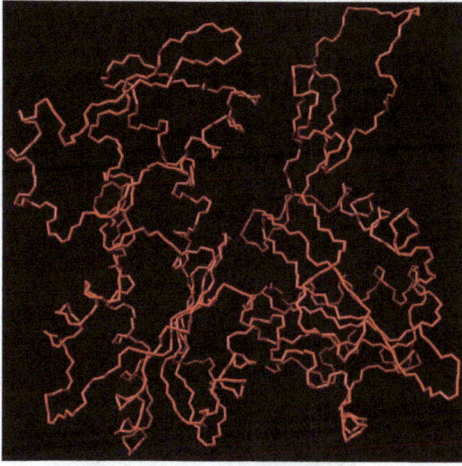

Fig. 5.3. Longest path of excitation propagation in $\mathcal{A}_0$ displayed by bonds.

5.3.2 (+)-stimulation

When we stimulate more than one node, the automaton $\mathcal{A}_0$ exhibits several 'epicentres' of excitation, the patterns of excitation propagate away from their origins (Fig. 5.4), and populate the graph. This stage is manifested in an increasing number of excited states at each step of the evolution (Fig. 5.5(a)). Eventually, depending on distances between sources of excitation, the graph becomes filled with waves and localisations, e.g. in illustration Fig. 5.5(a), a peak is reached in 7–8 steps. Then patterns of excitation start colliding with each other. They annihilate in the results of the collisions. The number of excited nodes decreases over time (Fig. 5.5(a)). The graph returns to the totally resting state. The larger the portion of the initially excited nodes, the quicker the evolution halts in the resting state (Fig. 5.5(b)). 'Quicker' can be quantified by a polynomial function $p = 4.7 \cdot \rho^{-0.6}$, where p is a length of transient period and ρ is a ratio of initially excited nodes.

5.3.3 (+−)-stimulation

In a 'classical' two-dimensional discrete excitable medium, stimulation of the medium with an excited node neighbouring with a

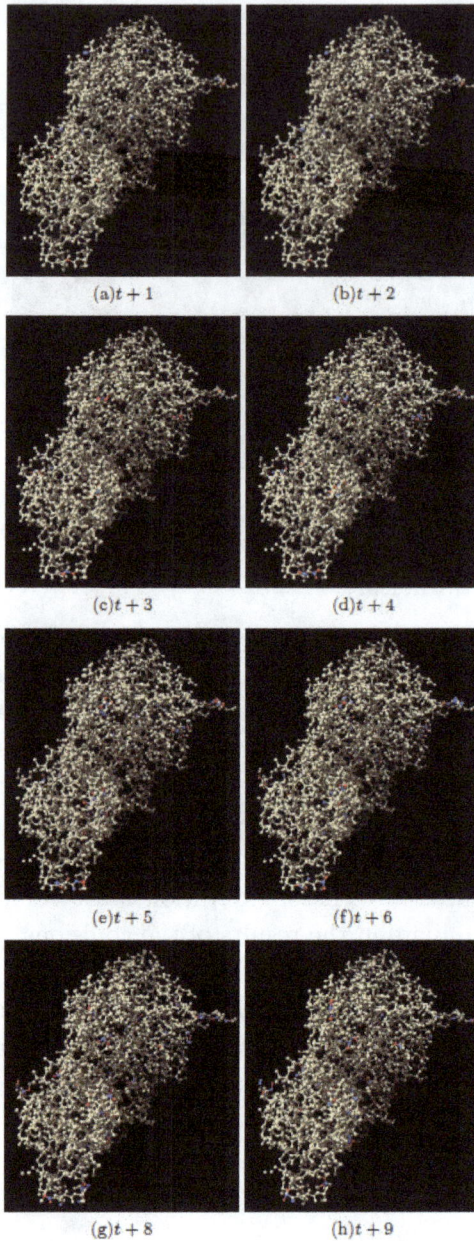

Fig. 5.4. Exemplar dynamics of $\mathcal{A}_1$. In the initially resting configuration, 1% of nodes is excited. Edges of $\mathcal{A}$ are shown by grey colour, excited nodes are red, refractory nodes are blue, resting nodes are not shown.

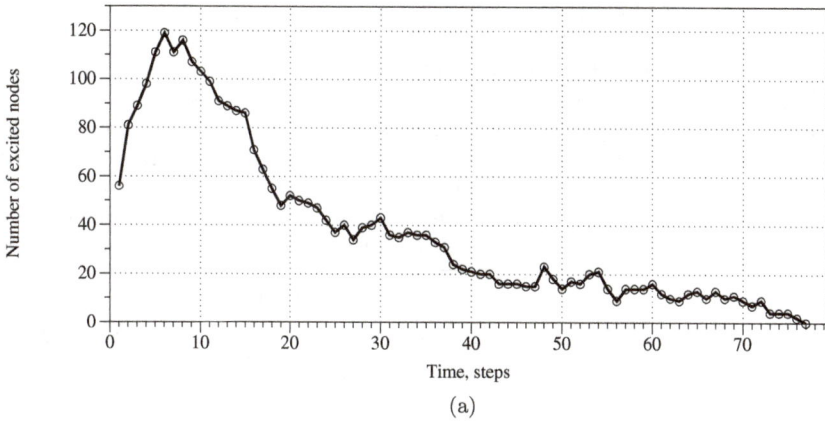

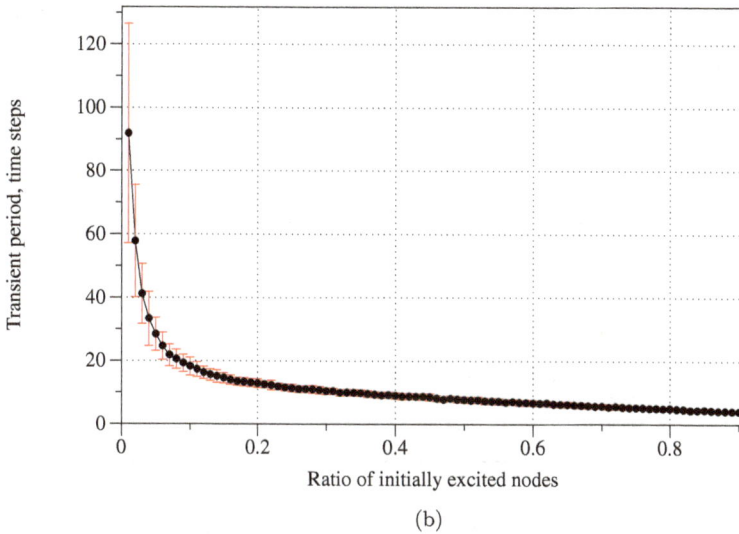

Fig. 5.5. Integral dynamics of $\mathcal{A}_0$ in scenarios when a portion of nodes is excited initially. (a) A number of excited nodes vs. time: initially, 1% of nodes are assigned excited states. (b) Ratio of initially excited nodes vs. average transient period. Standard deviation is shown as error bars.

refractory node leads to the formation of a spiral wave. Due to the spiral waves, excitation can persist in a modelled medium indefinitely. F-actin automata follow this principle. When we stimulate nodes of $\mathcal{A}_0$ such that some of the nodes get excited states and some refractory states, we evoke the excitation patterns. The average

Table 5.1. Characterisation of excitation dynamics in actin automata $\mathcal{A}$. (abc) Dependence of the dynamic on ratios of stimulated nodes. (a) $\mathcal{A}_0$, initially excited ratio ρ of nodes. (b) $\mathcal{A}_1$, initially excited ratio ρ of nodes. (c) $\mathcal{A}_1$, initial ratio ρ of nodes assigned excited and refractory states at random. Data are collected in 10 experiments for each value of ρ. For 10 ratios ρ of initially stimulated nodes, $\rho = 0.1, 0.2, \ldots, 0.9$, we calculated lengths of transient period p, lengths of cycles c, numbers e of excited nodes in a cycle, and their standard deviations, $\sigma(p)$, $\sigma(c)$, $\sigma(e)$. Values are rounded to integers. (d) Characteristics averaged all over stimulation ratios for each rule and stimulation scenario.

| (a) $\mathcal{A}_0$, (+−)-start | | | | | | | (b) $\mathcal{A}_1$, (+)-start | | | | | | |
ρ	p	c	e	$\sigma(p)$	$\sigma(c)$	$\sigma(e)$	ρ	p	c	e	$\sigma(p)$	$\sigma(c)$	$\sigma(e)$
10	1613	6	535	1820	1	55	0.1	1154	13	570	1251	12	39
20	1432	5	562	988	0	52	0.2	1388	13	553	961	12	50
30	1984	7	626	1275	8	108	0.3	893	11	575	362	10	8
40	2536	14	598	3064	15	15	0.4	920	24	590	487	11	51
50	1583	13	786	610	8	206	0.5	996	16	575	832	13	11
60	2614	14	719	3322	14	191	0.6	746	16	594	238	12	24
70	2052	9	805	2236	4	207	0.7	891	16	625	455	11	77
80	1311	5	705	521	1	180	0.8	1354	16	639	408	12	109
90	2850	16	651	2835	15	157	0.9	1729	15	577	1368	13	13

| (c) $\mathcal{A}_1$, (+−)-start | | | | | | | (d) | | | | |
| ρ | p | c | e | $\sigma(p)$ | $\sigma(c)$ | $\sigma(e)$ | | $\mathcal{A}_0$ | | $\mathcal{A}_1$ | |
								(+)-start	(+−)-start	(+)-start	(+−)-start
0.1	893	118	496	934	355	172	p	840	1997	1118	1667
0.2	2328	5	583	1525	0	6	c	1	10	15	21
0.3	1953	13	599	1791	12	16	e	0	665	588	615
0.4	1957	8	591	2400	8	5					
0.5	1785	14	636	2567	12	130					
0.6	976	6	706	345	3	179					
0.7	1342	11	709	464	13	175					
0.8	1170	13	625	620	14	109					
0.9	2599	5	593	2452	1	14					

level of excitation over trials is proportional to a number of nodes stimulated (see row e in Table 5.1(a)). The automaton enters a limit cycle (Fig. 5.6). The limit cycle's length varies between 5 to 14 time steps (see row c in Table 5.1(a)). Apparently, the automaton falls into longest limit cycles when nearly half of the nodes are stimulated; however, due to high deviation of the results (see row $\sigma(c)$ in Table 5.1(a)), we would not state this as a fact. Lengths of transient periods, from stimulation to entering the limit cycle, are over a half of the number of nodes in $\mathcal{A}$.

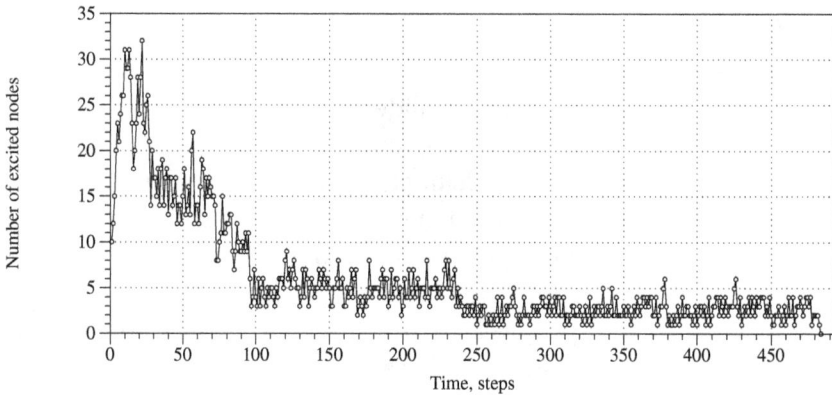

Fig. 5.6. Integral dynamics of $\mathcal{A}_0$ in scenarios when 1% of nodes are assigned excited or refractory states initially.

5.4 Dynamics of $\mathcal{A}_1$

5.4.1 Single node stimulation

When a single node is excited initially, the automaton $\mathcal{A}_1$ always evolves to a globally resting state. In sampling of 70 trials, we found that average length of the transient period is 862 time steps (standard deviation 230) and median transition period is 869. The average transient period to the resting state is 22 steps longer than the one in the automaton $\mathcal{A}_0$.

5.4.2 (+)-stimulation

In contrast to automaton $\mathcal{A}_0$, automaton $\mathcal{A}_1$ does not show a pronounced sensitivity to a ratio ρ of initially excited nodes. Transition periods for all values of ρ are grouped around 1112 (Table 5.1(b)). The automata always evolve to limit cycles. Cycle lengths are around 15 time steps with excitation level (number of excited nodes) of just below 600 nodes. The system shows a high degree of variability in lengths of transition periods and cycles, as manifested in large values of standard deviations $\sigma(p)$ and $\sigma(c)$. Level of excitation typically remains preserved.

5.4.3 (+−)-stimulation

A_1 behaves similarly to the scenario of (+)-start: there are many travelling localisations, which collide and, mostly, annihilate each other. Few localisations survive by finding a cyclic path to travel: if no other localisation enters their path, the remaining localisations can cycle indefinitely. The surviving localisations are responsible for A_1 falling into the limit cycle. Automaton starting with a mix of randomly excited and refractory states usually travels one-and-half times longer to its limit cycle than same automaton starting only with randomly excited states (compare Table 5.1(b) and 5.1(c)).

5.5 Stability of the dynamics

How does repeated stimulation affect excitation dynamics of A_0 and A_1? (+)-stimulation of A_0 at any stage of its evolution raises level of excitation by an amount equivalent to that of the stimulated resting automaton (Fig. 5.7). Thus, repeated stimulation prolongs

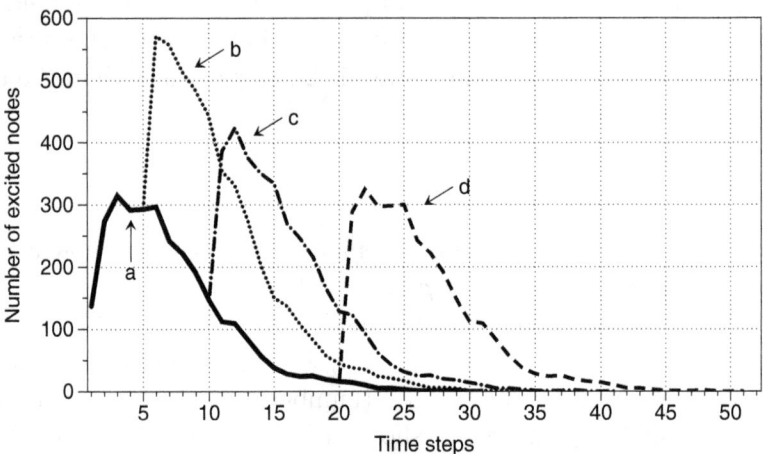

Fig. 5.7. Dynamics of A_0 automaton under repeated stimulation. In each trial, 5% of nodes were initially excited. (a) No more stimulation was applied (apart from the initial stimulation). (bcd) Automaton was stimulated by exciting 5% of nodes at (b) 5th step, (c) 10th step, and (d) 20th step of the evolution.

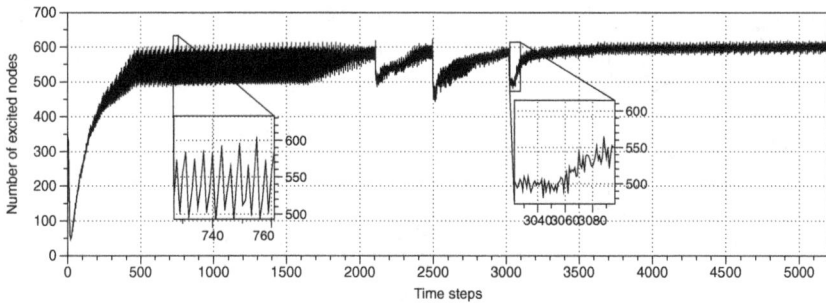

Fig. 5.8. Dynamics of $\mathcal{A}_0$ automaton under repeated stimulation. Initially 10% of nodes were assigned excited or refractory state at random. When the automaton reached limit cycle the stimulation was repeated.

return of the automaton to its resting state. In scenario of $(+-)$-stimulation, $\mathcal{A}_0$ evolves to a limit cycle. Repeated $(+-)$-stimulation of the automaton while it is in the limit cycle causes the automaton to change its trajectory in a state space. This change is characterised by the initially reduced level of excitation. Typically, excitation level drops by 100–150 nodes at the moment of stimulation. The level of excitation returns to its 'pre-stimulation' value in 400–500 time steps.

5.6 Implementation of memory

F-actin automata entering limit cycles could play a role of information storage in actin filaments. The minimal length of a limit cycle detected is 5 time steps. Thus, aromatic rings could be a substrate responsible for some patterns of cycling excitation dynamics. Let an aromatic ring automaton be stimulated such that a node is assigned an excited state and one of its neighbours is assigned refractory state. The wave of excitation (comprising one excited and one refractory states) propagates into the direction of its excited head (Fig. 5.9(a)). The excitation running along the aromatic ring cannot be extinguished by stimulation of one resting node (Fig. 5.9(b–e)) or two resting nodes (Fig. 5.9(f–h)). This is because an excited node surrounded by two resting neighbours excites both resting

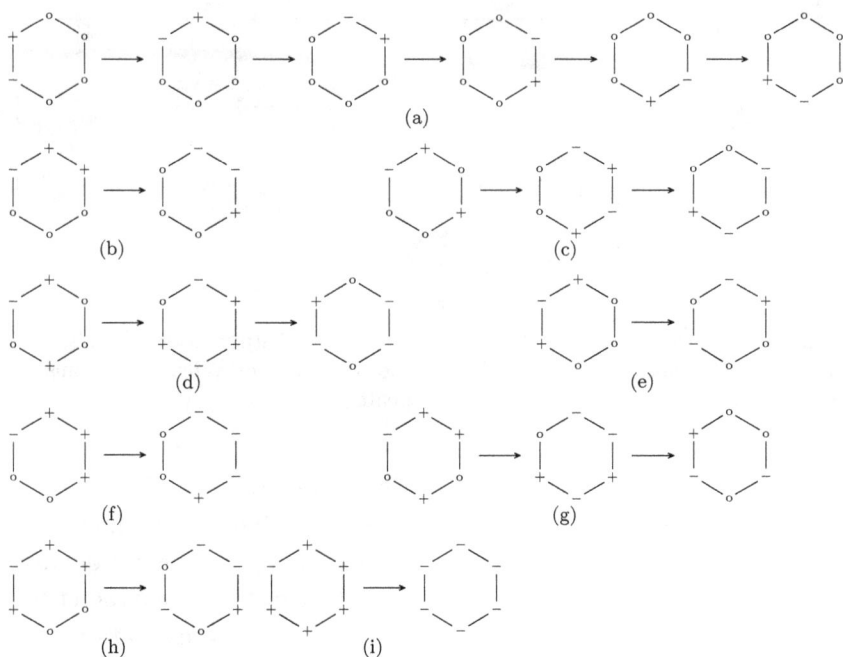

Fig. 5.9. Excitation dynamics of aromatic ring automaton governed by $\mathcal{A}_0$. (a) Propagation of excitation wave on undisturbed ring automaton. (b–i) Stimulation of ring automaton.

neighbours. Thus, excitation waves propagate along the ring in both directions. Therefore, even if the original excitation wave is cancelled by external stimulation, then similar running wave will emerge. To extinguish the excitation in an aromatic ring, we must externally excite all four resting nodes or force them into a refractory state.

The excited aromatic rings act as generators of excitation in F-actin automata. Let us consider an example. In Fig. 5.10, we see a histidine's aromatic ring stimulated: one node is assigned excited state and its neighbour, refractory state. The wave of excitation travels along the ring clockwise (Fig. 5.11(a–c)). When excitation reaches a node linked to the rest of the graph, the excitation propagates along the 'bridge' (Fig. 5.11(d)). The excitation then propagates further inside the graph (Fig. 5.11(e, f)) splitting into two compact excitation patterns at the junction (Fig. 5.11(g, h)). The overall pattern of

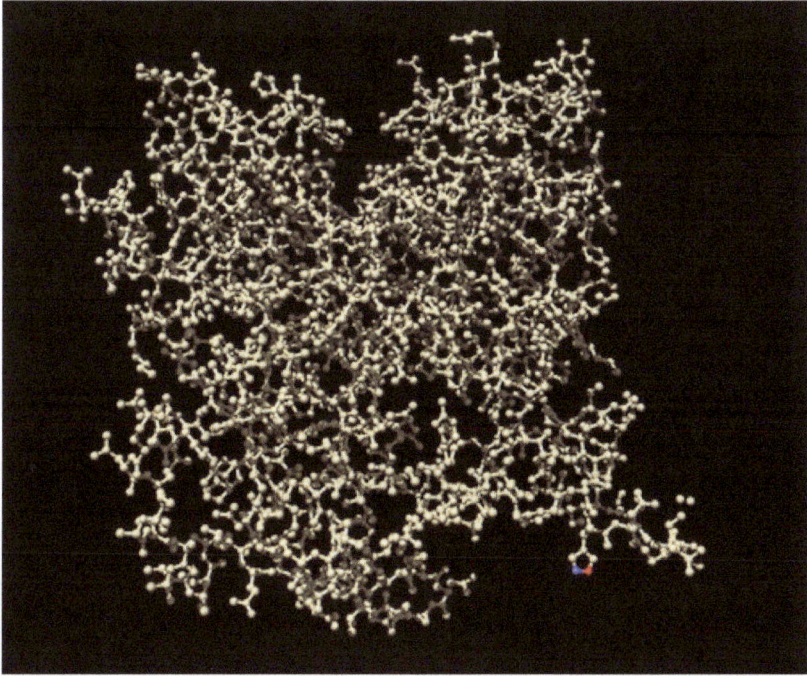

Fig. 5.10. Configuration of resting $\mathcal{A}_0$ at the moment of external stimulation of histidine's aromatic ring. Resting nodes are light-grey, excited nodes are red, refractory nodes are blue.

excitation in $\mathcal{A}_0$ recorded at the 90th step of evolution is shown in Fig. 5.12.

5.7 Discussion

The automaton model representing the F-actin unit serves as a valuable tool for rapid prototyping and studying the dynamics of excitation in actin filaments, allowing for controlled propagation of localisations at the atomic level. In our analysis, we specifically focused on two excitation rules. The first rule, denoted as $\mathcal{A}_0$, employs a classical threshold excitation criterion, where a resting node becomes excited if it has at least one excited neighbour. The second rule, $\mathcal{A}_1$, exhibits nonlinear excitation behaviour, as it requires exactly one excited neighbour to trigger excitation in a resting

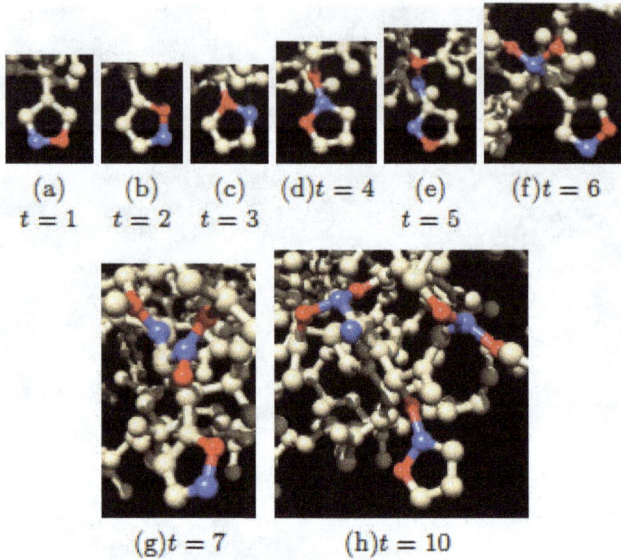

Fig. 5.11. Excitation dynamics of $\mathcal{A}_0$ triggered by excitation of histidine's aromatic ring. First 10 steps of the automaton evolution are shown. Only part of the graph adjacent to excitation is displayed. Resting nodes are light-grey, excited nodes are red, refractory nodes are blue.

node. We did not explore other ranges of thresholds or excitation intervals, as they consistently resulted in the extinction of excitation at the early stages of evolution. Both rules support the propagation of travelling patterns of excitation. However, automata governed by rule $\mathcal{A}_0$ demonstrate longer transient periods, smaller limit cycles, and higher average levels of excitation compared to automata governed by rule $\mathcal{A}_1$ (see Table 5.1(d)). When a resting automaton following rule $\mathcal{A}_0$ receives external excitation at specific nodes, the excitation patterns spread throughout the automaton graph, but the overall activity eventually diminishes, returning to a global resting state. Furthermore, when actin automata are stimulated with a combination of excited and refractory states, the excitation dynamics exhibit longer transient periods and give rise to repetitive patterns of excitation, reminiscent of oscillatory structures. These limit cycles remain stable, as an automaton subjected to repeated stimulation consistently returns to its pre-stimulation activity level.

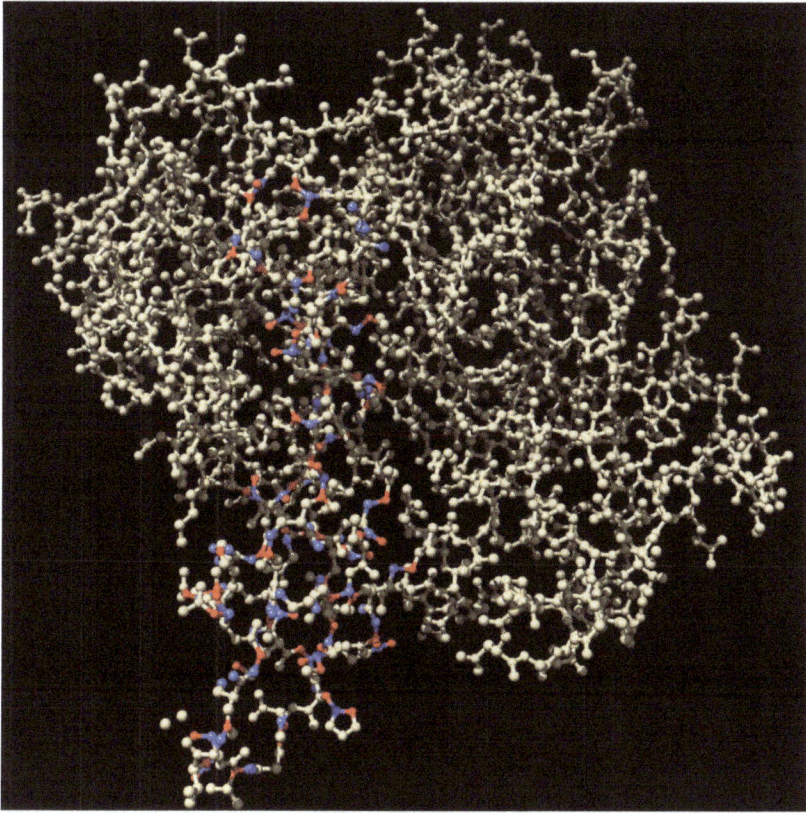

Fig. 5.12. Pattern of excitation of $\mathcal{A}_0$ triggered by excitation of histidine's aromatic ring as shown in Fig. 5.11. The pattern is recorded at 90th step of evolution. Resting nodes are light-grey, excited nodes are red, refractory nodes are blue.

The noise-tolerance of excitation waves propagating in aromatic rings suggests that these rings could serve as memory devices in a hypothetical actin computer. In this scenario, an excited aromatic ring represents one bit of information. To write a bit, we excite one node and inhibit (force into a refractory state) one of its neighbours. To erase a bit, we must either excite or inhibit all resting nodes. An F-actin unit consists of 40 aromatic rings, including 8 histidine rings, 12 phenylalanine rings, 4 tryptophan rings, and 16 tyrosine rings (see configuration in Fig. 5.13). Therefore, an F-actin unit can store

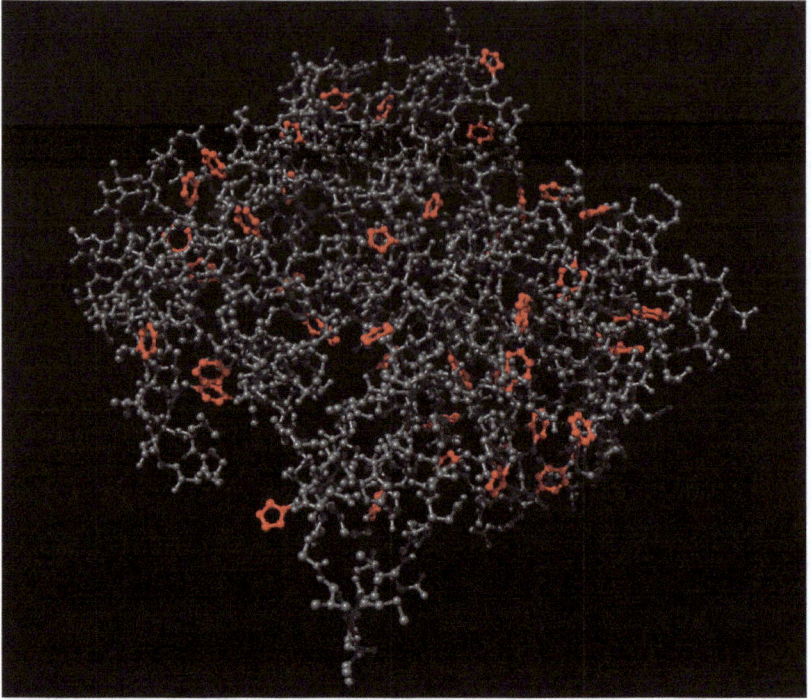

Fig. 5.13. F-actin molecule with aromatic rings highlighted in red.

40 bits of information. The maximum diameter of an actin filament is approximately 8 nm, and a single unit of F-actin has a diameter of approximately 4 nm (Fig. 11.1(a)). Considering that the persistent length of an F-actin polymer is 17 μm, it is reasonable to assume that it is feasible to store $32 \cdot 10^4$ bits on a double-stranded actin filament. If we have the necessary tools to read and write the dynamics of excitation in specific parts of the F-actin molecule, we can achieve a memory density of 64 Petabits per square inch ($6.452 \cdot 10^{16}$ per square inch). This estimate demonstrates the potential memory capacity offered by the actin polymer in terms of data storage.

Chapter 6

Actin Networks Voltage Circuits

Stefano Siccardi[*,ǁ], Andrew Adamatzky[†,**], Jack Tuszyński[‡,††],
Florian Huber[§], and Jörg Schnauß[¶,‡‡]

[*]*Consorzio Interuniversitario Nazionale per l'Informatica, Italy and
Unconventional Computing Laboratory, UWE Bristol, UK*
[†]*Unconventional Computing Laboratory, UWE Bristol, UK*
[‡]*Department of Oncology, University of Alberta, Edmonton, Canada;
DIMEAS, Politecnico di Torino, Turin, Italy*
[§]*Netherlands eScience Center, Amsterdam, The Netherlands*
[¶]*Soft Matter Physics Division, Peter Debye Institute for Soft Matter
Physics, Faculty of Physics and Earth Science, Leipzig University,
Germany; Fraunhofer Institute for Cell Therapy and Immunology (IZI),
DNA Nanodevices Group, Leipzig, Germany*
[ǁ]*stefano.siccardi@unimi.it*
[**]*andrew.adamatzky@uwe.ac.uk*
[††]*jack.tuszynski@gmail.com*
[‡‡]*joerg.schnauss@uni-leipzig.de*

Abstract
We treat the bundles of actin filaments as electrical wires, where the
filament densities can vary, allowing us to analyse and discuss the
key electrical parameters of the system. In our analysis, we solve a
set of equations that describe the network, considering different initial
conditions. We apply input voltages, representing information bits, at
specific locations within the system and calculate the corresponding
output voltages at other designated positions. We investigate two
scenarios: an idealised situation where point-like electrodes can be
inserted at any desired points within the actin bundles, and a more
realistic scenario where electrodes are placed on a surface, conforming
to typical dimensions used in the industry. Through our investigations,
we make the remarkable discovery that in both cases, the system is
capable of implementing essential logical gates and can function as a
finite-state machine. This implies that the actin bundle network can

perform logical operations and exhibit memory capabilities. Our findings demonstrate the tremendous potential of actin bundle structures as a basis for computing circuits. By harnessing the unique properties of actin filaments, we can realize the implementation of logical gates and create systems with the capability to process and store information. This research opens up exciting new avenues for the development of innovative computing architectures based on self-assembled actin bundles.

6.1 Introduction

We present a comprehensive and physics-based model that describes the behaviour of voltage solitons on bundles of actin filaments (AF). Our model considers AF networks as wire-like structures, with specific node locations connecting the bundles. Each bundle is characterised by its unique electrical parameters and enables the transport of ions along its length. By employing this approach, our objective is to enhance the comprehension of AF networks and their potential as computing devices through the incorporation of a robust physics-based framework.

6.2 The model

The fundamental model utilised in this study is presented in Ref. [254], which focuses on elucidating the characteristics of actin filaments (AFs) and provides the derivation of pertinent equations. The model is built upon several key assumptions, which we will highlight. First, it assumes that each monomer within the filament carries 11 negative excess charges, taking into account the presence of charged residues within the AF structure. The double helical nature of the filament introduces regions with non-uniform charge distribution, resulting in areas of higher and lower charge density. These variations in charge density are incorporated into the model. Additionally, the concept of the Bjerrum length (λ_B) is integrated into the model. This length represents a specific distance beyond which the influence of thermal fluctuations becomes more significant than the electrostatic attraction or repulsion between charges in the surrounding solution. By considering these assumptions and factors, the model aims to provide a comprehensive understanding of AF behaviour and their interaction with the surrounding ionic environment. The derivation of formulas and calculations presented in Ref. [254] serve as the fundamental basis for further advancements and investigations conducted in this study. It is inversely proportional to temperature and directly proportional to the ions' valence z:

$$\lambda_B = \frac{ze^2}{4\pi\epsilon\epsilon_0 k_B T},$$ (6.1)

where e is the electrical charge, ϵ_0, the permittivity of the vacuum, ϵ, the dielectric constant of the solution with AFs immersed in (estimated to be similar to $\epsilon_{\text{water}} \approx 80$), k_B, Boltzmann's constant, and T, the absolute temperature. If δ is the mean distance between charges, counterion condensation is expected when $\lambda_B/\delta > 1$. Considering the temperature is $T = 293$ K and the ions are monovalent, Ref. [254] finds $\lambda_B = 7.13 \times 10^{-10}$ m and Ref. [129], $\lambda_B = 13.8 \times 10^{-10}$ m for Ca^{2+} at $T = 310$ K. Considering actin filaments, δ is estimated to be 0.25 nm because assuming an average of 370 monomers per μm, there are approximately $4e$/nm. Each monomer behaves like an electrical circuit with inductive, capacitive, and resistive components. The model is based on the transmission line analogy.

The capacitance C is computed considering the charges contained in the space between two concentric cylinders, the inner with radius half the width of a monomer ($r_{\text{actin}} = 2.5$ nm) and the outer with radius $r_{\text{actin}} + \lambda_B$; both cylinders are one monomer high (5.4 nm). Thus,

$$C_0 = \frac{2\pi\epsilon l}{\ln\left(\frac{r_{\text{actin}}+\lambda_B}{r_{\text{actin}}}\right)},$$ (6.2)

where $l \approx 5.4$ nm is the length of a monomer.

The charge on this capacitor is assumed to vary in a nonlinear way with voltage, according to the following formula:

$$Q_n = C_0(V_n - bV_n^2).$$ (6.3)

Nonlinear voltage dependence of electrochemical capacitance for nano-scale conductors is due to finite density of states of the conductors. Details can be found in Refs. [159, 267]. We did not try to evaluate this parameter; instead, we used some trial values in our equations and found that, as long as b is reasonably small, the solutions converge to the constant ones in the cases that we considered. So we focused on constant solutions, and we can say that nonlinearity is not needed for our results.

The inductance L is computed as

$$L = \frac{\mu N^2 \pi (r_{\text{actin}} + \lambda_B)^2}{l}, \tag{6.4}$$

where μ is the magnetic permeability of water and N is the number of turns of the coil, that is the number of windings of the distribution of ions around the filament. It is approximated by counting how many ions can be lined up along the length of a monomer as $N = l/r_h$, and it is supposed that the size of a typical ion is $r_h \approx 3.6 \times 10^{-10}$ m.

The resistance R is estimated considering the current between the two concentric cylinders, obtaining

$$R = \frac{\rho \ln((r_{\text{actin}} + \lambda_B)/r_{\text{actin}})}{2\pi l}, \tag{6.5}$$

where resistivity ρ is approximately given by

$$\rho = \frac{1}{\Lambda_0^{K^+} c_{K^+} + \Lambda_0^{Na^+} c_{Na^+}}. \tag{6.6}$$

Here, c_{K^+} and c_{Na^+} are the concentrations of sodium and potassium ions, which were considered in previous papers to be 0.15 M and 0.02 M, respectively; $\Lambda_0^{K^+} \approx 7.4(\Omega m)^{-1} M^{-1}$ and $\Lambda_0^{Na^+} \approx 5.0(\Omega m)^{-1} M^{-1}$ are positive constants that depend only on the type of salts but not on the concentration [254]. With this formula, R_1 is computed and R_2 is taken as $\tilde{1}/7R_1$. Here, R_1 accounts for viscosity.

Figure 6.1 illustrates the circuit schema, where an actin monomer unit in a filament is delimited by the dotted lines.

The main equation for filaments is the following, derived from Ref. [254] (see Fig. 6.1 for the meaning of R_1, etc.).

$$LC_0 \frac{d^2}{dt^2}(V_n - bV_n^2)$$

$$= V_{n+1} + V_{n-1} - 2V_n - R_1 C_0 \frac{d}{dt}(V_n - bV_n^2)$$

$$- R_2 C_0 \left\{ 2 \frac{d}{dt} (V_n - b V_n^2) - \frac{d}{dt} (V_{n+1} - b V_{n+1}^2) \right.$$

$$\left. - \frac{d}{dt} (V_{n-1} - b V_{n-1}^2) \right\}. \tag{6.7}$$

In Ref. [235], we used this equation to compute the evolution of some tens of monomers in a filament.

It must be observed that the Bjerrum length will probably not be constant, but may vary both from point to point and with time. Also, one could consider the effects described in leading to charge density waves. However, the effects reported in that work refer mainly to electrostatically condensed bundles, while the bundles in our experimental setting were built using different network formation processes, that is via depletion forces. The depletion forces are a fundamental, entropic effect and do not rely on counterion condensation [41, 42]. Thus, the situations, i.e. the charge distributions, are completely different. Moreover, the present work is a computational analysis to prepare real experiments and to speculate about potential solutions, and will use the simplest possible stimuli, i.e. constant ones. Therefore, we did not consider phenomena that, even if they happened, in this context might be considered transient.

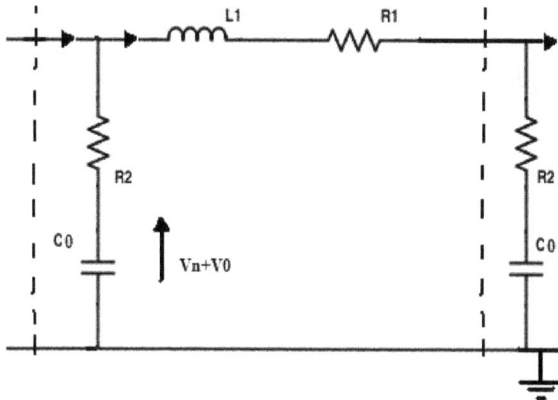

Fig. 6.1. A circuit diagram for the nth unit of an actin filament. From Ref. [254].

6.3 Extension to bundle networks

To extend the model to bundle networks, we must compute suitable electrical parameters. The actin filaments are made of elements, the actin monomers. We will model bundles as made of elements of the same height of a single monomer, and width depending on the bundle density. We will consider two possibilities:

(1) The filament density in the bundle is so low that each filament stands at a distance greater than twice λ_B from all the others. In this situation, we assume that filaments do not interact and that each one behaves as if it would not be in the bundle.
(2) The inner-bundle density is high enough that areas closer than λ_B to the filaments intersect. In this situation, we will conservatively assume that the influences of the filaments' ions cancel out.

In Case 1, we can either consider the parameters for a filament and solely multiply results by the number of filaments in the bundle or compute C, L, and R using the standard formulas for electrical parallel circuits. In Case 2, we only use the bundle radius instead of the filament one in the above formulas.

Considering a Bjerrum length $\lambda_B = 7.13 \times 10^{-10}$ m [254], results for high-density bundles at different bundle widths are displayed in Table 6.1. Results for low-density bundles made of varying filament numbers are shown in Table 6.2.

In the following, we will define equations for nodes. Equation (6.7) applies to elements **inside** the bundle, so we will use Equation (6.8) instead, where n is the index of the element, M is the number of elements linked to it, and the suffix n_k ranges in the set of such linked elements.

Table 6.1. C_0, L, and R_1 for high-density bundles.

Width	200 nm	450 nm	700 nm
C_0 in pF	$33.8 \cdot 10^{-4}$	$76 \cdot 10^{-4}$	$11.8 \cdot 10^{-3}$
L in pH	1668	8378	20227
R_1 in MΩ	0.173	0.077	0.049

Table 6.2. C_0, L, and R_1 for low-density bundles.

Filaments	1	25	50	75
C_0 in pF	$102.6 \cdot 10^{-6}$	$4.1 \cdot 10^{-6}$	$2 \cdot 10^{-6}$	$1.4 \cdot 10^{-6}$
L in pH	1.92	$7.66 \cdot 10^{-2}$	$3.83 \cdot 10^{-2}$	$2.56 \cdot 10^{-2}$
R_1 in MΩ	5.7	0.23	0.11	0.08

The term F_n represents an input voltage, which is supposed to be non-zero only for some values of n.

$$\frac{d^2}{dt^2}(V_n - bV_n^2) = \frac{1}{LC_0}$$

$$\times \left\{ \sum_{k=1}^{M} V_{n_k} - M \times V_n + F_n - R_1 C_0 \frac{d}{dt}(V_n - bV_n^2) \right.$$

$$\left. - R_2 C_0 \left\{ M \times \frac{d}{dt}(V_n - bV_n^2) - \sum_{k=1}^{M} \frac{d}{dt}(V_{n_k} - bV_{n_k}^2) \right\} \right\}. \quad (6.8)$$

These equations can represent any type of element in the network. When $M = 2$, they coincide with (6.7) and represent internal elements of a bundle. When $M = 1$, they refer to a free terminal element of a bundle that is not connected to anything else. We note that in Ref. [235] we used a slightly different equation for this case, namely, we always kept $M = 2$. The present form is more consistent with the model and its generalisation. Other values of M represent generic nodes.

6.4 The network

We used a stack of low-dimensional images of the three-dimensional actin network, produced in experiments on the formation of regularly spaced bundle networks from homogeneous filament solutions [123]. The network was chosen because it resulted from a protocol that reliably produces regularly spaced networks due to self-assembly effects [98, 123] and thus could be used in prototyping of cytoskeleton computers. From the stack of images we extracted a network description, in terms of edges and nodes, and used it as a substrate to compute the electrical behaviour. The extracted structure takes into account the main bundles in each image, with their intersections, and an estimate of bundles that can connect nodes in two adjacent images. It is not an accurate portrait of all the bundles, but it captures the main characteristics of the network.

The main steps to compute the network structure were as follows:

(1) After some preprocessing of the images (e.g. thresholding, contour finding, distance transform), we looked at the points placed at the local maxima of distance from the background. We considered that these are the nodes of the network. Each node found in this way has a centre and a radius (corresponding to the circle that can be inscribed in the foreground).

(2) We then tried to link nodes to each other with straight lines or elliptical arcs, checking that they do not go out of the bundles (with some tolerance, as the bundles are often bent). For this, starting from, e.g. Node 1, we considered the point spaced about 16 pixels along the line from Node 1 to, e.g. Node 2. If its colour was above the threshold, we went on to the next point 16 pixels farther. If not, we considered the points in a neighbourhood 4 pixels wide: if at least one was above the threshold, we considered that the edge is still in the bundle and went on, if not, we stopped.

(3) If we could not find any straight line, we tried some elliptical arcs, with big axis = distance between nodes and a range of small axes, using the same procedure.

(4) When we were able to reach, e.g. Node 2 from Node 1, we added the edge to the network, with its length (distance of the linked nodes for straight edges or approximate ellipse arc length for the others) and width = the average of the radii of its nodes; when we were not able to reach Node 2, we did not add the edge; if a node was not connected to anything, we did not consider it anymore.

(5) We also tried to detect edges between images. For this, we merged the bundles of two consecutive ones, shrunk them a bit, and applied the method of Step 2 to link nodes. We used a lower tolerance in this step and considered straight lines only.

All computations described here and in the other sections were performed using Python including its libraries Matplotlib [127] and SciPy [264].

At the end, we got a table of pictures — nodes; for each node we got its radius as well as position and the list of nodes that we were able to reach starting from there; and another table with edges and their characteristics. Based on this, we derived the data of Table 6.3.

These figures can be compared to typical characteristics found in other experiments [123].

The bundles are formed by depletion forces and neighbouring filaments will maximise their overlap region; the bundles will be at least as long as the longest filaments, experimentally it is even hard to form super long bundles. A typical length distribution of actin filaments has mean value of about 9–10 μm, which is already rather long, and a range between 10 and 50 μm for bundles. In our model, the length distribution is quite skewed, with 80% of edge lengths in the range 10–30 μm and another 7% in the range 30–50 μm.

About bundle sizes, 84% of radii in the model are in the range of 1.22–3.42 μm. This can be considered a reasonable estimation and compares well to the experiments [123].

Table 6.3. Parameters of the actin network used in the modelling.

Parameter	Value
Number of nodes in the main connected graph	2968
Number of edges in the main connected graph	7583
Max number of nodes linked to a node	13
Average number of nodes linked to a node	5.07
Standard deviation of nodes linked to a node	2.14
Average radius of edges in pixels	8.48
Max radius of edge in pixels	20
Min radius of edge in pixels	3
Standard deviation of radii of edges in pixels	2.62
Average edge radius, one pixel is 244.14 nm	2.07 μm
Average length of edges in pixels	70.11
Max length of edge in pixels	465.40
Min length of edge in pixels	4.12
Standard deviation of lengths of edges in pixels	41.54
Average edge length, one pixel is 244.14 nm	17.12 μm

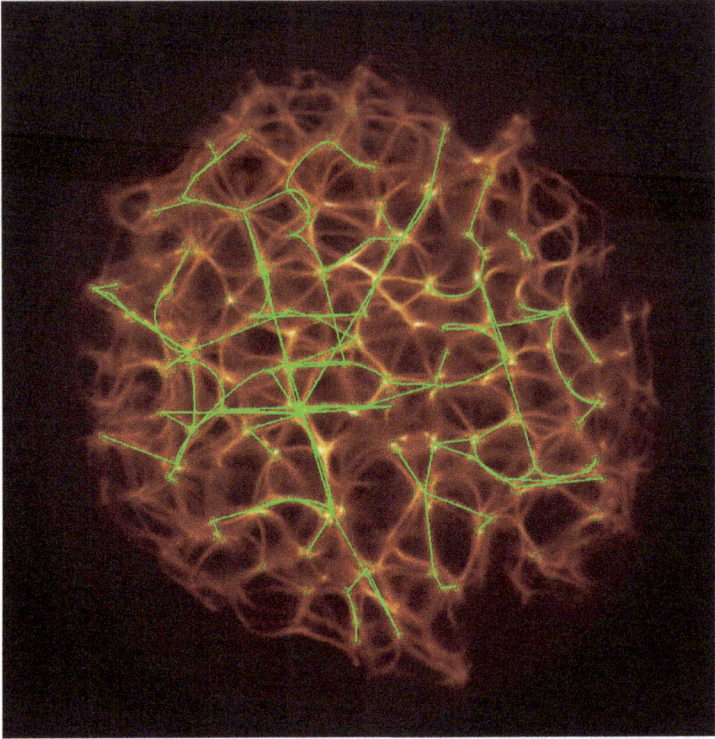

Fig. 6.2.   Two-dimensional centres and edges of a Z-slice.

The average number of filaments in bundles is quite difficult to measure experimentally. An estimation based on comparing the fluorescence intensity of a bundle against a single filament yielded a result of 45 (±25) filaments per bundle [246]. Considering even small bundles having 100 nm radius, we find that they can accommodate more than 100 filaments, considering a radius of 5.4 nm plus $2\lambda_B$. Hence, the low-density model for bundles is probably appropriate, and we decided not to consider possible interactions between actin filaments within the bundles.

As shown in Fig. 6.2, the white lines drawn on the original image represent the computed edges.

In a realistic experiment, one would have to set a support with electrodes in contact with the network. In a typical configuration, we

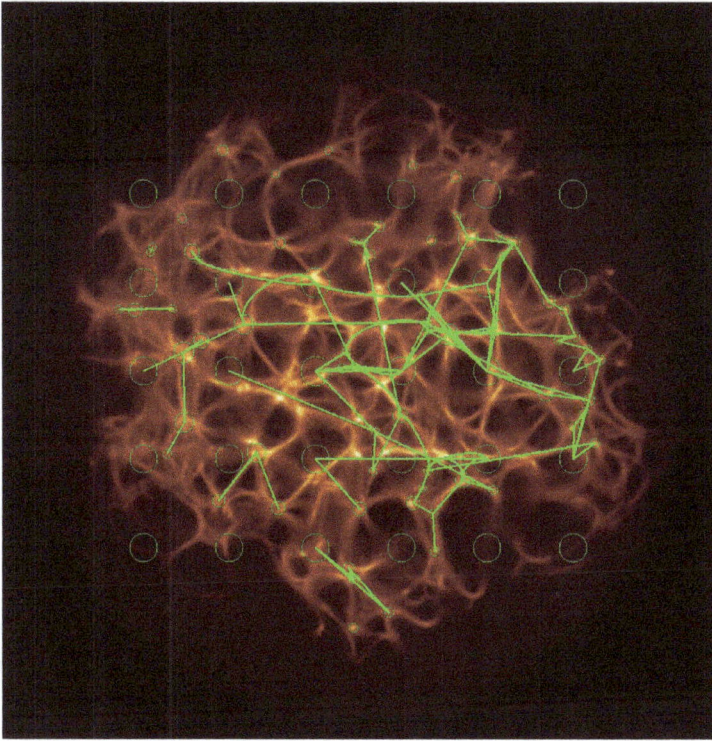

Fig. 6.3. The grid of electrodes as it appears when the network forms on top of the supporting glass.

will consider a grid of 5×6 electrodes, for instance, on a thin glass; their diameter is 10 μm and centre-to-centre distances are 30 μm. We considered two situations: (1) the network is grown in droplets sitting on the glass surface that holds the electrodes. This is actually very close to the experimental setup described in Refs. [122, 123]. And (2) the array of electrodes is set inside the network, i.e. in the middle of the actin droplet along its vertical axis. This might be the case if the networks were grown around the electrode layer or if it were placed in the network later. Figure 6.3 is an illustration of the network grown on top of the electrode-containing glass.

The general features of the network are listed in Table 6.3.

6.5 Preliminary results

We used the simplest possible form for the input functions F_n, which are constant functions:

$$\text{for } 0 < t < t_1 \; F_n \equiv \begin{cases} 1 & \text{if } n \in N_1 \\ 0 & \text{if } n \in N_0, \\ -1 & \text{if } n \in N_{-1} \end{cases} \tag{6.9}$$

where N_1, N_0, and N_{-1} are three sets of indices and t_1, the duration of the input stimuli, which can be equal to or less than the whole experiment time.

Numerical integration has been performed for bundles consisting of some tens of elements using various stimuli and electrical values. We considered both open bundles with free extreme elements, and closed ones where every element is connected to two others. Examples can be found in Fig. 6.4 (high-density open bundle) and Fig. 6.5 (low-density closed bundle).

These numerical experiments demonstrate that in all the cases considered, the solutions become constant after a transient time.

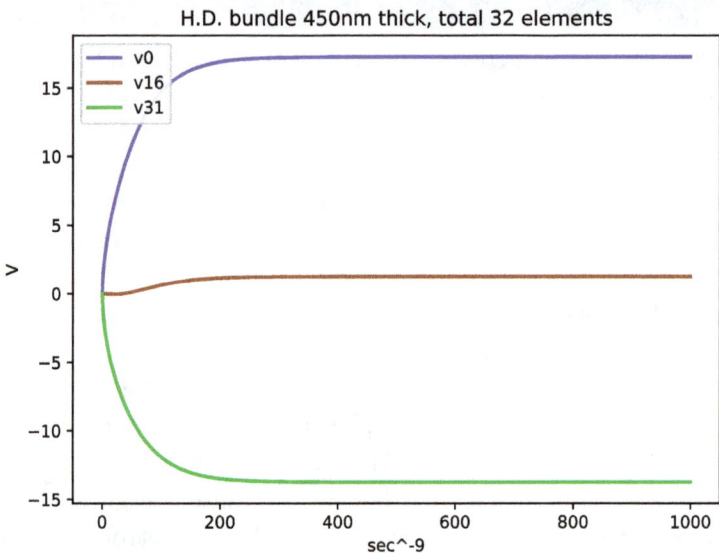

Fig. 6.4. Evolution of the potential of first, middle, and last elements of an open high-density (HD) bundle 32 elements long, 450 nm thick. Input was set at 1 and −1 at the first and last elements.

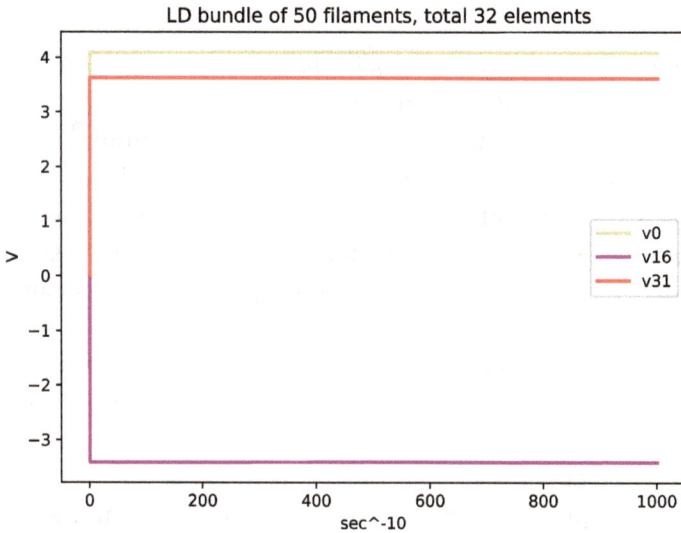

Fig. 6.5. Evolution of the potential of some elements of a closed low-density (LD) bundle 32 elements long, made of 50 filaments. Input was set at 1 and −1 at elements with index 0 and 15, respectively.

Moreover, when the inputs are blocked, all the solution converges to the same constant value, so that no currents can be detected.

We therefore considered constant stimuli lasting for all the experiment time and searched for constant solutions. Input bits are defined as a pair of points of the network, so that a +1 potential (in arbitrary units) is applied at one of them and −1 at the other to encode a value 1 of the bit; when no potential is applied, the bit value is zero.

Analogously, we chose pairs of points and measured the difference of their potential to read an output bit. A suitable threshold has been defined to distinguish the 1 and 0 values.

6.6 Results

6.6.1 Ideal electrodes

In this section, we consider ideal electrodes that (1) can be placed in any point on the network surface or inside it and (2) are so small that they would be in contact with one element of a single bundle only. Moreover, we use a slightly idealised network of spherical shape.

6.6.1.1 Boolean gates

We have randomly chosen 8 sites in the network and considered them as 4 pairs to represent 4 input bits.

Then we applied in turn all the possible input states from (0000) to (1111) and solved the system (6.8). It reduces to a linear algebraic structure and simplifies finding the values of the potential in the **nodes**. Then we checked, for all the sets of input states that correspond to a logical input, whose output bits correspond to the expected results for a gate.

For instance, to find the NOT gates we considered input state sets ((0000),(0001)), ((1000),(1001)) etc.; then we looked for all the output bits that are 1 for the first state and 0 for the second of one of the input sets.

The same procedure was used to find OR, AND, and XOR gates. We used three values for the output threshold: 2, 1, and 0.5.

The results of three runs are shown in Tables 6.4–6.7 revealing that once an input position is chosen, it is possible to find a suitable number of edges that behave as output for the main gate types.

6.6.1.2 Time estimates

We also computed the time that the network would need to converge to the constant solutions taking the time into account needed for an

Table 6.4.   Number of possible NOT gates.

Run	Thresh. 2	Thresh. 1	Thresh. 0.5
1	8266	8944	8409
2	3688	4660	4682
3	5730	7043	7455

Table 6.5.   Number of possible OR gates.

Run	Thresh. 2	Thresh. 1	Thresh. 0.5
1	4385	8191	12494
2	6360	8188	11336
3	5835	8260	11063

Table 6.6. Number of possible AND gates.

Run	Thresh. 2	Thresh. 1	Thresh. 0.5
1	3600	3562	2577
2	4506	43842	3119
3	4954	5076	3726

Table 6.7. Number of possible XOR gates.

Run	Thresh. 2	Thresh. 1	Thresh. 0.5
1	1543	2155	3749
2	584	986	1799
3	1009	1499	3083

element to discharge. As a first estimate, we used the value $R_1 C_0$, that is the discharge time of a pure RC circuit. Using the parameters for a single filament (or for low-density bundles made of independent filaments), we got $2.248 \cdot 10^{-3}$ sec to travel the 3843876 elements of the whole network. Parameters for a high-density network, adjusted for the estimated width of each bundle, gave a time of $2.25 \cdot 10^{-3}$ sec. In both cases, the velocity is of the order of 4.7 m/sec, two orders of magnitude larger than the estimate found in Ref. [129] with a different model (pure RC), but in the range estimated in Ref. [254] using the presented method.

6.6.2 Realistic electrodes

In this section, we consider electrodes that could be actually available, with their supporting glass. Moreover, we use the real network dimensions (the confocal images are 250 μm × 250 μm and they are spaced 110 μm in depth).

We considered both the case with the network being on top of the glass holding the electrodes, and the case when the electrodes are placed inside the network, along the middle plane of the confocal image stack.

6.6.2.1 *Boolean gates*

In the case of the network on top of the glass, we have randomly chosen 8 electrodes and considered them as 4 pairs to represent 4 input bits.

We applied in turn all the possible input states from (0000) to (1111) and solved the system (6.8). Then we computed the potential differences for all the pairs of electrodes that were not used as input and applied a suitable threshold to distinguish 0 and 1 bits. The threshold we used was the median of the differences.

As 10 electrodes out of the 18 connected to the network were not used as input, we had 45 potential output bits. We found that, considering all the possible input and output bits, we have 101 NOT gates, 113 OR gates, 46 AND gates, and 13 XOR gates.

It must be noted that the same pair of output electrodes may have been counted many times in these numbers. For instance, the potential difference of electrodes between 46th and 32nd electrode (electrodes in row 4 column 6 and in row 3 column 2) were considered a possible NOT gate for all the cases listed in Table 6.8.

In the case of the network with electrodes placed in the interior of the network, we have randomly chosen 12 electrodes and considered them as 6 couples to represent 6 input bits.

We applied in turn all the possible input states from (000000) to (111111) and solved the system (6.8). Then we computed the potential differences for all the pairs of electrodes that were not used as input and applied a suitable threshold to distinguish 0 and 1 bits. The threshold we used was the median of the differences.

As 15 electrodes out of the 27 connected to the network were not used as input, we had 105 potential output bits. We found that, considering all the possible input and output bits, we have 1885 NOT gates, 1279 OR gates, 783 AND gates, and 467 XOR gates.

6.7 Finite state machine

The actin network implements a mapping from $\{0,1\}^k$ to $\{0,1\}^k$, where k is a number of input bits represented by potential difference in pairs of electrodes, as described above. Thus, the network can

Table 6.8. Possible NOT with a single edge.

Input state	Output value	NOT on bit
1100	1	4
1101	0	
1010	1	4
1011	0	
1000	1	4
1001	0	
0110	1	4
0111	0	
0100	1	4
0101	0	
0111	0	2
0011	1	
0101	0	2
0001	1	
1011	0	1
0011	1	
1001	0	1
0001	1	

be considered as an automaton or a finite-state machine, $\mathcal{A}_k = \langle \{0,1\}, C, k, f \rangle$. The behaviour of the automaton is governed by the function $f : \{0,1\}^k \to \{0,1\}^k$, $k \in \mathbf{Z}_+$. The structure of the mapping f is determined by exact configuration of electrodes $C \in \mathbf{R}^3$ and geometry of the AF bundle network.

6.7.1 Using two values of $k = 4$ and $k = 6$

The machine $\mathcal{A}_4$ represents the actin network placed onto an array of electrodes. In this case, having at our disposal 45 potential output bits, the number of combinations of 4 of them is 148995. We therefore limited the study at the output positions that assume a 1 value more than 6 and less than 11 times for the 16 input states. In this way we found 11 output bits and computed the state transitions for the 330 machines that one can obtain choosing 4 out of them, $k = 4$.

The machine $\mathcal{A}_6$ represents the actin network where the array of electrodes is inside the network. In this case, having at our disposal 105 potential output bits, the number of combinations of 6 of them is quite large. We therefore limited the study at the output positions that assume a 1 value 32 times for the 64 input states. In this way we found again 11 output bits and computed the state transitions for the 462 machines that one can obtain choosing 6 out of them, $k = 6$.

We derived structures of functions f_4 and f_6, governing behaviour of automata $\mathcal{A}_4$ and $\mathcal{A}_6$, as follows. There is potentially an infinite number of electrode configurations from $\mathbf{R}^3$. Therefore, we selected 330 and 462 configurations C for machines $\mathcal{A}_4$ and $\mathcal{A}_6$, respectively, and calculated the frequencies of connections of input to output states, obtaining two probabilistic state machines $= \langle \{0,1\}, p, k, f \rangle$, where $p : \{0,1\}^{k\{0,1\}} \to [0,1]$, the p assigns a probability to each mapping from $\{0,1\}^k$ to $\{0,1\}$. Thus, a state transition of $\mathcal{A}_k$ is a directed weight graph, where weight represents a probability of the transition between states of $\mathcal{A}_k$ corresponding to nodes of the graph. The weighted graph can be converted to a non-weighted directed graph by removing all edges with weight less than a given threshold θ. In the following, we perform trimming for several thresholds with 0.1 increment.

The graph remains connected for θ till 0.1 (Fig. 6.6). The graph for $\mathcal{A}_4$ is characterised by having no unreachable nodes and several absorbing states (Fig. 6.6(a)), while the graph for $\mathcal{A}_6$ has a number of unreachable nodes (Garden-of-Eden states) and less, than $\mathcal{A}_4$, absorbing states (Fig. 6.6(b)).

The state transition graph of $\mathcal{A}_6$ becomes disconnected for $\theta = 0.2$ (Fig. 6.7(b)) and the graph of $\mathcal{A}_4$ remains connected (Fig. 6.7(a)).

Another way of converting weighted, probabilistic, state transition graphs into non-weighted graphs is by selecting for each node x a successor y such that the weight of the arc (xy) is the highest among all arcs outgoing from x. These graphs G_4 and G_6 of most likely transitions are shown in Fig. 6.8. The graph G_4 (Fig. 6.8(a)) has two disconnected sub-graphs, eight Garden-of-Eden states, and two absorbing states corresponding, respectively

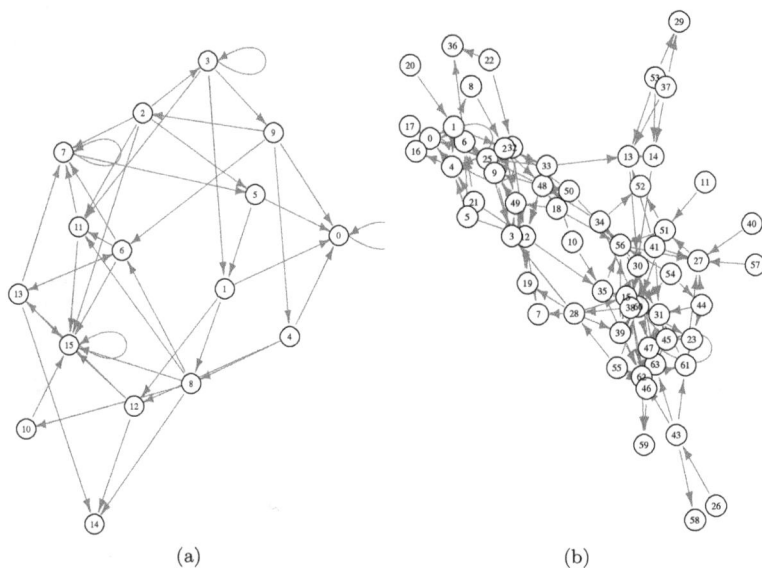

Fig. 6.6. State transitions graphs for (a) $\mathcal{A}_4$ and (b) $\mathcal{A}_6$, trimming threshold is $\theta = 0.1$. Nodes are labelled by digital representation of 4-bit (a) and 6-bit (b) states.

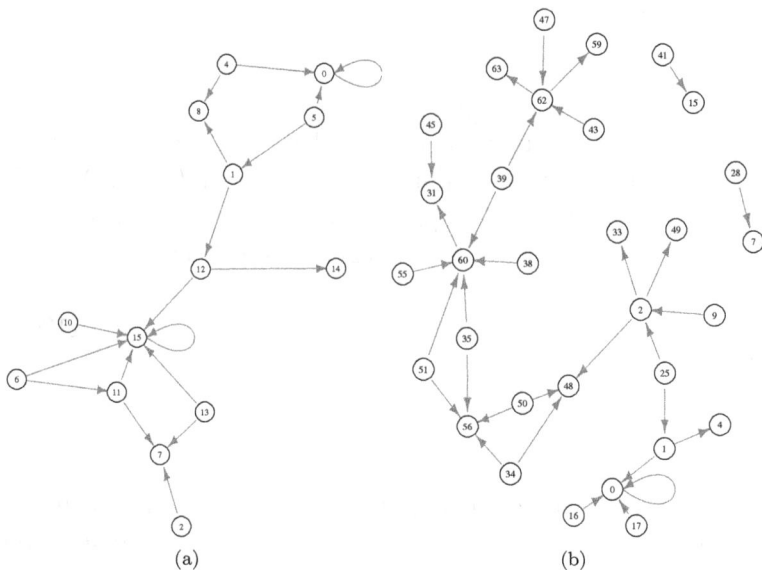

Fig. 6.7. State transitions graphs for (a) $\mathcal{A}_4$ and (b) $\mathcal{A}_6$, trimming threshold is $\theta = 0.2$. Nodes are labelled by digital representation of 4-bit (a) and 6-bit (b) states, respectively.

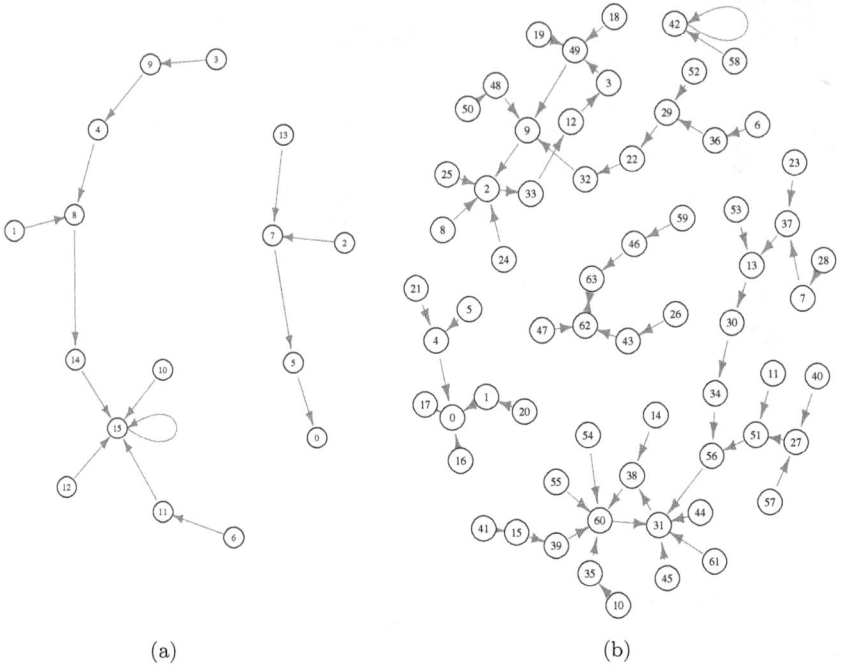

(a)   (b)

Fig. 6.8. Graphs representing most likely transitions G_4 (a) and G_6 (b) of $\mathcal{A}_4$ (a) and $\mathcal{A}_6$ (b).

to (1111) and (0000); the graph has no cycles. The graph G_6 has five disconnected sub-graphs (Fig. 6.8(b)). Two of them have only absorbing states, corresponding to (00000) and (101010), and no cycles. Three of the sub-graphs do not have an absorbing state but have cycles: (111110) $\rightarrow$ (111111) $\rightarrow$ (111110), (001111) $\rightarrow$ (001111) $\rightarrow$ (100110) $\rightarrow$ (111100) $\rightarrow$ (001111) and (000011) $\rightarrow$ (110001) $\rightarrow$ (001001) $\rightarrow$ (000010) $\rightarrow$ (100001) $\rightarrow$ (001100) $\rightarrow$ (000011).

6.8 Discussion

In this study, we employed a physical model of ionic currents on a nonlinear transmission network to demonstrate the realisation of Boolean functions on actin networks and investigate the distribution of achievable Boolean gates. The geometry of the three-dimensional actin bundle network used in the model was derived from experimental data obtained in the laboratory. The outcomes of our study serve as a feasibility assessment for potential future experimental prototypes of cytoskeleton computing devices.

Furthermore, we derived finite-state machines that can be implemented on actin networks. These machines hold significant importance in two regards. First, the state transition graphs of these machines can serve as unique identifiers of actin networks formed under different experimental or physiological conditions, offering valuable insights into structural variations within actin networks. Second, the study of machine structures can advance our understanding of the computational capabilities of actin networks in the realm of formal language recognition.

Overall, our research contributes to the exploration of actin networks as a promising platform for computation, paving the way for further investigations and potential applications in the field of cytoskeleton computing.

Probabilistic finite-state machines are computational models that are capable of modelling and generating distributions over sets of strings, sequences, words, phrases, terms, and trees. Unlike deterministic finite-state machines, probabilistic finite-state machines incorporate probabilistic transitions, allowing for a broader range of language recognition and generation capabilities. These machines have been shown to be capable of recognising an uncountable number of languages, providing a powerful framework for probabilistic modelling and language processing tasks.

Indeed, the actin structures studied in our research may not directly lead to the development of powerful computing machines comparable to modern computers. However, our findings provide a proof of principle that these structures have the potential to perform basic processing tasks. One of the notable advantages of such structures is their biodegradability, which eliminates concerns regarding recycling and environmental impact. This aspect opens up possibilities for the development of green technologies that align with sustainable practices.

Chapter 7

Logical Gates Implemented by Solitons at the Junctions

Stefano Siccardi[*,‡] and Andrew Adamatzky[†,§]

*Consorzio Interuniversitario Nazionale per l'Informatica,
Italy and Unconventional Computing Laboratory,
UWE Bristol, UK*
†*The Unconventional Computing Lab,
University of the West of England, Bristol, UK*
‡*stefano.siccardi@uwe.ac.uk*
§*andrew.adamatzky@uwe.ac.uk*

Abstract

We investigate the propagation of solitons on one-dimensional lattices of nodes. The neighbouring nodes interact with each other through Morse interaction. Through numerical integration, we demonstrate that solitons colliding at the junctions of two lattices can function as Boolean logic gates, specifically the AND, OR, and NOT gates. The logic gates are determined by the geometry of the junctions and the phase difference of the colliding solitons. By analysing the behaviour of solitons at the junctions, we identify the conditions under which the solitons can implement the desired logic operations. The geometry of the junctions and the relative phase of the colliding solitons play crucial roles in determining the output of the logic gates. Furthermore, we assess the feasibility of using these soliton-based logic gates in practical applications by designing a NOR gate. The NOR gate is an important component in digital logic circuits and is known for its versatility. By utilising the properties of solitons at the junctions, we evaluate the potential of implementing the NOR gate using the proposed design. Overall, this study explores the capabilities of solitons in implementing Boolean logic gates through their interactions at lattice junctions. The findings provide insights into the feasibility of employing solitons for practical applications in digital logic circuit design.

7.1 Introduction

The concept of computing with solitons has been of interest to engineers, computer scientists, and physicists since the late 1980s [245] and has gained momentum with the emergence of collision-based computing [9]. Logical gates can be implemented using solitons in two ways.

In the first approach, logical variables are encoded in the states of solitons, which include characteristics such as shape, phase, speed, and energy level. When two solitons collide, their states can potentially change. The states of the solitons before the collision represent input logical values, while the states after the collision represent the output logical variables. In the second approach, logical variables are encoded based on the presence or absence of solitons at specific locations in space. Certain locations are designated as inputs, while others serve as outputs. If a soliton is present at an input location, the input is considered 'True'; otherwise, it is considered 'False'. When two solitons collide, their velocity vectors can change, and they may reflect, annihilate, or merge. The resulting trajectories of the solitons after the collision are interpreted in terms of logical operations. Significant progress has been made in collision-based computing with solitons [37, 132–134], including the construction of functionally complete sets of logical gates [244], information transfer between colliding solitons [131], computation via phase coding [245], and even quantum computing using the entanglement of solitons in the Frenkel–Kontorova model [161]. Gliders in cellular automata serve as discrete analogies of solitons [91, 149, 165, 168, 191, 249, 282], which means that designs of glider-based computing devices implemented in cellular automata [29, 170, 214, 243] can also be considered in the context of soliton-based computing. We became interested in solitons due to our ongoing studies in computing potential of actin cytoskeleton networks. While designing experimental laboratory prototypes of computing devices from living slime mould *Physarum polycephalum* [5], we found that actin networks might play a key role in distributed sensing, decentralised information processing, and parallel decision-making in a living cell [172–174]. We proposed a finite state machine, or automaton network, model of an actin filament [27], and found that the actin automata fibres exhibit a wide range of mobile and stationary patterns. We later exploited these patterns to design computational models of quantum [230] and Boolean [235] gates implementable on actin fibres, including universal computation with cyclic tag systems [166]. Automata network models are easy to work with and are convenient for fast prototyping of computing circuits based on collisions between travelling localisations. However, they might not grasp certain key features of the physical substrate. Thus, we decided to represent logical signals with solitons. Solitons in protein chains have been relatively well studied since the pioneer works by Davydov [71–73]. Moreover, there is evidence of soliton-like travelling localisations: electrical charges and ionic clouds along actin filaments [254, 255].

To assess the potential of developing computing circuits using programmable polymerisation of branching actin networks, we simplify the actin filaments by representing them as one-dimensional lattices composed of nodes with nearest neighbour interactions. We focus on studying the propagation of localised excitations, known as solitons, and their interactions at the junctions or branching sites of the lattices. In this study, we employ the Morse interaction potential between nodes, which was proposed by Velarde *et al.* [261–263]. One of the advantages of this approach is that it allows us to consider the movement of both electrons and lattice excitations (solitons) simultaneously. For instance, previous research [261,262] has revealed that when two solitons with opposite vector vectors collide, they exchange electron probability density between each other. However, it is important to note that in the present work, we will not be considering electrons as part of our analysis.

7.2 Model

We consider one-dimensional lattices of nodes. Each node has a potential energy. The potential energy of a node depends on the positions of the neighbouring nodes. Nodes oscillate around their equilibrium positions, a minimum of the potential energy. When a few nodes are excited, the oscillation travels along the lattice as a solitary wave.

The Hamiltonian of the lattice nodes is as follows [262]:

$$H_{Mo} = \sum_{i=1}^{N} \left[\frac{1}{2} \frac{p_i^2}{M} + D(1 - \exp[-b(q_i - q_{i-1})])^2 \right], \quad (7.1)$$

where N is the number of nodes, M is the mass of a lattice node (all nodes have the same mass), q_i is the displacement of node i from its equilibrium position, p_i is the momentum of node i, D is the depth of the well defined by the potential (dissociation energy), b is the stiffness of the spring constant in the Morse potential.

The dynamics of the motion of nodes is as follows:

$$\frac{d^2 q_i}{dt^2} = (1 - \exp[-(q_{i+1} - q_i)])\exp[-(q_{i+1} - q_i)]$$
$$- (1 - \exp[-(q_i - q_{i-1})])\exp[-(q_i - q_{i-1})]. \quad (7.2)$$

Following Refs. [261, 262], we take $(2Db^2/M)^{-1/2}$ as a unit of time, $(2D)$ as unit of energy, b^{-1} as displacement. The values used in Refs. [261, 262] are $b = 4.45\text{Å}^{-1}$, $D = 0.1\,\text{eV}$, and the time unit is $3.29\,10^{-13}$ sec. For actin monomers, we can use $b = 4.89\,\text{nm}^{-1}$, $D = 7.67\,\text{eV}$ (see [39]), and $M = 6.7\,10^{-23}\,\text{kg}$ corresponding to a time unit of $2.5\,10^{-11}$ sec.

The one-dimensional lattices cross each other and form junctions (Fig. 7.1). Geometry 1 (Fig. 7.1(a)) consists of two chains forming a junction at an internal point and of a third chain starting at the junction. Geometry 2 (Fig. 7.1(b)) consists of three lattices forming a Y-shape, the junction is formed by the extreme points of the lattices.

At the junction, a node of a lattice interacts also with its nearest neighbours in complementary lattices. This interaction is shown by grey lines in Fig. 7.1(c). We use the same potential for the junctions as for the lattices, but we introduce an attenuation parameter to account for a weaker interaction between nodes of different lattices.

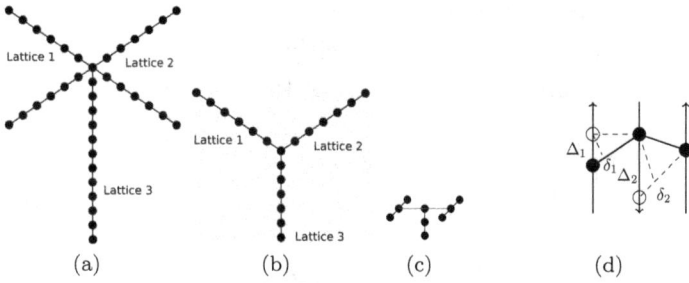

Fig. 7.1. Junctions of lattices. Nodes are black discs. Interactions between nodes along the same lattice is shown by solid lines. (a) Geometry 1. (b) Geometry 2. (c) Perspective view of details of a generic junction of Geometry 1. (d) Neighbourhood structure at the junction, filled circles refer to equilibrium positions, empty circles and dotted lines to displaced positions.

When nodes in the junction oscillate around their equilibrium positions, they move along their lattices. Their distances from their nearest neighbours in the complementary filaments changes and their interactions are affected. This is shown in Fig. 7.1(c), where the filled circles represent nodes in their equilibrium positions, empty circles are nodes in displaced positions, and dotted lines are the corresponding distance changes. We approximate the variations of the distances between the three nodes of a junction with the variations of their positions in their respective lattices.

We note that the coordinate q_i of the node i is relative to the internal reference of its filament. If the node is part of a junction, an increase of its q_i may result in an increase or a decrease of its distance from the other nodes in the junction, depending on the positions and directions of the filaments. For example, given the orientations and positions as shown in Fig. 7.1(d), an increase Δ_1 of the coordinate of the leftmost node will result in a decrease δ_1 of its distance from the central node (discordant variations). An increase Δ_2 of the coordinate of the central node will result in an increase δ_2 of the central node's distance from the rightmost node (concordant variations). We assume that a displacement q_i of the ith node along its lattice corresponds to a fraction of the displacement along the junction. We incorporate this effect in the exploratory parameter α in the expression of the potential acting between the nodes of the junction. We assume all structures have concordant variations of the q_i unless discordant variation is explicitly stated.

The Hamiltonian for the complex of three lattices is

$$
\begin{aligned}
H_{\text{lattice}} = \sum_{i=1}^{N_1} & \left[\frac{1}{2}\frac{p_i^2}{M} + D(1 - \exp[-b(q_i - q_{i-1})])^2 \right] \\
+ & \sum_{i=N_1+1}^{N_1+N_2} \left[\frac{1}{2}\frac{p_i^2}{M} + D(1 - \exp[-b(q_i - q_{i-1})])^2 \right] \\
+ & \sum_{i=N_1+N_2+1}^{N_1+N_2+N_3} \left[\frac{1}{2}\frac{p_i^2}{M} + D(1 - \exp[-b(q_i - q_{i-1})])^2 \right] \\
+ & \alpha D(1 - \exp[-b(q_{i_1} - q_{i_2})])^2] \\
+ & \alpha D(1 - \exp[-b(q_{i_2} - q_{i_3})])^2].
\end{aligned}
\tag{7.3}
$$

A set of equations is written for each lattice. The first and last equations of each set lack the terms with q_{i-1} and q_{i+1}, respectively, and a term is added for the nodes in the junctions. where N_1, N_2, N_3 are the numbers of nodes in the Lattices 1, 2, and 3, q_{i_1}, q_{i_2}, q_{i_3} are the nodes in the junction, α parametrises the strength of the interaction in the junction. This Hamiltonian refers to structures with concordant variations of the coordinates; if discordant variations were considered, some of the q_{i_k} would have the opposite sign.

We derive equations of motion analogous to the equations of motion in Equation (7.2). They have been numerically solved for Geometries 1 and 2. The conservation of energy has been checked at every integration step.

7.3 AND gate

Inputs are Lattices 1 and 2. Output is Lattice 3. Presence of a soliton at the top end of Lattice 1 or 2 means logical input TRUE. Presence of a soliton at the bottom end of Lattice 3 means logical output TRUE. Absences of the solitons corresponds to logical values FALSE.

In all conditions, the excitation spreads in part to all the lattices, see e.g. Fig. 7.4, which shows an example of the evolution of the structure. Therefore, by presence of a soliton we mean that the lattice energy is above the specified threshold; absence, if the energy is below the threshold.

The AND gate is implemented in Geometry 1 with three lattices of 40 nodes each. Propagation and interaction of solitons is tested for $\alpha = 0.1$ and 0.4. Runs with initial conditions with no solitons, one at the beginning of Lattice 1 or Lattice 2, show that a single soliton's energy will basically stay in its lattice, with just a small amount passing to the others (Fig. 7.2). If there is a soliton only in one of the input lattices, no soliton will appear in the output lattice.

When both inputs of the gate are TRUE, two solitons generated in input Lattices 1 and 2 transfer a considerable amount of energy

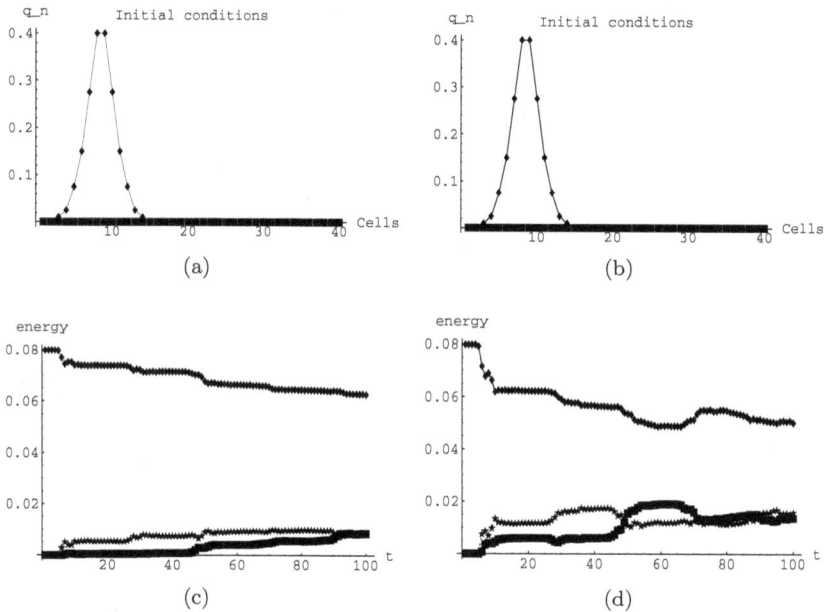

Fig. 7.2. Gate AND in Geometry 1 (Fig. 7.1(a)): only one input is TRUE, either soliton is excited in Lattice 1 or in Lattice 2. (a, b) Initial energy distribution along lattices. (c, d) Energy evolution with time. (a, c) $\alpha = 0.1$. (b, c) $\alpha = 0.4$. Energy for Lattice 1 is shown by solid line with rhomb markers, Lattice 2, dashed line with square markers, Lattice 3, dotted line with start markers. Values q_n are measured in b^{-1}, energy in $2D$, and time in $(2Db^2/M)^{-1/2}$ units.

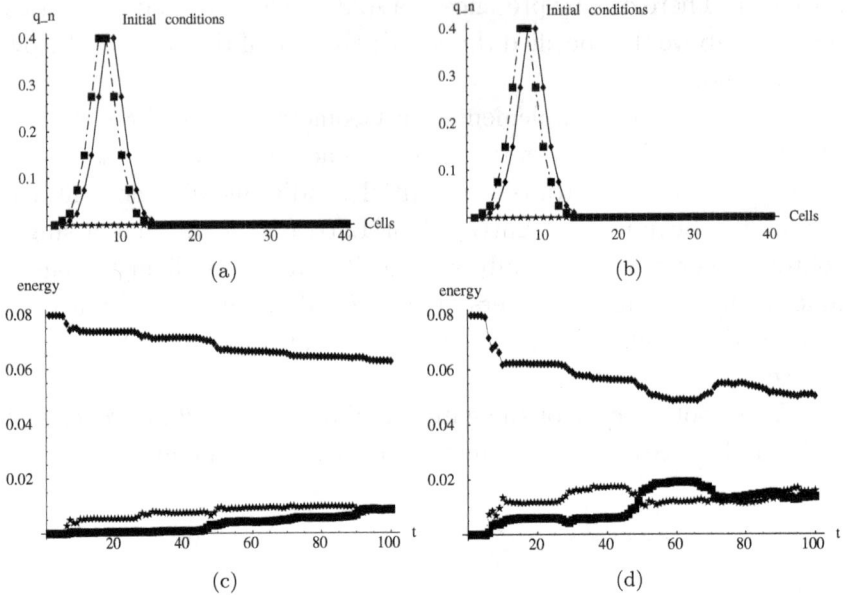

Fig. 7.3. Gate AND in Geometry 1 (Fig. 7.1(a)): both inputs are TRUE, solitons are excited in Lattice 1 and 2. (a, b) Initial energy distribution along lattices. (c, d) Energy evolution with time. (a, c) $\alpha = 0.1$. (b, c) $\alpha = 0.4$. Values q_n are measured in b^{-1}, energy in $2D$ and time in $(2Db^2/M)^{-1/2}$ units. Energy for Lattice 1 is shown by solid line with rhomb markers, Lattice 2 dashed line with square markers, Lattice 3 dotted line with start markers.

to output Lattice 3 if their motions along the filaments are in phase (that is they start at the same positions or very near). The amount of the energy transferred from input lattices to output lattice depends on the strength of the junction (Fig. 7.3). Thus, when solitons in input lattices are in phase, the Geometry 1 implements AND gate.

As an illustration, in Fig. 7.4 we show a three-dimensional plot of the displacement of nodes around their equilibrium positions of the lattices for the case $\alpha = 0.1$ for a single input TRUE: a single initial soliton in Lattice 1. The excitation is transmitted to Lattices 2 and 3. Two solitons with opposite directions of motion are excited in Lattice 2, even if most of the energy stays in its original lattice.

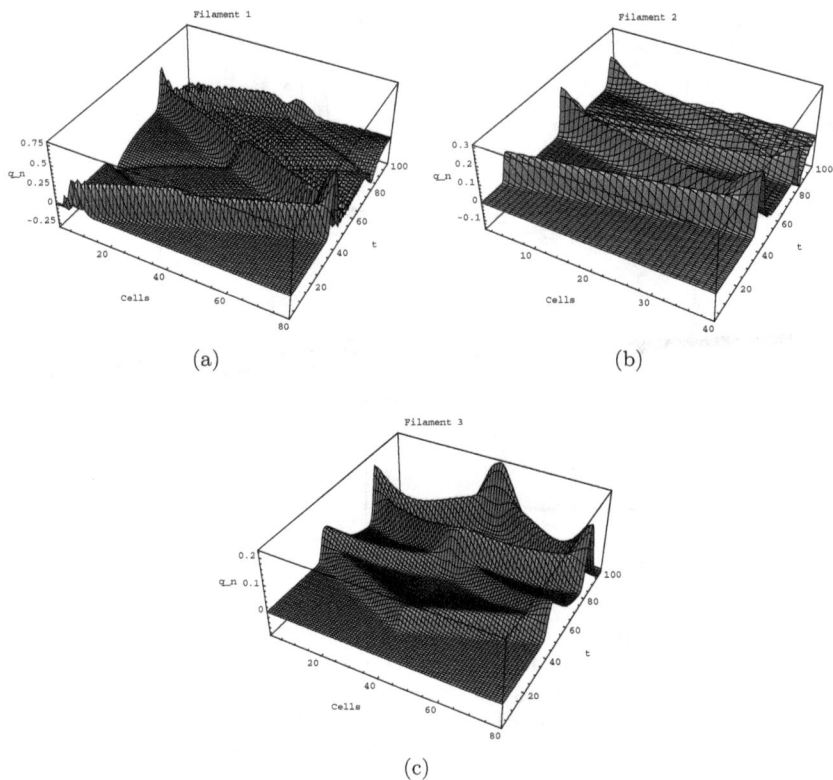

(a)

(b)

(c)

Fig. 7.4. Evolution of a single soliton starting at the beginning of Lattice 1, Gate AND in Geometry 1. (a) Evolution of Lattice 1; (b) Evolution of Lattice 3; (c) Evolution of Lattice 2; $\alpha = 0.1$.

The excitations are reflected at the ends of the filaments and they meet again and again. As they are soliton-like but not perfect solitons, they interfere and may change or lose their shapes. Changing the geometry of the filament structures, that is making them cross at different points, may change the relative phases of the excitation periodic motions along the filaments, so that they meet at different points with different consequences on their changes. In case of segments of infinite length, this phenomenon would not occur.

If the solitons are not in phase, but shifted by some nodes from each other, a much less amount of energy is transferred to the inner

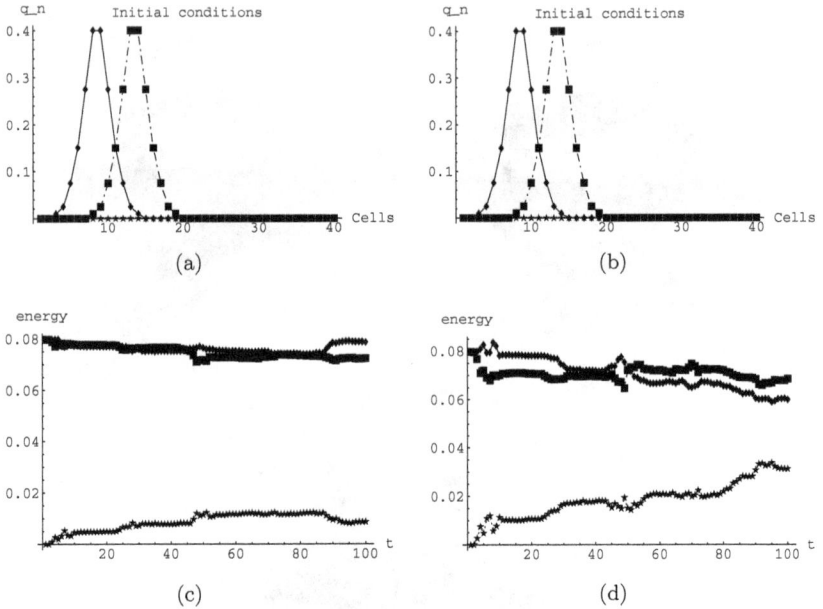

Fig. 7.5. Evolution of two solitons starting at different points of Lattices 1 and 2 (a, b) Initial conditions. (d, c) Energy evolution with time. (a, d) $\alpha = 0.1$. (b, d) $\alpha = 0.4$. Values q_n are measured in b^{-1}, energy in $2D$, and time in $(2Db^2/M)^{-1/2}$ units. Energy for Lattice 1 is shown by solid line with rhomb markers, Lattice 2, dashed line with square markers, Lattice 3, dotted line with start markers.

lattice: see Fig. 7.5, where the starting point of the solitons differs by 6 nodes. In this case, no logical operation can be performed by the structure.

7.4 OR gate

The OR is not implemented in Geometry 1, but in Geometry 2 only (Fig. 7.1(b)). Figure 7.6 shows the evolution of excitation in scenarios where at least one input is TRUE. The left column represents the evolution of an excitation starting at the beginning of the lattice, the central column, the evolution of two excitations starting at the beginning of Lattices 1 and 2, and the right column, a single excitation in Lattice 2. The interaction parameter is $\alpha = 0.4$. The

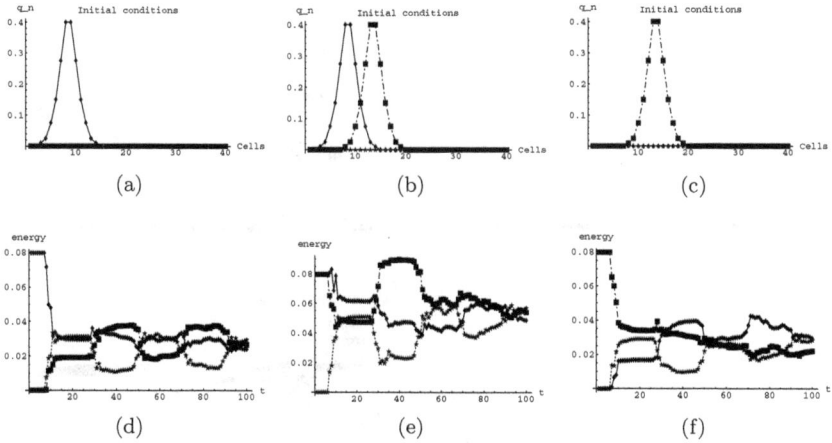

Fig. 7.6. OR gate in Geometry 2 (Fig. 7.1(b)), not implemented in Geometry 1. (a–c) Initial distribution of energy; (d–f) Energy evolution with time. (a, d) Input 1 is TRUE. Single soliton in Lattice 1. (b, e) Inputs 1 and 2 are TRUE. Two solitons at the beginning of Lattices 1 and 2. (c, f) Input 2 is TRUE. Single soliton in Lattice 2. Values q_n are measured in b^{-1}, energy in $2D$ and time in $(2Db^2/M)^{-1/2}$ units. Energy for Lattice 1 is shown by solid line with rhomb markers, Lattice 2, dashed line with square markers, Lattice 3, dotted line with start markers.

energy evolutions show that the excitations are always transferred from input Lattices 1 and 2 to output Lattice 3. Thus, the gate OR is implemented.

7.5 NOT gate

The NOT gate is not implemented in Geometry 1, but in Geometry 2 only (Fig. 7.1(b)) with discordance of nodes' coordinates in the junction. The Lattices 2 and 3 are in such a position that the variations of their coordinates are so that an increase of the internal coordinate q_i (Fig. 7.1(d)) of the node of the Lattice 2 in the junction results in a decrease of the node's distance from the central node in the junction.

Figure 7.7 shows the evolution of excitations. We keep the input Lattice 2 excited. The evolution with the resting input Lattice 1 is shown in Fig. 7.7(a, c). The excitation is transmitted to the output

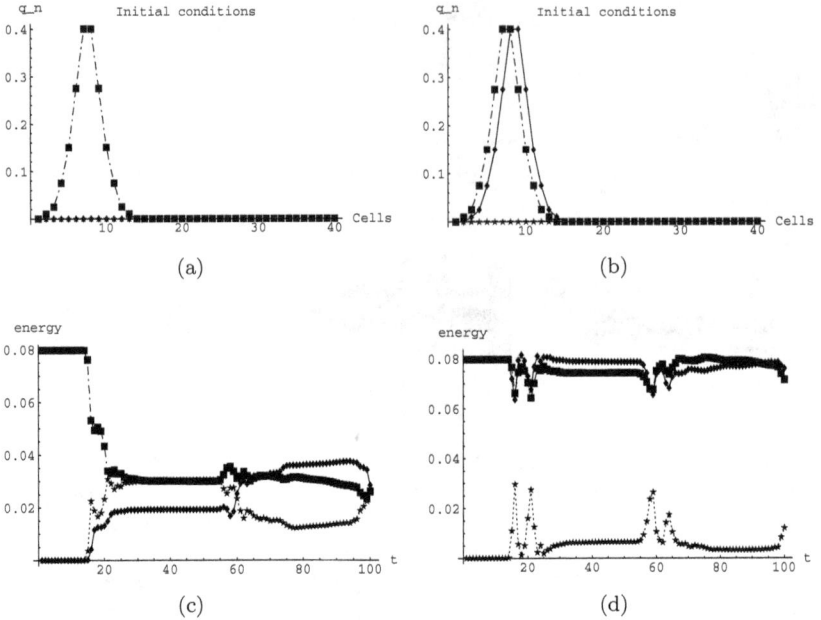

Fig. 7.7. NOT gate in Geometry 2 with discordant coordinates, not implemented in Geometry 1. (a, b) Initial distribution of energy; (c, d) Energy evolution with time. (a, c) Single soliton in Lattice 2. (b, d) Two solitons at the beginning of Lattices 1 and 2. Values q_n are measured in b^{-1}, energy in $2D$, and time in $(2Db^2/M)^{-1/2}$ units. Energy for Lattice 1 is shown by solid line with rhomb markers, Lattice 2, dashed line with square markers, Lattice 3, dotted line with start markers.

Lattice 3. The evolution with the excited input Lattice 1 is shown in Fig. 7.7(b, d). The excitation is not transmitted to the output Lattice 3. Thus, the negation operation is implemented.

7.6 Cascading gates

To make a NOR gate, we connect NOT gate to OR gate as shown in Fig. 7.8(a). There five filaments are connected in two junctions. The coordinates of the first junction (black circle) are concordant. The coordinates of the second junction (empty circle) are discordant. All lattices have 40 nodes, except the bottom right lattice. The bottom right lattice carries the auxiliary signal necessary to implement negation. This lattice has 80 nodes. The input signals, starting at the

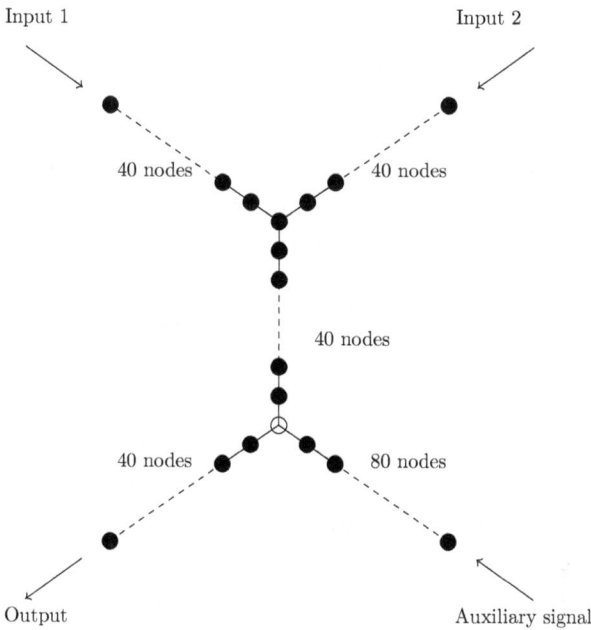

(a)

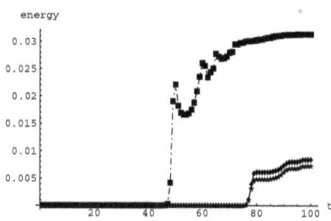

(b)

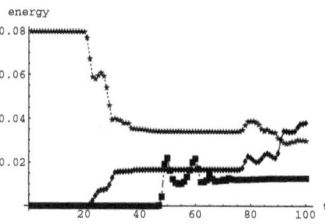

(c)

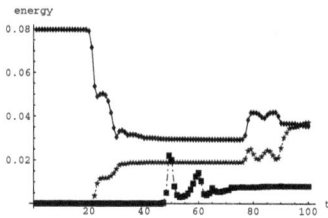

(d)

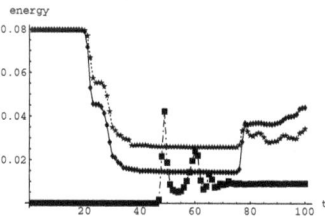

(e)

Fig. 7.8. A NOR gate. (a) Structure. (b) Energy evolution with input (FALSE, FALSE). (c) Energy evolution with input (FALSE, TRUE). (d) Energy evolution with input (TRUE, FALSE). (e) Energy evolution with input (TRUE, TRUE). Energy for Input 1 is shown by solid line with rhomb markers, Input 2, dotted line with start markers, Input 3, dashed line with square markers.

top lattices, travel along the same number of nodes as the auxiliary
signal before they collide with each other.

The energy diagrams (Fig. 7.8(b–e)) show that the circuit works
correctly. If no input signals are applied, the output energy is high. If
one or two input signals are applied, the output energy is low. From
a quantitative point of view, the values of input energies are 0.0797
(in units $2D$). We consider an output represents TRUE when energy
is 0.031 (39% of the input energy), see Fig. 7.8(b). Logical FALSE
corresponds to energy 0.01 (15% of the input), see Fig. 7.8(c–e).

7.7 Signal immunity to noise

In logical circuits discussed signals are solitons. We consider now
what happens with the signals if the circuits operate in the presence
of noise or if there is a dissipation of energy in the medium. Following
solitonic signals standardisation approach by Rand *et al.* [209] we
evaluated the signal immunity to noise and friction. We added the
following terms:

$$- \gamma_1 \frac{dq_i}{dt} + \gamma_2 \xi(t) \tag{7.4}$$

to the equations of motions. The first term describes the standard
friction between the nodes and the surrounding environment. The
second term represents stochastic forces acting on lattice nodes. One
might also consider nonlinear friction, which models active Brownian
particles that carry refillable energy depots, e.g. see [61,260]), but we
here limit ourselves to the basic mechanism that can subtract energy
and make the signals more difficult to detect.

In Fig. 7.9, we show the variations of the energies for increasing
values of friction and noise. The friction parameters, γ_1 in Eq. (7.4),
increase from 0 to 0.2 in Fig. 7.9(a–c). The corresponding changes
in the ratio of the energies during the last steps of the computation,
when solitons have propagated into the output lattice, lead to the
expected energy values for FALSE or TRUE. The values of γ_2 range
from ≈1% to 4% of the maximum of the order of magnitude of
Eq. (7.2) for the initial conditions used for the computations and
from ≈3% to 20% of the average of the same value. This seemed

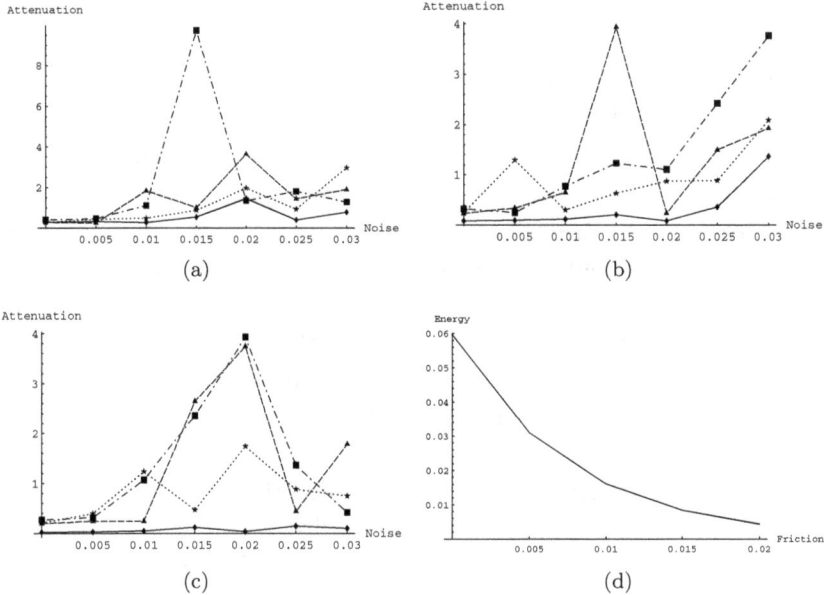

Fig. 7.9. Effects of friction and noise on signals. Evolution of energy of the auxiliary filament in NOR gate and distinguishability of logical values. (a) No friction. (b) Friction parameter $\gamma_1 = 0.1$. (c) Friction parameter $\gamma_1 = 0.2$. (d) Energy damping in function of friction, without noise. Energy for Input 1 is shown by solid line with rhomb markers, Input 2, dotted line with start markers, Input 3, dashed line with square markers.

an appropriate range of values to study the noise influence on the performance of the gates.

As a measure of the change in energy, we take the mean ratio of the initial to final energy of the auxiliary filament, which is always excited, whatever the input values of the gate. The friction lowers the energy, a phenomenon that can in part be compensated by the noise. When no noise is considered, the energy is clearly damped, see panel Fig. 7.9(d), which shows the behaviour of the absolute value of the auxiliary filament energy as a function of friction and absent noise. The noise can make the differences between the logical zero and one in the output not reliable, in general, starting with $\gamma_2 \approx 0.1$, even if in some cases a good separation is found also for bigger values of the noise. The friction does not make the signals less separated.

7.8 Discussion

Building upon our previous research on the computational potential of actin filaments, including studies such as Adamatzky et al. [27], Siccardi et al. [231,238], and Mayne et al. [173,175], we embarked on investigating the feasibility of implementing logical circuits through interconnected protein chains. In this novel approach, Boolean values are represented by travelling localisations that propagate along the filaments, and computation occurs through collisions and interactions between these travelling localisations at the junctions between chains. Our research demonstrated successful implementations of AND, OR, and NOT gates within these setups, and further showed the cascading of these gates to create hierarchical circuits, exemplified by the NOR gate.

However, it is important to acknowledge the limitations inherent in our approach, which should be addressed in future studies. Presently, we represent each globular protein in an actin filament as an abstract node, but expanding the model to consider the distribution of electrical charges on the surface of these globular units and the interactions between neighbouring charges would provide a more detailed representation. Additionally, the precise architecture of the junctions between actin filaments must be taken into account. Several families of auxiliary proteins, such as Hartwig et al. [119], Mullins et al. [184], Weaver et al. [271], Derivery and Gautreau [78], and Korobova and Svitkina [147], play significant roles in these junctions, and their unique electrical characteristics influence signal propagation between filaments. Incorporating these characteristics into the model would enhance its accuracy.

Furthermore, when considering the experimental fabrication of these architectures, it is crucial to evaluate the feasibility of interfacing protein chain-based logical circuits with conventional silicon hardware. One potential avenue is exploring optical inputs and outputs by connecting the actin filaments with light-sensitive molecular structures, as demonstrated in studies such as Kisselev et al. [144], Palczewski et al. [190], and Archer et al. [39]. Additionally, if soliton-based computing is implemented in vivo, an interface between the host hardware and cytoskeleton-based gates could be achieved by utilising carbon nanotubes, as shown in experimental works on interfacing tubes with living cells by Liopo *et al.* [155], Shaibani *et al.* [232], and Bianco *et al.* [49]. The size compatibility between actin filaments (with a diameter of approximately 6 nm) and single-wall carbon nanotubes (with a diameter around 0.3 nm) suggests the feasibility of such interfacing.

Chapter 8

Boolean Gates on Actin Filaments

Stefano Siccardi[*,§], Jack A. Tuszynski[†,¶], and Andrew Adamatzky[‡,||]

*Consorzio Interuniversitario Nazionale per l'Informatica, Italy and
Unconventional Computing Lab, UWE, Bristol, UK
†Department of Oncology, University of Alberta, Edmonton, Alberta,
Canada
‡Unconventional Computing Lab, UWE, Bristol, UK
§ssiccardi@2ssas.it
¶jack.tuszynski@gmail.com
||andrew.adamatzky@uwe.ac.uk

Abstract
We employ a coupled nonlinear transmission line model to investigate the interactions between voltage pulses along actin filaments. By assigning the logical value of 'True' to the presence of a voltage pulse and 'False' to its absence, we represent digital information transmission along these filaments. When two pulses, representing the Boolean values of input variables, interact, they can either facilitate or inhibit each other's further propagation. By exploring this phenomenon, we are able to construct Boolean logical gates and a one-bit half-adder using the interacting voltage pulses. These findings have implications for understanding cellular processes and may also find applications in various technological and computational domains.

8.1 Background

Previously we proposed a model of actin filaments as two chains of one-dimensional binary-state semi-totalistic automaton arrays for signalling events, and discovered local activity rules that support travelling or stationary localisations [26]. This finite state machine model has been further extended to a quantum cellular automata (QCA) model in Ref. [230]. We have shown that quantum actin automata can perform basic operations of Boolean logic, and implemented a binary adder [230], and three valued logic operations [4]. These models were implemented in general algorithmic terms, without describing any specific physical mechanisms that could be used to implement cell excitations and interactions. They also did not employ interactions between propagating localisations in a spirit of collision-based computing [7].

We aim to rectify these omissions and present a model of actin in terms of RLC (resistance, inductance, capacitance) nonlinear electrical transmission wires that can implement logical gates via interacting voltage impulses. These models take inspiration from the idea of a cellular automaton, but do not fit exactly in its definition, as cells do not possess discrete states, there are no intrinsic time steps, and no rules are defined for state transitions. However, the states can be digitised by defining a suitable threshold, a time step can be defined by convenience, and transitions are governed by Kirchhoff's circuit laws so that a deterministic evolution takes place even in absence of explicit transition rules.

Our starting point is the usual model of electrical wires as sequences of circuits composed by resistors, capacitors, and inductors. Our model is based on the Tuszyński et al. [254] model of actin monomers in terms of electrical components. This latter model was developed to explain experimental observations of ionic conductivity along actin filaments [153]. This model exhibits, in a continuous limit, solitons (standing waves and travelling impulses).

The aim of the present work is to use the same equations as in Ref. [254], but without invoking the continuous limit approximation and to study the behaviour of several types of solutions for tens of monomers. We will show how an excitation moves along the actin filament and how it collides with another excitation coming from elsewhere; we will show that these collisions can be used to implement logical operations. This could provide a physical description of signal propagation and processing in the cellular milieu and also possibly in hybrid bio-nano technological applications.

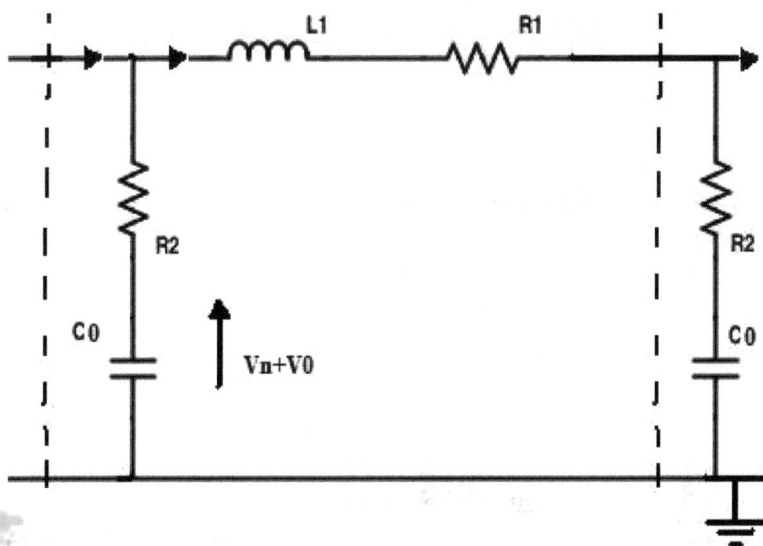

Fig. 8.1. A circuit diagram for the nth unit of an actin filament (from Ref. [254]).

8.2 Actin filaments as nonlinear RLC transmission lines

In this section, we recall the basic model that will be used; more details can be found in the original paper [254].

Referring to Fig. 8.1, where an actin monomer unit in a filament is delimited by the dotted lines, we assume that capacitors are nonlinear (see discussion in Refs. [158, 268]):

$$Q_n = C_0(V_n - bV_n^2). \tag{8.1}$$

The main equation is

$$LC_0 \frac{d^2}{dt^2}(V_n - bV_n^2)$$

$$= V_{n+1} + V_{n-1} - 2V_n - R_1 C_0 \frac{d}{dt}(V_n - bV_n^2)$$

$$- R_2C_0 \left\{ 2\frac{d}{dt}(V_n - bV_n^2) - \frac{d}{dt}(V_{n+1} - bV_{n+1}^2) \right.$$

$$\left. - \frac{d}{dt}(V_{n-1} - bV_{n-1}^2) \right\}. \tag{8.2}$$

The above is the basic equation that will be used to compute the cells' behaviour.

In Ref. [254], formulas are derived to obtain values for the relevant parameters. The Bjerrum length is defined as follows:

$$\lambda_B = \frac{e^2}{4\pi\epsilon\epsilon_0 k_B T}, \tag{8.3}$$

where e is the electrical charge, ϵ_0, the permittivity of the vacuum, ϵ, the dielectric constant of the solution where the actin molecule is placed, estimated similar to $\epsilon_{\text{water}} \approx 80$, k_B, Boltzmann's constant, and T, the absolute temperature, which is taken $\approx 293\,\text{K}$. The Bjerrum length is the distance beyond which thermal fluctuations are stronger than electrostatic interactions between charges. The parameter used in Ref. [254] is $\lambda_B = 7.1 \cdot 10^{-10} m$.

The formula for the capacitance is as follows:

$$C_0 = \frac{2\pi\epsilon\epsilon_0 l}{\ln\left(\frac{r_{\text{actin}}+\lambda_B}{r_{\text{actin}}}\right)}, \tag{8.4}$$

where l is the length of a monomer, taken $\approx 5.4\,\text{nm}$; and $r_{\text{actin}} \approx 2.5\,\text{nm}$. With the parameters given, the capacitance per monomer is estimated $C_0 \approx 96 \cdot 10^{-6}\,pF$.

For the inductance, we have

$$L = \frac{\mu N^2 \pi (r_{\text{actin}} + \lambda_B)^2}{l}, \tag{8.5}$$

where μ is the magnetic permeability and N is the number of turns of the coil, that is the number of windings of the distribution of ions around the filament. It is approximated by counting how many ions can be lined up along the length of a monomer as $N = l/r_h$, and it is supposed that the size of a typical ion is $r_h \approx 3.6 \cdot 10^{-10}\,\text{m}$. With these assumptions, we have for a monomer $L \approx 1.7\,pH$.

For the resistance, we have

$$R = \frac{\rho \, ln((r_{\text{actin}} + \lambda_B)/r_{\text{actin}})}{2\pi l}, \tag{8.6}$$

where resistivity ρ is approximately given by

$$\rho = \frac{1}{\Lambda_0^{K^+} c_{K^+} + \Lambda_0^{Na^+} c_{Na^+}} \tag{8.7}$$

where c_{K^+} and c_{Na^+} are the concentrations of sodium and potassium ions, considered, respectively, 0.15 M and 0.02 M; $\Lambda_0^{K^+} \approx 7.4(\Omega m)^{-1} M^{-1}$ and $\Lambda_0^{Na^+} \approx 5.0(\Omega m)^{-1} M^{-1}$ depends only on the type of salts. Accordingly, $R_1 \approx 6.11 \, M\Omega$; R_2 is taken $= R_1/7 \approx 0.9 M\Omega$ following Ref. [254].

We note that resistance, inductance, and capacity values could be tuned by changing the temperature and the ion concentrations. Further sections deal with solutions of Eq. (8.2) and their interactions.

8.3 Evolution of pulses

We recap the parameters that will be used in almost all computations:

$$L = 1.7 \text{pH}, \quad C_0 = 96 \cdot 10^{-6} \text{pF},$$
$$R_1 = 6.11 \cdot 10^6 \Omega, \quad R_2 = 0.9 \cdot 10^6 \Omega. \tag{8.8}$$

With these parameters, we measure time in nanoseconds. Let us consider the case when an initial voltage is set at one end of the filament for some time as, for example, could take place when an actin filament is in close proximity of an ion channel. We use for voltage the form

$$V_0 = \frac{1}{2} - \frac{e^{t-t_0} - e^{-t+t_0}}{2(e^{t-t_0} + e^{-t+t_0})} \tag{8.9}$$

In all the following computations, we consider $t_0 = 3$ ns.

We find that pulses actually travel, but after a few nanoseconds they are damped (Fig. 8.2(a)). From the diagram we see that the

Cells

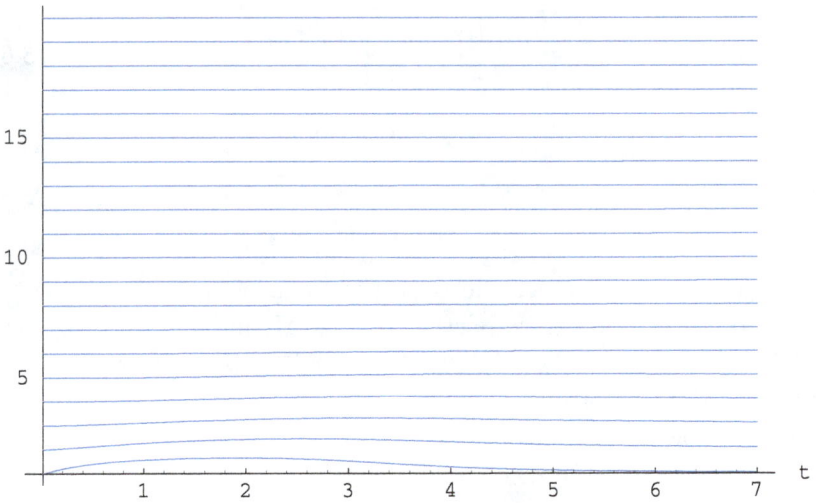

(a)

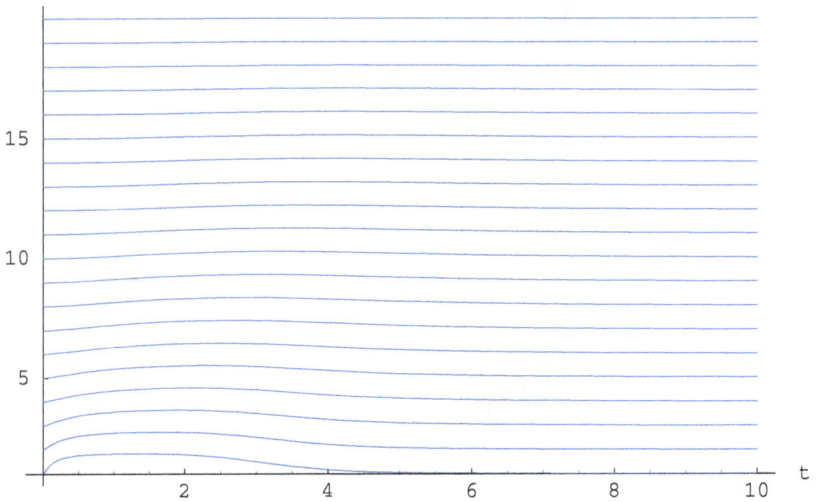

(b)

Fig. 8.2. An input pulse travelling along the actin filament: (a) parameters (8.8), (b) the same parameters, but $R_1 = 61.1\,10^5\,\Omega$. The pictures represent the cell evolutions, each cell voltage is represented by a diagram with time on the horizontal axis and voltage V on the vertical one. The diagrams are tiled vertically with Cell 1 at the bottom and Cell 20 at the top, with a unitary displacement.

12th unit is reached in about 4 ns, so that the speed is approximately 5.4 nm · 12/4 ns, that is about 16 m/sec. That is consistent with the estimates in Ref. [254]. If we decrease the resistance R_1 by an order of magnitude, the speed doubles. Then we can obtain a much longer lasting pulse (Fig. 8.2(b)).

Figure 8.3(a) shows the case where two identical impulses are applied at the ends of the actin filament. In Fig. 8.3(b), we see the interaction of two pulses of opposite signs applied to ends of the filament.

In the next two sections, we consider two types of inputs: forced and unforced. An unforced input is an impulse applied just once at input sites of actin-based gates. A forced input is an impulse applied continuously during evolution of the gate.

8.4 Gates with unforced pulses

Let us consider interactions and possible gates implemented using only initial conditions of cells, without any forced pulses applied. They are based on the assumption that we can initialise any cells with a predetermined electrical potential value and to reliably measure output potentials at given units of actin filaments.

Using parameters (8.8) and $b = 0.1$, and initial condition $V = V_0 \neq 0$, we look at cells having voltages greater than a fixed percentage of initial potential V_0. We find that some logical operations are possible to implement, if we consider signals above $\approx 0.1V_0$ as output, to read outputs of the logical gates.

Figure 8.4 shows the AND operation obtained with two input cells at Positions 8 and 15. The output is in Cells 11 and 12, which are excited at the end of the evolution if both the input cells were excited initially. Figure 8.5 shows the OR operation obtained with two input cells at Positions 9 and 11. The output is in Cell 10, which is excited at the end of the evolution if one of the input cells or both of them were excited initially.

To obtain a NOT gate, we initialise an auxiliary cell at Position 9 at $-V_0$ and place the input cell at Position 11; the output is found in Cell 10 as shown in Fig. 8.6.

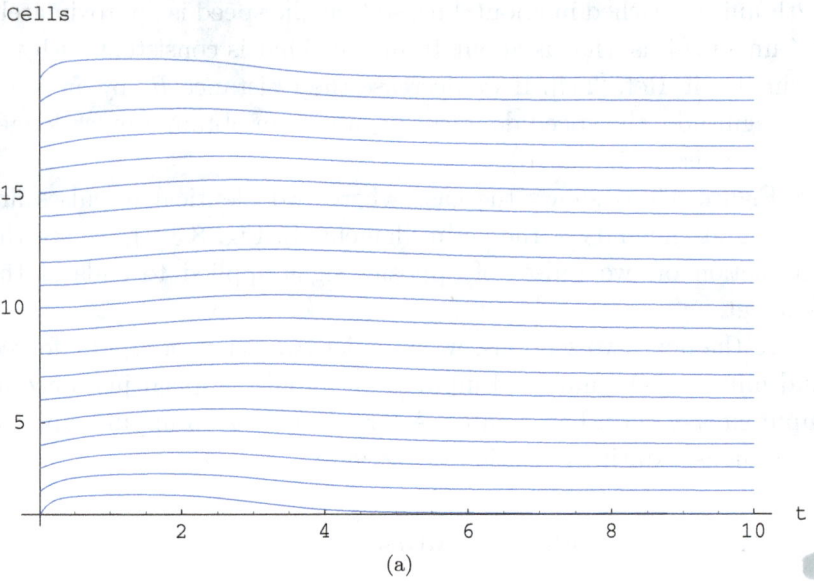

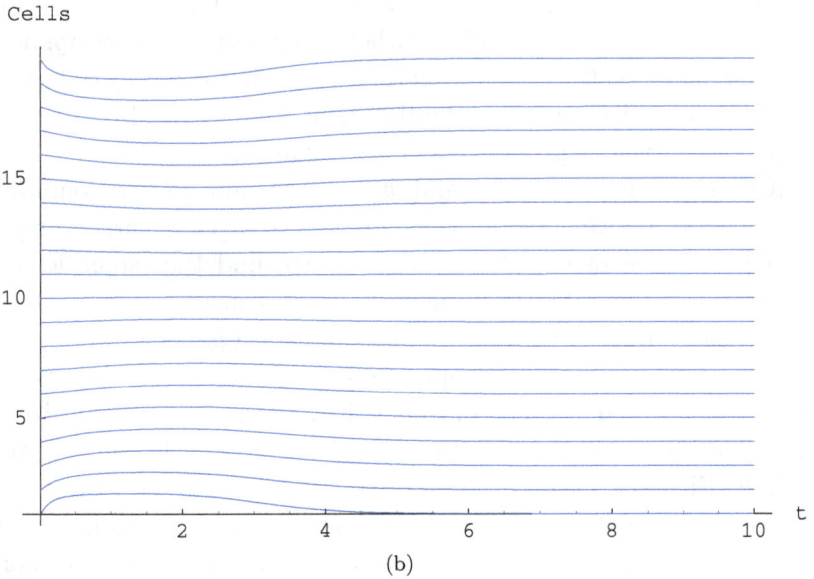

Fig. 8.3. Two input pulses travelling along actin monomers from the opposite ends, with parameters (8.8), but $R_1 = 61.1 \, 10^5 \, \Omega$. (a) Identical pulses. (b) Opposite voltage.

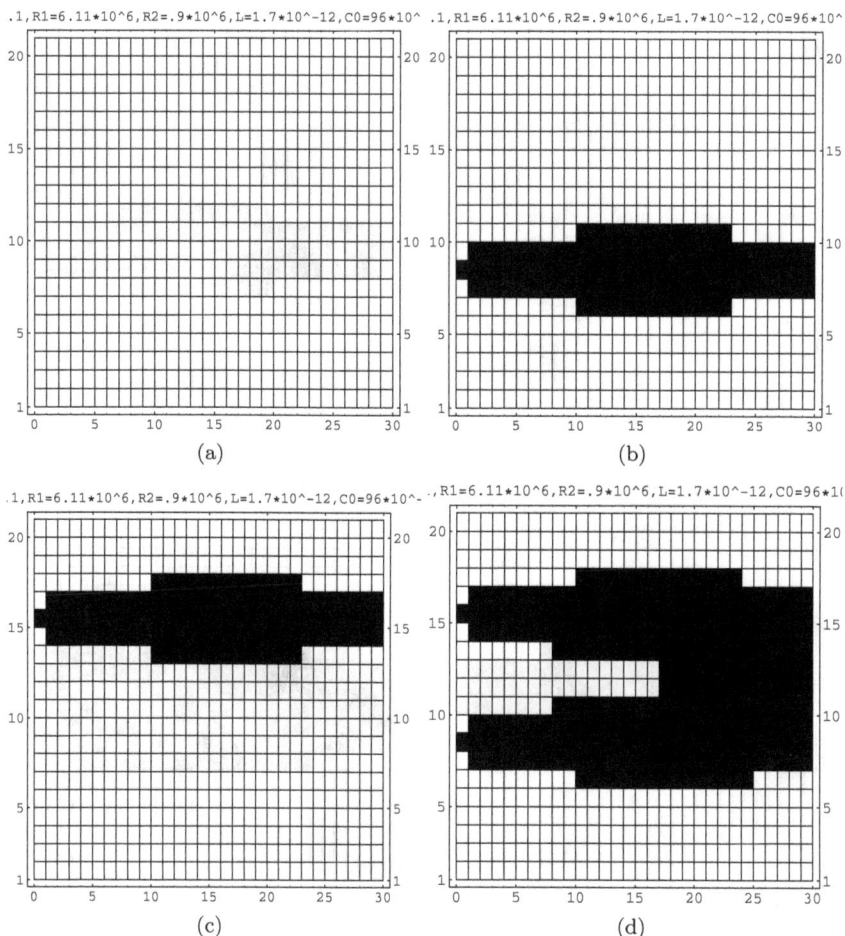

Fig. 8.4. Unforced pulses, AND operation. Inputs are (a) 00, (b) 10, (c) 01, and (d) 11 in Cells 8 and 15; output in Cells 11 and 12. Cells excited above $0.1\,V_0$ are filled in black; time evolves along the horizontal axis and cells are aligned along the vertical axis.

To build an XOR gate, we use the formula

$$a \oplus b = (\neg(a \wedge b)) \wedge (a \vee b) \tag{8.10}$$

as shown in Fig. 8.7. Each run is composed by four gates, and it is supposed that some mechanism reads the output of each gate, amplifies the signal, and passes the signal to the next gate. In the

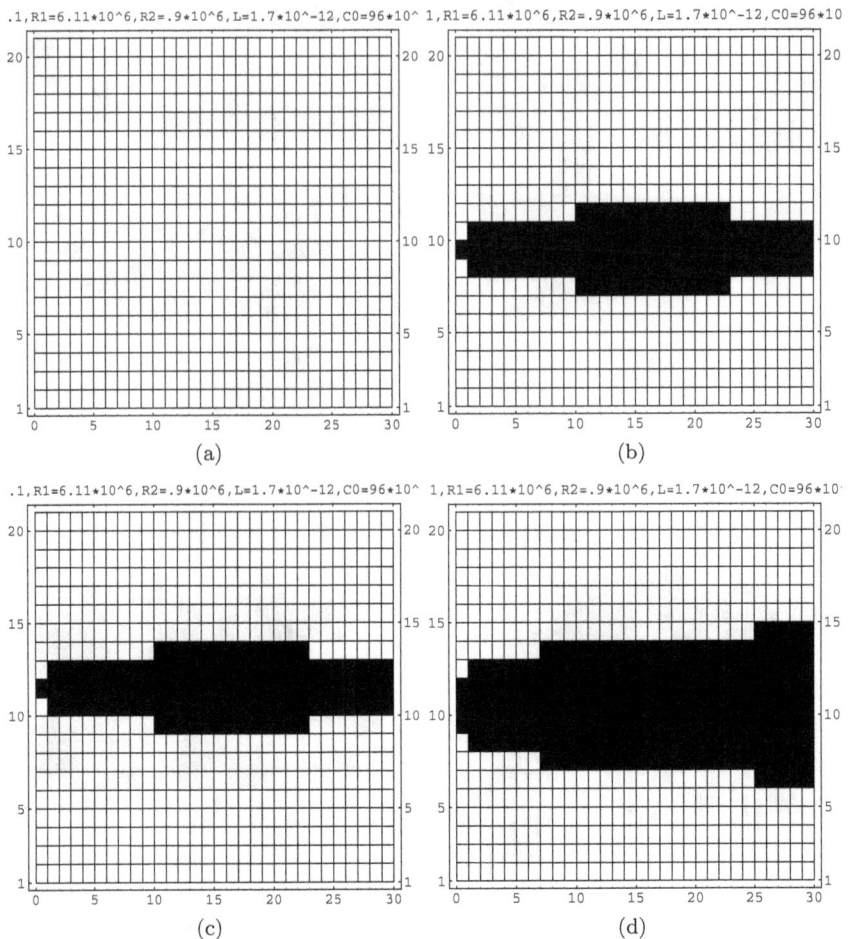

Fig. 8.5. Unforced pulses, OR operation. Inputs are (a) 00, (b) 10, (c) 01, and (d) 11 in Cells 9 and 11; output in Cell 10. Cells excited above $0.1\,V_0$ are filled in black; time evolves along the horizontal axis and cells are aligned along the vertical axis.

figure, the arrays represent signals travelling between gates. The first from the left is an AND gate with input Cells 8, 15 and output Cell 12; its output goes to the next gate: the NOT gate having input in Cell 11 and output in Cell 10. The original inputs, coming out at Cells 8 and 15 of the first gate, are sent to input Cells 9 and 11 of the OR gate in the third section. Finally, the output of the OR gate in Cell 10

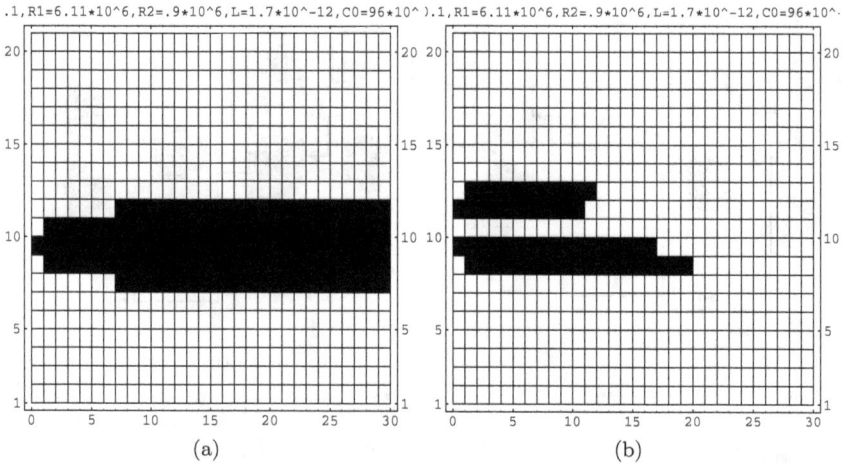

Fig. 8.6. Unforced pulses, NOT operation. Input is (a) 0 and (b) 1 in Cell 11; output in Cell 10. Cells excited above $0.1V_0$ are filled in black; time evolves along the horizontal axis and cells are aligned along the vertical axis.

and the output of the NOT gate are sent to input Cells 8 and 15 of the final AND gate. The final output of the XOR operation is found in Cells 11 and 12 at the right end.

8.5 Gates with forced pulses

Let us consider interactions and possible gates implemented by applying input pulses continuously at input sites. Namely, we send a sinusoidal pulse to input during the whole evolution of the system, with the intensity of the input pulse remaining constant.

We use slightly different values for resistances: $R_1 = 9.23 \, 10^6 \, \Omega$, $R_2 = 1.32 \, 10^6 \, \Omega$, that can be obtained changing concentrations of K^+ and Na^+; excitations are sinusoidal with a 1 ns period.

Figure 8.8 shows the AND operation obtained with two input cells at Positions 8 and 14. The output is found in Cells 11 that is excited at the end of the evolution if both the input cells are excited.

Figure 8.9 shows the XOR operation obtained with two input cells at Positions 10 and 14. The output is found in Cell 12, which is excited at the end of the evolution if only one of the input cells is.

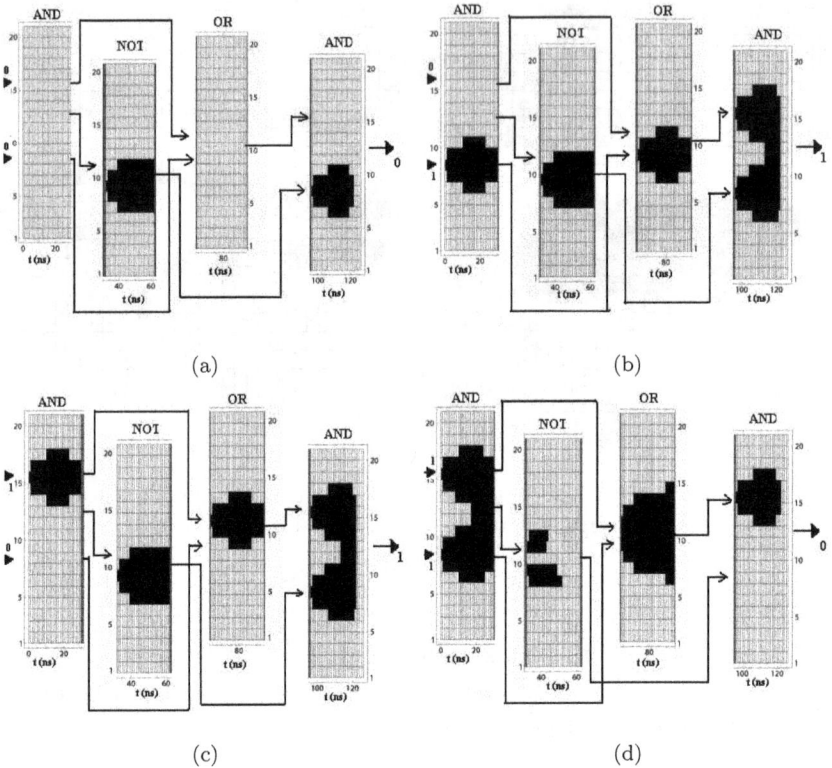

(a)   (b)

(c)   (d)

Fig. 8.7. Unforced pulses, XOR operation. Inputs are (a) 00, (b) 10, (c) 01, and (d) 11 in Cells 8 and 15 of the leftmost gate; output in Cells 11 and 12 of the rightmost. Cells excited above $0.1V_0$ are filled in black; time evolves along the horizontal axis and cells are aligned along the vertical axis.

The input pulses are sinusoidal with opposite phases. The other logical gates can be implemented in a similar way.

We also show in Fig. 8.10 a one-bit half-adder: it is composed by an AND gate followed by an XOR. The AND gate has input Cells 8 and 14 and outputs the most significant bit (the carrier) in Cell 11; the XOR has input Cells 10 and 14 and outputs the less significant bit in Cell 12. It is supposed that the input in Cell 14 is common to both gates, while the input in Cell 8 of the AND gate travels to Cell 10 of the XOR changing its phase.

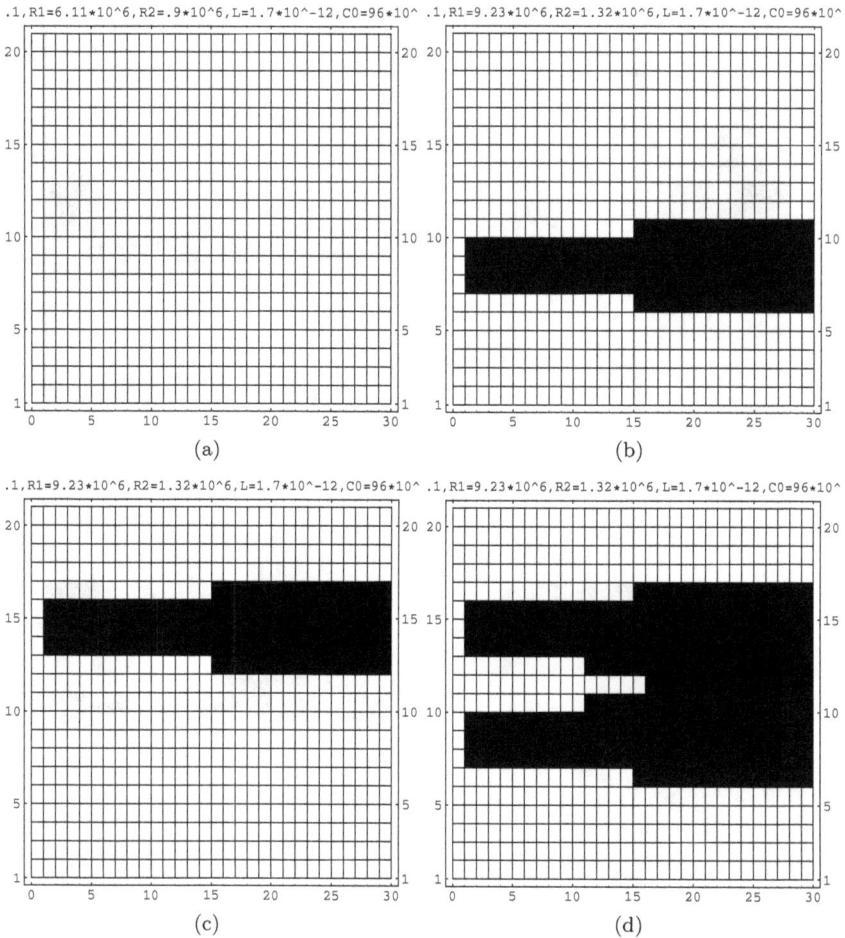

Fig. 8.8. Forced pulses, AND operation. Inputs are (a) 00, (b) 10, (c) 01, and (d) 11 in Cells 8 and 14; output in Cell 11.

One could argue that it could be difficult in practice to send an input pulse to exactly one monomer (or to initialise it in an excited status), which requires a localisation precision of under 5 nm. We therefore compute some evolutions supposing that the pulse hits a number N of monomers. Under these conditions, the values of resistance, inductance, and capacitance must be computed with

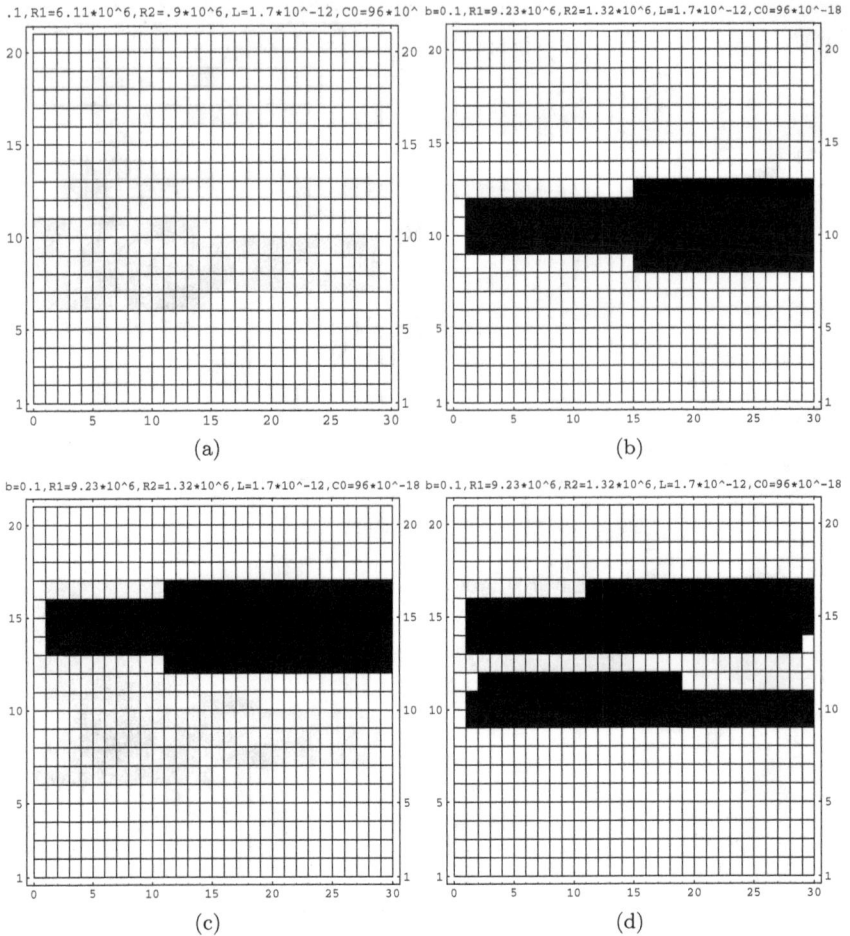

Fig. 8.9. Forced pulses, XOR operation. Inputs are (a) 00, (b) 10, (c) 01, and (d) 11 in Cells 10 and 14; output in Cell 12.

appropriate addition rules. For resistance, we have

$$R_{2\,\mathrm{tot}} = \left(\sum_{i=1}^{N} R_{2,i}^{-1} \right)^{-1}, \quad R_{1\,\mathrm{tot}} = \sum_{i=1}^{N} R_{1,i}. \qquad (8.11)$$

Total inductance and capacitance are obtained summing those of individual monomers. Figure 8.11 shows the XOR gate with input

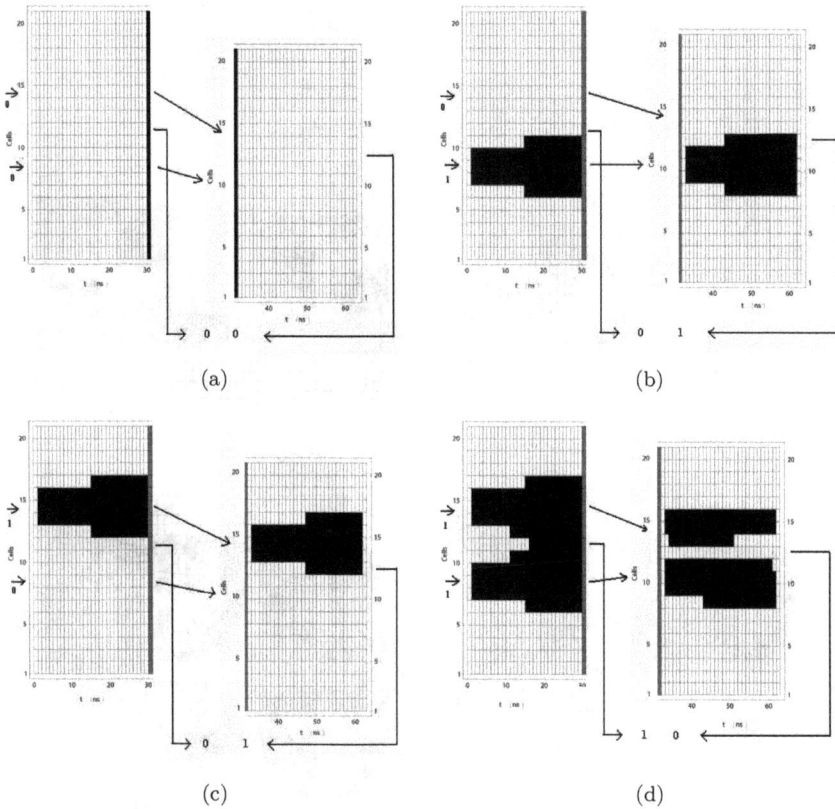

Fig. 8.10. Forced pulses, half-adder. Inputs are (a) 00, (b) 10, (c) 01, and (d) 11 in Cells 8, 10, and 14; outputs in Cells 11 and 12.

sent to groups of 10 monomers in Positions 8–17 and 23–32. As in Fig. 8.9, the input pulses are sinusoidal with opposite phases. Output is in Cell 20. In our approximation, we do not consider the detailed evolution inside the groups of monomers receiving the input, they are treated as a single cell, with R, L, C parameters computed summing on 10 monomers.

We note that the results are robust: pulses fade slowly and a threshold of 0.3 of the input signal can be used, instead of 0.1 as in the previous computations.

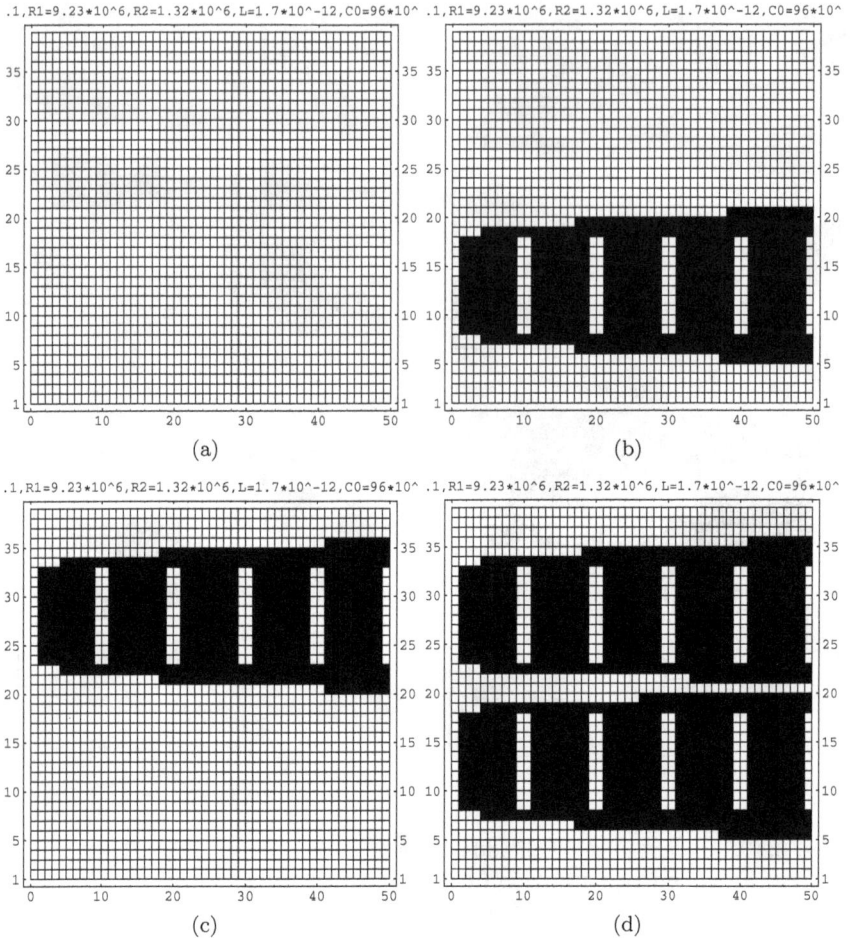

Fig. 8.11. Forced pulses, XOR operation. Inputs are (a) 00, (b) 10, (c) 01, and (d) 11 in Cells 8–17 and 23–32; output in Cell 20.

8.6 Conclusion

In our numerical experiments, we have successfully demonstrated the construction of basic logical gates using actin filaments modelled as RLC (resistor–inductor–capacitor) circuits. This implementation relies on the ability to send input pulses to specific locations and set initial voltage values in actin monomers. By leveraging interactions between these input pulses, we were able to realize several logical gates. For example, to implement an AND gate, we directed identical phase pulses to the input cells. These pulses combined with each other, and the output threshold was set in such a way that if only one input was active, the output potential remained below the threshold. However, if two input potentials were summed, the threshold was exceeded, indicating an active output. To implement an XOR gate, we sent pulses with opposite phases, causing them to subtract from each other when they met.

While our numerical experiments provide promising results, it is important to consider further aspects for a real implementation. Specifically, we need to address how the output can be measured with sufficient sensitivity, such as achieving measurements at 0.1 or 0.3 of the initial voltage. Additionally, ensuring reliable transmission of output pulses from one gate to another as input needs to be carefully studied. These points will be thoroughly examined in future works as we aim to advance our understanding of cascading gates and develop practical strategies for their implementation. By addressing these considerations, we can enhance the reliability and scalability of actin filament-based logical circuits, paving the way for their potential applications in information processing and computing.

Actin filaments commonly exist in networks and can form cross-links with different shapes, such as parallel and orthogonal arrangements, facilitated by various binding proteins. One example is filamin A, which constructs orthogonal networks by cross-linking actin filaments. In our model, we assume that electric pulses generated as output in one filament can propagate to another filament through the filamin molecule, effectively acting as a connecting wire. Additionally, actin filament networks can exhibit branching, initiated by the ARP2/3 protein, which leads to the propagation of two actin filaments at acute angles.

To simplify the analysis, we can consider similar electrical mechanisms observed in actin filaments and compute the parameters R_1, R_2, C_0, and L for a filamin based on its dimensions. This allows us to model the electrical behaviour of a filamin as an electrical component in our circuit. In principle, any number of actin gates could be connected through the filamin cross-links, providing a means to create more complex circuits.

However, a challenge arises in maintaining reliable signal transmission within the actin network over longer distances. Without amplification, the intensity of the pulses will diminish rapidly, resulting in weakened signals. To address this challenge, an amplification method must be implemented to sustain the signal strength and ensure reliable signal transmission within the actin network. This amplification strategy is crucial to overcoming the attenuation of signals and maintaining the integrity of the information being transmitted through the network. Future research efforts should focus on developing and integrating appropriate amplification mechanisms into the actin-based circuitry to address this challenge effectively.

Indeed, additional studies, such as Ref. [213], have shed light on the diverse roles of filamin A in signal transduction. Furthermore, research, such as Ref. [223], has highlighted the regulation of voltage-gated ion channels by actin filaments. These findings suggest the intriguing possibility of a system where the output voltage of one gate can activate an ion channel, which, in turn, triggers the activation of the next gate, creating a cascade of signal propagation within the actin network.

However, to draw practical conclusions and assess the feasibility of such a system, a more comprehensive analysis is necessary. Several factors need to be carefully evaluated, including the spatial distribution of ion channels within the actin network, their dimensions, the required electrochemical gradients, the suitability of specific ion types for signal propagation, and their diffusion rates. These considerations are crucial in assessing the practical viability of developing a logical device based on this concept.

Future research efforts should aim to investigate these aspects in more detail to gain a deeper understanding of the interactions between actin filaments, ion channels, and signal propagation within the network. Such investigations will help determine the feasibility and practical challenges associated with developing a logical device based on this concept. By addressing these research gaps, we can further advance our knowledge and explore the potential applications of actin filaments and ion channels in developing novel computing architectures.

Indeed, tubulin, a significant structural protein, forms cylindrical microtubule structures that have been shown to exhibit electrical amplification properties [203]. Actin filaments can be connected to microtubules through microtubule-associated proteins (MAPs) [75, 106]. If the electrical properties of MAPs are similar to those of actin units, it is conceivable to achieve cascading Boolean gates into arithmetic circuits by utilising actin-MAP-microtubule-MAP-actin connections.

Recent research has also proposed a link between microtubule phosphorylation, mediated by the calmodulin kinase II enzyme complex, and cellular memory formation [68]. Moreover, it has been demonstrated that signals propagating along microtubules can interact through logical gates [68]. These findings suggest the potential integration of signal storage and processing capabilities within the cytoskeleton, composed of both actin filaments and microtubules that interact with each other.

In the context of *in vitro* design of actin-based computing circuits, the insertion of tailored mono-molecular amplifiers [140] between actin strands in cascaded gates could enhance the signal amplification process. This approach could potentially address the challenge of signal attenuation over distance, improving the reliability and efficiency of information transmission within actin-based circuits.

Further research and experimentation are required to explore the precise electrical properties of MAPs, the mechanisms of signal propagation between actin filaments and microtubules, and the feasibility of integrating signal storage and processing capabilities within the cytoskeleton. Investigating the potential integration of actin filaments, microtubules, and tailored amplification techniques in computational systems holds promise for advancing the field of cytoskeleton-based computing and developing novel computing architectures.

Chapter 9

Quantum Actin Automata

Stefano Siccardi[*,‡] and Andrew Adamatzky[†,§]

*Consorzio Interuniversitario Nazionale per l'Informatica, Italy and
Unconventional Computing Lab, UWE, Bristol, UK
†Unconventional Computing Lab, UWE, Bristol, UK
‡ssiccardi@2ssas.it
§andrew.adamatzky@uwe.ac.uk

Abstract

In our study, we employ a model that represents actin filaments as
two chains of one-dimensional quantum automata arrays. This model
provides a framework for exploring hypothetical signalling events that
propagate along these chains. Our primary focus is to investigate the
functions of automaton state transitions and analyze their behaviour,
with particular attention to the role of superposition in the initial
states. By conducting a thorough analysis, we are able to observe and
study the propagation of localised particles along the actin chains. One
significant outcome of our research is the discovery that logical gates
can be realised through collisions between these travelling particles.
This finding demonstrates that the interaction of particles along the
actin chains enables the implementation of logical operations. Building
upon this insight, we leverage these particle collisions to successfully
create a binary adder, which serves as a compelling demonstration
of the computational capabilities of the system. By adopting this
model and uncovering the potential for logical gate implementation and
more complex computations, our study contributes to advancing our
understanding of the computational properties of actin filaments. This
work opens up new avenues for exploring the role of actin filaments in
information processing and lays the groundwork for the development of
novel computing architectures inspired by quantum-inspired models.

9.1 Quantum Cellular Automata

Quantum cellular automata (QCA) present a variant of cellular automata where the states of individual cells are represented by qubits or cubits, rather than classical bits. The formal definitions of QCA have been proposed by several researchers, including notable references such as Delgado [77], and Aoun and Tarifi [37]. To transition classical automata into their quantum counterparts, various methods have been explored, as documented in works like Inokuchi and Mizoguchi [129] and Schumacher and Werner [225]. We will initially examine two classical definitions of one-dimensional QCA (1QCA) put forth in the existing literature, specifically in the works of Watrous [311] and Kondacs and Watrous [312].

Definition 9.1. A 1QCA $\mathcal{A} = \langle Q, \lambda, N, \delta \rangle$ is determined by a finite set Q of states, a *quiescent state* λ, a neighbourhood $N = \{n_1, ..., n_r\} \subseteq Z$, with $n_1 < n_2 < \cdots < n_r$ and a local transition function

$$\delta : Q^{r+1} \to C_{[0,1]}$$

satisfying the following three conditions:

(1) Local probability: for any $(q_1, ..., q_r) \in Q^r$,

$$\sum_{q \in Q} |\delta(q_1, ..., q_r, q)|^2 = 1.$$

(2) Stability of the quiescent state: if $q \in Q$, then

$$\delta(\lambda, ..., \lambda, q) = 1 \, if \, q = \lambda, \quad 0 \text{ otherwise.}$$

(3) To state the third condition, we introduce the automaton's mapping as follows. A configuration $c : Z \to Q$ is a mapping such that $c(i) \neq \lambda$ only for finitely many i. Let $C(\mathcal{A})$ denote the set of all configurations. Computation of $\mathcal{A}$ is done in the space $H_{\mathcal{A}} = l_2(C(\mathcal{A}))$ with the basis $\{|c\rangle \, | \, c \in C(\mathcal{A})\}$. In one step, $\mathcal{A}$ transfers from one basis state $|c_1\rangle$ to another $|c_2\rangle$ with

the amplitude

$$\alpha(c_1, c_2) = \prod_{i \in Z} \delta(c_1(i + n_1), ..., c_1(i + n_r), c_2(i)).$$

A state in H_A, in general, has the form

$$|\phi\rangle = \sum_{c \in C(A)} \alpha_c |c\rangle$$

with normalised α_c. The evolution operator E_A of A maps any state $|\phi\rangle$ into $|\psi\rangle = E_A|\phi\rangle$ such that

$$|\psi\rangle = E_A|\phi\rangle = \sum_{c \in C(A)} \beta_c |c\rangle,$$

where

$$\beta_c = \sum_{c' \in C(A)} \alpha_{c'} \alpha(c', c).$$

Now we can state the third condition: unitarity. The mapping E_A must be unitary.

In general, to decide whether the unitarity condition is satisfied is a non-trivial problem [1, 83, 107]. Partitioned 1QCA, that are well suited to represent cells and their states, are easier to deal with.

Definition 9.2. A partitioned 1QCA (P1QCA) is a 1QCA $A = \langle Q, \lambda, N, \delta \rangle$ that satisfies the following conditions:

(1) The set of states Q is the cartesian product $Q = Q_1 \times \cdots \times Q_r$ of $r = |N|$ nonempty sets.
(2) The local transition function δ is the composition of two functions:

$$\delta_c : Q^r \to Q \quad \text{a classical mapping}$$

and

$$\delta_q : Q \to C^Q \quad \text{a quantum mapping}$$

where $\delta_c((q_{1,1}, ..., q_{1,r}), ..., (q_{r,1}, ..., q_{r,r})) = (q_{r,1}, q_{r-1,2}, q_{1,r})$.

(3) The function δ_q defines a 1QCA with operator

$$U_{\mathcal{A}_q}(q_2, q_1) = [\delta_q(q_1)](q_2)$$

$U_{\mathcal{A}_q}$ is also called a local transition matrix of $\mathcal{A}$. It can be shown that the evolution of a P1QCA is unitary if and only if this operator is unitary.

9.2 Application to actin-like structures

Our next step is to translate the actin automaton defined in Ref. [6] to a quantum automaton based on the P1QCA definition. The original actin automaton is composed of two layers of cells, labelled x_k and y_k. In our model, we use only one type of cells, q_i, identifying $q_{2k+1} = x_k$ and $q_{2k} = y_k$ like in Fig. 9.1. With this convention, the neighbourhood of any q_i is $(q_{i-2}, q_{i-1}, q_i, q_{i+1}, q_{i+2})$, which corresponds both to the neighbours of cells x_k and y_k in the original notation. Moreover, the state of the generic cell q_i is a vector $(q_{i1}, q_{i2}, q_{i3}, q_{i4}, q_{i5})$ and the first step of the evolution includes building of the intermediate state

$$\tilde{q}_i = (q_{i+2,1}, q_{i+1,2}, q_{i,3}, q_{i-1,4}, q_{i-2,5})$$

In the classical actin automaton [6], the node/cell transition rule has the following form $((a_{00}, a_{01}, a_{02}, a_{03}, a_{04}), (a_{10}, a_{11}, a_{12}, a_{13}, a_{14},)) = (X_0, X_1)$ with $a_{ik} \in \{0, 1\}$ and X_i is a 5-elements array.

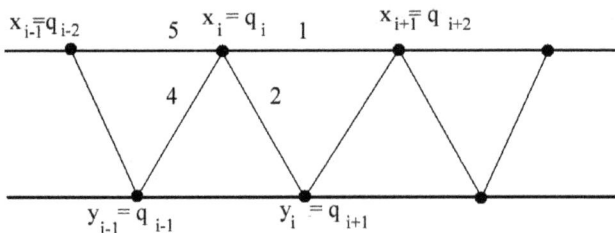

Fig. 9.1. A scheme of quantum actin automata.

To compute the state of cell q_i at time $t + 1$, the sum

$$S = q_{i-2} + q_{i-1} + q_{i+1} + q_{i+2}$$

is computed at time t, and the new state of q_i takes the value $a_{q_i(t),S}$ where $q_i(t)$ is the state of q_i at time t.

In the quantum case, we must specify the matrix (of dimension 32) $U(q_1, q_2) = [u_{a_1a_2a_3a_4a_5,b_1b_2b_3b_4b_5}]$ whose elements give the probability that a state $\tilde{q} = (a_1a_2a_3a_4a_5)$ evolves into another state $q = (b_1b_2b_3b_4b_5)$.

For the generic element of U, we have:

$$u_{a_1a_2a_3a_4a_5,b_1b_2b_3b_4b_5} = 0 \quad \text{if } b_3 \neq X_{a_3,S} \text{ and}$$

$$u_{a_1a_2a_3a_4a_5,b_1b_2b_3b_4b_5} = 1 \quad \text{if } b_3 = X_{a_3,S} \text{ and } b_{i \neq 3} = B_i(S),$$

(9.1)

where $B_i(S)$ is an array or a rule, possibly dependent on S, that must be specified. In the P1QCA case, we have to specify also how the four 'extra' states evolve. Several choices are possible, so that this kind of automata is quite flexible and suitable to model different situations. In other words, when using quantum automata we have three choices to make: (1) the specific definition of the automaton, e.g. a partitioned quantum automaton; (2) the transition matrix type; (3) the rule for the evolution, like in the classical case.

Note that:

(1) The value $\lambda = (0,0,0,0,0)$ must be stable, so the rules $(1, x, y, z, w), (\text{any})$ are excluded.
(2) We have to check if the matrix U is unitary for specific rules.
(3) The above automaton can be very similar to a classical one: we can use a mechanism that generates superposition of the states or does not; in the latter case, we can start with cells in superposition or without. If we start with classical — not superposed — states, the evolution is classical.
(4) We used the following definition to manage the states transition:

$$u_{a_1a_2a_3a_4a_5,b_1b_2b_3b_4b_5} =$$

$$1 \, if \, b_3 = X_{a_3,S} \quad \text{and} \quad b_{i \neq 3} = a_i$$

(9.2)

that is, the four 'extra' substates of a node/cell are copied from the neighbours of the node/cell.

9.3 Implementation

9.3.1 Unitarity

Unitarity check must be performed, as explained in the above discussion, for the 512 rules considered in the classical actin automata (excluding rules 1xxxx because of quiescent states stability). Using Definition (9.2), where a cell inherits the substates of its neighbours, one obtains 16 unitary matrices, for rules $(i, 31 - i)$ and $i = 0, ..., 15$. It is worth noting that for all the rules, conditions that excite a quiescent cell will deactivate an excited one.

9.3.2 Evolution

In the first runs, we used seeds like those shown in our previous paper [6], five x_i and five y_i cells, all the others being in the quiescent state. The automata had a total of 60 cells and the automaton was evolved for 60 steps. As our aim at this stage was a qualitative comparison of the classical and quantum models, we tried to obtain comparable results, so we prepared pictures in terms of x and y (that is q_{2n+1} and q_{2n}) cells.

Figure 9.2 shows the evolution of the five substates of the x cells; time is the horizontal axis and x, the vertical axis. We recall that the states are actually qubits, that is they are a linear combination of the two basis elements $|0\rangle$ and $|1\rangle$. Accordingly, the picture represents

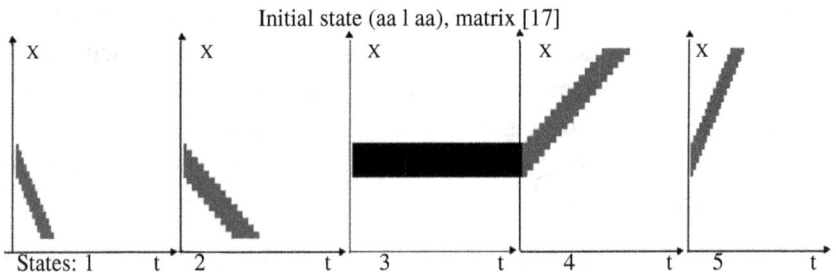

Fig. 9.2. Rule $(0, 31)$, initial state is $(aa1aa)$ as explained in the text, Matrix 9.2.

the module of the coefficient of $|1\rangle$ in grey, black meaning 1, and white, 0. Of course, it is expected that the identity rule (0, 31) will leave the cells in their initial condition, and it is indeed the case starting with five x_i and five y_i cells in the state (00100).

It is worth to note that, if we start with cells in a superposition state, e.g. $1/4 \sum (a_j a_k 1 a_i a_l)$, with all the combinations of $a = 0$ and $a = 1$, we get an evolution for the 'lateral' substates, even with the Rule (0, 31), as shown in Fig. 9.2.

We did not aim at a systematic analysis of seeds, but looked for seeds that are particles (gliders, localisations) moving to the right or to the left and that could be used as building blocks for some logical functions. Examples of evolutions starting with one seed generating a particle moving to the right and another seed generating a particle moving to the left are shown in Fig. 9.3. In this and in the subsequent pictures, we make no distinctions between x_i and y_i, and the pictures show the evolution of all the cells together, which will be labelled generically x.

Seeds are just two sets of three adjacent cells each, at Positions (6, 7, 8) and (12, 13, 14). Cell 7 is initialised at 00010, Cells 8 and 12 at 00100, Cell 13 at 01000. Cells 6 and 14 carry in their rightmost

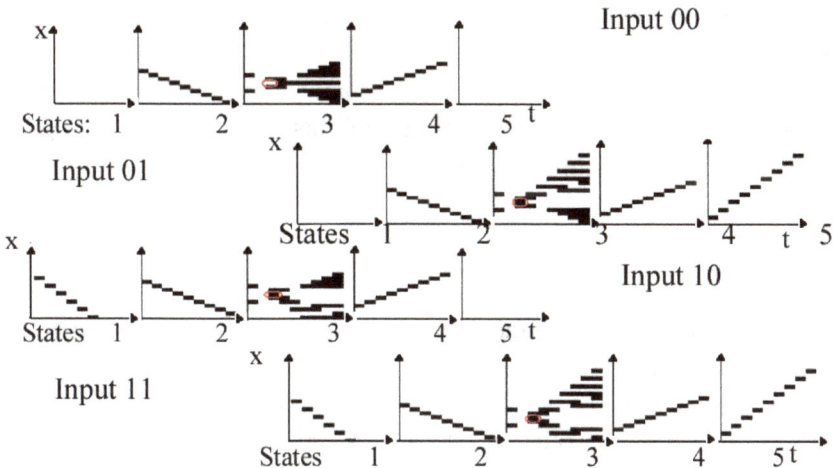

Fig. 9.3. Rule (14,17), Matrix 9.2 of two cells at Positions 6 and 14 in initial right and left states 00, 01, 10, 11. The 'output' of the logical OR gate is the red circles.

and leftmost states, respectively, the bits that we want to use in an operation. In the four runs, we have the four combinations, 00, 01, 10, and 11.

When the seeds meet at the third step, the resulting central state of Cell 10 (circled in red in the picture) is 1 unless both Cells 8 and 12 are initialised at 0. This means that this automaton is acting like a classical gate OR: if it could be 'prepared' in the initial states containing the two seeds at specified positions, and if it could be 'read' at the middle position after the specified number of steps, it could be used as a building block for a classical computer.

The same automaton — when initialised in a superposition of states — propagates superposition. For instance, one could initialise Cells 7 and 13 at a superposition of excited and idle states, e.g. $1/\sqrt{2} \cdot 00010 + 1/\sqrt{2} \cdot 00000$ and $1/\sqrt{2} \cdot 01000 + 1/\sqrt{2} \cdot 00000$, respectively, and let Cells 8 and 12 start at 00100 like in the previous case. Cells 6 and 14 that carry the qubits used in the operation are initialised at $1/2 \cdot 00000 + \sqrt{3}/2 \cdot 00001$ and $1/2 \cdot 00000 + \sqrt{3}/2 \cdot 10000$. What one obtains is that the output state (State 3 of Cell 10 at Step 3) is 0 if both inputs are 0, it is $1/2|0\rangle + \sqrt{3}/2|1\rangle$ if one of the inputs is in the active state $1/2 \cdot 00000 + \sqrt{3}/2 \cdot 00001$, and it is $1/4|0\rangle + \sqrt{15}/4|1\rangle$ if both input cells are active.

So we see that although the automaton is perfectly working with quantum states, the translation of classical to quantum logical operations is not completely straightforward.

9.4 Gates

We now discuss all the logical gates we can build in the evolution of given consecutive cells.

We recall that, considering two input qubits, we have 16 logical gates as in Table 9.1. With five cells, each carrying an active substate, which is a qubit, we can choose the input qubits in 10 different ways,

Table 9.1. Logical gates definition, columns contain the output value, given the input values of the headings.

(0,0)	(0,1)	(1,0)	(1,1)	Gate
0	0	0	0	Constant 0
0	0	0	1	$x \wedge y$
0	0	1	0	$x \wedge \neg y$
0	0	1	1	$x \forall y$
0	1	0	0	$\neg x \wedge y$
0	1	0	1	$y \forall x$
0	1	1	0	$x \oplus y$
0	1	1	1	$x \vee y$
1	0	0	0	$\neg x \wedge \neg y$
1	0	0	1	$x \oplus \neg y$
1	0	1	0	$\neg y \forall x$
1	0	1	1	$x \vee \neg y$
1	1	0	0	$\neg x \forall y$
1	1	0	1	$\neg x \vee y$
1	1	1	0	$\neg(x \wedge y))$
1	1	1	1	constant 1

and for each choice we can choose each of the given qubits as the output. Moreover, for each choice of the input qubits, we can choose in eight different ways the initial state of the other three 'ancillary' qubits.

So, for each rule, we have $10 \cdot 5 \cdot 8 = 400$ possible gates, and we run the evolution for the values $(0, 0)$, $(0, 1)$, $(1, 0)$, $(1, 1)$ to check which logical gates it is possible to obtain. In experiments, we considered three time steps, to check if the results are stable, or if we could have different gates, just waiting one or two more time steps.

As expected, the first rules like $(1, 30)$ give less interesting results than others, that can excite states starting with a lower number of cells.

Results for Rule $(4, 27)$ are reported in Tables 9.2 and 9.3; the first time step of the collision is reported, similar results have been found for second and third steps later.

Table 9.2. Logical gates depending on the choice of 2 input qubits, 1 output qubit, and the initial state of the qubits other than the 2 input ones. Rule (4, 27), time step 1, ancillary states 000-011.

inp1	inp2	out	000	001	010	011
1	2	3	$x \wedge y$	$x \oplus y$	$x \oplus y$	$\neg x \wedge \neg y$
1	3	3	$y \forall x$	$x \oplus y$	$x \oplus y$	$x \oplus \neg y$
1	4	3	$x \wedge y$	$x \oplus y$	$\neg(x \wedge y))$	$x \oplus \neg y$
1	5	3	$x \wedge y$	$x \oplus y$	$\neg(x \wedge y))$	$x \oplus \neg y$
2	3	3	$y \forall x$	$x \oplus y$	$x \oplus y$	$x \oplus \neg y$
2	4	3	$x \wedge y$	$x \oplus y$	$\neg(x \wedge y))$	$x \oplus \neg y$
2	5	3	$x \wedge y$	$x \oplus y$	$\neg(x \wedge y))$	$x \oplus \neg y$
3	4	3	$x \forall y$	$x \oplus y$	$x \oplus y$	$x \oplus \neg y$
3	5	3	$x \forall y$	$x \oplus y$	$x \oplus y$	$x \oplus \neg y$
4	5	3	$x \wedge y$	$\neg(x \wedge y))$	$x \oplus y$	$x \oplus \neg y$

Table 9.3. Logical gates depending on the choice of two input qubits, one output qubit and the initial state of the qubits other than the two input ones. Rule (4, 27), Time step 1, ancillary states 100-111.

inp1	inp2	out	100	101	110	111
1	2	3	$\neg(x \wedge y))$	$x \oplus \neg y$	$x \oplus \neg y$	$x \vee y$
1	3	3	$x \oplus y$	$x \oplus \neg y$	$x \oplus \neg y$	$y \forall x$
1	4	3	$x \oplus y$	$\neg x \wedge \neg y$	$x \oplus \neg y$	$x \vee y$
1	5	3	$x \oplus y$	$\neg x \wedge \neg y$	$x \oplus \neg y$	$x \vee y$
2	3	3	$x \oplus y$	$x \oplus \neg y$	$x \oplus \neg y$	$y \forall x$
2	4	3	$x \oplus y$	$\neg x \wedge \neg y$	$x \oplus \neg y$	$x \vee y$
2	5	3	$x \oplus y$	$\neg x \wedge \neg y$	$x \oplus \neg y$	$x \vee y$
3	4	3	$x \oplus y$	$x \oplus \neg y$	$x \oplus \neg y$	$x \forall y$
3	5	3	$x \oplus y$	$x \oplus \neg y$	$x \oplus \neg y$	$x \forall y$
4	5	3	$x \oplus y$	$x \oplus \neg y$	$\neg x \wedge \neg y$	$x \vee y$

9.5 Adders

Using the gates shown in Tables 9.2 and 9.3 for Rule (4,27), we can build a simple two-bit half-adder.

To solve this problem, we use a slightly different matrix, that makes active Substates 1 and 2 swap to Substates 5 and 4, when they meet in a cell that has an active Substate 3. In this way, it is possible, e.g. to get a half adder whose output state for the carrier is

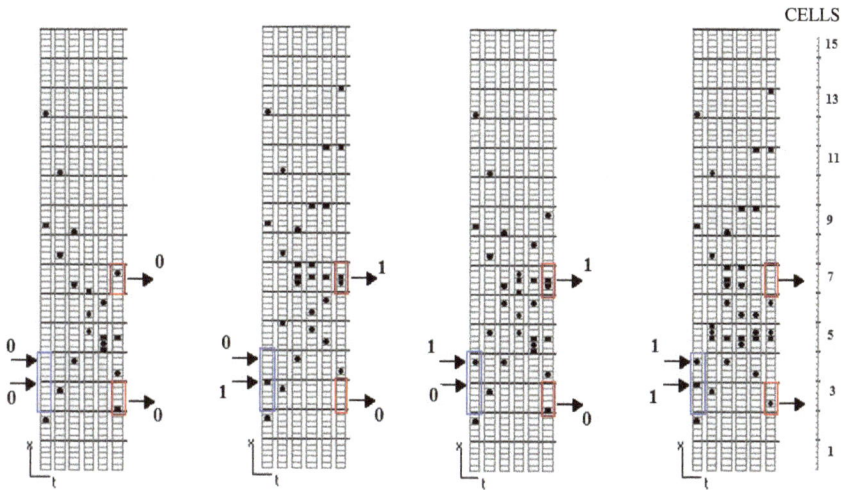

Fig. 9.4. A two-bit half adder with gates built using Rule (4,27). Blue squares, input cells, red square, output cells, Substate 3 for the sum, and Substate 2 for the carrier.

Substate 2 of Cell 3 at Step 6 (Fig. 9.4). Cells are represented along the vertical axe and time evolves on the horizontal one from left to right. In general, three or more time steps are shown in the pictures. Each cell is composed by 5 squares, representing its substates and a dark bar divides a cell from its neighbours. Note also that cells are drawn in a single column to make figures simpler to read, but they are actually arranged in two layers, like in Fig. 9.1, so that, e.g. Cell 3 interacts with its neighbours 1, 2, 4, and 5. Empty squares represent 0 values for the states, a black dot represents 1.

In such a way, it is possible, in principle, to send the output of a gate to another one in an arbitrarily long chain. However, as states go on moving without the possibility of being turned off, it is challenging to program even slightly more complex circuits, like, e.g. a full adder, because particles interact in a lot of unwanted ways.

To get usable devices, it is necessary to implement separate automata, each one performing a specific task. In such a way, we can program, for instance, a full adder. Example of the full adder states for $0 + 1 + 0 = 1$, $1 + 0 + 1 = 10$, and $1 + 1 + 1 = 11$ are shown in Figs. 9.5–9.7.

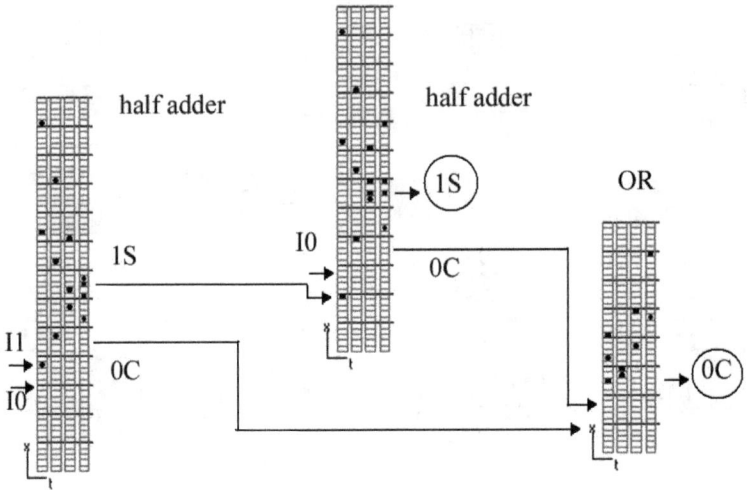

Fig. 9.5. A three-bit full adder with gates built using Rule (4,27): $0+1+0 = 1$. The final sum and carrier are encircled.

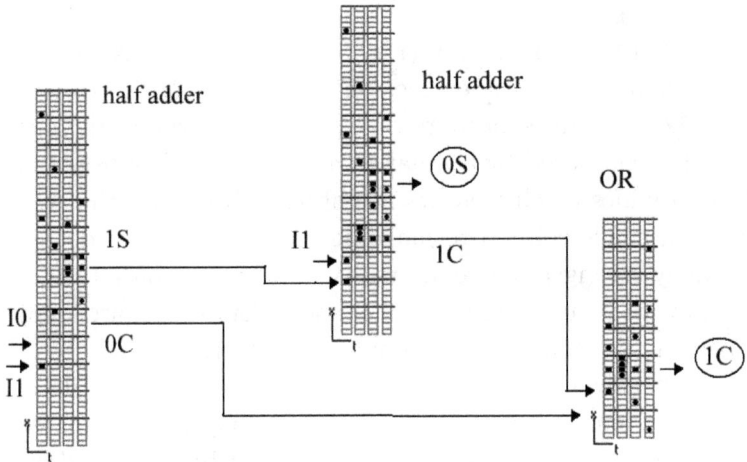

Fig. 9.6. A three-bit full adder with gates built using Rule (4,27): $1+0+1 = 10$. The final sum and carrier are encircled.

9.6 Discussion

We have presented a simple partitioned quantum automaton that can be implemented using a two-layer physical system. Each cell in the system has a principal internal state and four substates representing

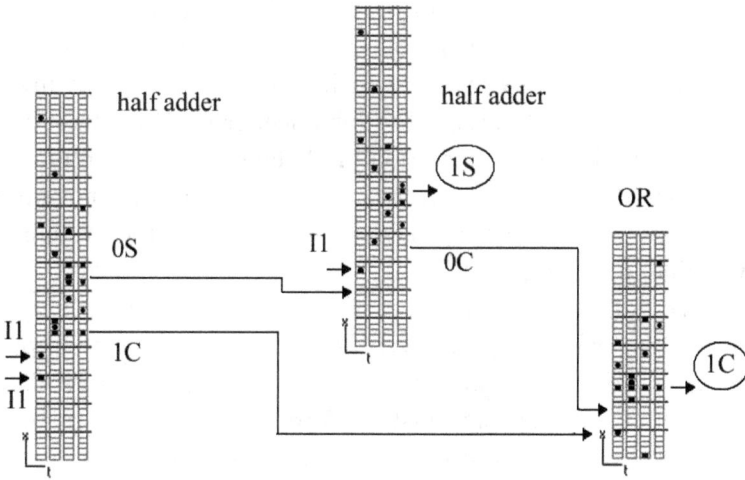

Fig. 9.7. A three-bit full adder with gates built using Rule (4,27): $1+1+1 = 11$. The final sum and carrier are encircled.

signals from neighbouring cells. The influence of the substates on the principal state is described by a unitary matrix, which can generate superposition or not, depending on the chosen matrix.

We extensively studied the behaviour of the automaton in a classical setting, using a matrix that does not generate superposition and classical input states. Despite its simplicity, we demonstrated that the system can construct a wide range of gates and even a full adder. We specifically used the (4, 27) rule for our computations, but other rules such as (6, 25), (7, 24), and (8, 23) exhibit similar computational power, as shown in tables like Tables 9.2 and 9.3.

The crucial next step is to investigate if existing circuit models of actin, proposed in previous research, can be interpreted in terms of pulses that activate and deactivate cell states according to our hypotheses. We need to determine a mechanism through which pulses can exhibit constructive interference to activate a state and destructive interference to deactivate it. The natural time step of the system is likely to play a significant role in this regard. It is possible that our simple automaton may not be sufficient when considering real systems, and a more complex model may be necessary.

Nonetheless, we believe that our partitioned quantum automaton and its behaviour can provide valuable guidance for further exploration.

Implementing a physical system based on this concept involves considering additional factors. For instance, errors may arise during signal transmission or composition, and the strength of interactions between cells may diminish over time, which can impact computations positively or negatively. Different sections of the system may follow different rules as well.

Finally, for the realisation of a practical computer using this system, it is crucial to develop efficient methods for preparing input states, reading output states, and defining routes for storing information. These aspects must be carefully considered to build a functional and reliable computing system based on this approach.

https://doi.org/10.1142/9789811285073_0010

Chapter 10

Multi-valued Logic in Quantum Actin Automata

Stefano Siccardi*,‡ and Andrew Adamatzky†,§

*Consorzio Interuniversitario Nazionale per l'Informatica, Italy
†Unconventional Computing Lab, UWE, Bristol, UK
‡ssiccardi@2ssas.it
§andrew.adamatzky@uwe.ac.uk

Abstract
To demonstrate the computational capabilities of actin filaments, we propose an innovative model that represents actin as a helix comprising two one-dimensional quantum automata arrays. This model extends our previous work and takes advantage of the inherent quantum properties of the automaton, specifically leveraging the concept of superposition. By carefully selecting functions for automaton state transitions, we are able to compute and observe the evolution of actin automata, successfully implementing three-valued logical gates within the actin automata framework. Through illustrative examples, we showcase the implementation of operators from diverse three-valued logical systems. These findings establish a strong basis for further theoretical explorations and open up possibilities for potential experimental realisations of multiple-valued logical circuits in laboratory settings.

10.1 Introduction

From their inceptions in polyvalence of logical statements by Peirce [294] and further development in three-valued logic by Łukasiewicz [295], imaginary logic by Vasil'ev [296], modal logics by Russell and MacColl [297], Bochvar's calculus [298], Kleene's [299] and Belnap Jr. [300] logical systems, the concepts of partial judgement, free the laws of excluded middle and non-contradiction formed a mathematical basis for artificial cognitive systems. The mapping of multiple-valued logical systems onto a substrate with topology similar to that of human nervous system has been first implemented research directions inspired by McCullock and Pitts neurons [301]. Further logical networks got a new lease of life with introduction of random Boolean networks, where nodes are Boolean logical functions, chosen at random, with arguments being Boolean states of the node's neighbours in the network [302,303]. The networks with Boolean logic nodes were adopted as models of memory and cognition [304–306]. We are unaware of any published results of implementing multiple-valued logics in molecular systems, polymer chains or any other components of the cytoskeleton.

A few authors have considered QCA for a multiple-valued logic or, more in general, the relationships between the logic and the quantum information. Thus [307] has proposed multiple-valued logic gates for quantum computing, and analysed their advantages for scaling up quantum computers. More specifically, [308, 309] and [310] considered extensions of QCA to deal with three-valued logics. In particular, [310] tackled the problem of the implementation of Łukasiewicz NOT, AND and OR functions with a QCA. What these approaches have in common is that they propose a version of the QCA, sometimes called an Extended QCA, where cells can take three states — qutrits — instead of two states as in the standard QCA models. This is achieved by using cells consisting of eight quantum dots with a specific structure.

In contrast, our approach differs from these previous works. We utilize a standard QCA, where the states are expressed in qubits. However, we take a different interpretation by considering superposed states as representing the third logical value ($\frac{1}{2}$). This approach is inspired by the research conducted by Dubois [285, 286], where the intermediate logical value is interpreted in an epistemic manner, indicating 'unknown' or 'controversial'. In our study presented in Ref. [287], we explicitly implemented logical operations of several three-valued logics, including Łukasiewicz, Gödel, Ja'skowski, Soboci'nski, and Sette logics. Our research delved into the programming and interconnection of automata, exploring their capabilities. Although it is evident that all three-valued operations can be represented using, for example, Lukasiewicz logic and the augmentation operator, we examined the programming and coupling aspects in greater depth.

We report here the classical definition of one-dimensional quantum cellular automata as proposed in Watrous [311–313].

Definition I. A one-dimensional quantum cellular automaton (1QCA) is a tuple $\mathcal{A} = \langle Q, \lambda, N, \delta \rangle$ with a finite set Q of states, that include one so-called *quiescent state* λ, a neighbourhood $N = \{n_1, \ldots, n_r\} \subseteq \mathbb{Z}$, with $n_1 < n_2 < \cdots < n_r$ and a local transition function $\delta \colon Q^{r+1} \to \mathbb{C}_{[0,1]}$ satisfying the following three conditions. First, local probability: for any $(q_1, \ldots, q_r) \in Q^r$, $\sum_{q \in Q} |\delta(q_1, \ldots, q_r, q)|^2 = 1$. Second, stability of the quiescent state: if $q \in Q$, then $\delta(\lambda, \ldots, \lambda, q) = 1$ *if* $q = \lambda$, 0 otherwise. To state the third condition we have to introduce the automaton's mapping.

A configuration $c : \mathbb{Z} \to Q$ is a mapping such that $c(i) \neq \lambda$ only for finitely many i. Let $C(\mathcal{A})$ denote the set of all configurations. Computation of $\mathcal{A}$ is done in the space $H_{\mathcal{A}} = l_2(C(\mathcal{A}))$ with the basis $\{|c\rangle \mid c \in C(\mathcal{A})\}$. In one step, $\mathcal{A}$ transfers from one basis state $|c_1\rangle$ to another state $|c_2\rangle$ with the amplitude $\alpha(c_1, c_2) = \prod_{i \in \mathbb{Z}} \delta(c_1(i + n_1), \ldots, c_1(i + n_r), c_2(i))$ A state in $H_{\mathcal{A}}$ has, in general, the following form $|\phi\rangle = \sum_{c \in C(\mathcal{A})} \alpha_c |c\rangle$ with normalised α_c. The evolution operator $E_{\mathcal{A}}$ of $\mathcal{A}$ maps any state $|\phi\rangle$ into $|\psi\rangle = E_{\mathcal{A}} |\phi\rangle$ such that $|\psi\rangle = E_{\mathcal{A}} |\phi\rangle = \sum_{c \in C(\mathcal{A})} \beta_c |c\rangle$ where $\beta_c = \sum_{c' \in C(\mathcal{A})} \alpha_{c'} \alpha(c', c)$ The third condition is unitarity: the mapping $E_{\mathcal{A}}$ must be unitary.

We will deal with a specific kind of QCA, a partitioned quantum one-dimensional cellular automaton:

Definition II. A partitioned one-dimensional quantum cellular automaton (P1QCA) is a 1QCA $\mathcal{A} = \langle Q, \lambda, N, \delta \rangle$ that satisfies the following restrictions: (1) The set of states Q is the Cartesian product $Q = Q_1 \times \cdots \times Q_r$ of $r = |N|$ nonempty sets; (2) The local transition function δ is a composition of two functions: $\delta_c : Q^r \to Q$ a classical mapping and $\delta_q : Q \to C^Q$ a quantum mapping where $\delta_c((q_{1,1}, \ldots, q_{1,r}), \ldots, (q_{r,1}, \ldots, q_{r,r})) = (q_{r,1}, q_{r-1,2}, \ldots, q_{1,r})$.

The function δ_q defines a 1QCA with an operator $U_{\mathcal{A}_q}(q_2, q_1) = [\delta_q(q_1)](q_2)$ The operation $U_{\mathcal{A}_q}$ is called the local transition matrix of $\mathcal{A}$. It can be shown that the evolution of a P1QCA is unitary if and only if the operator is unitary.

In the models used in the present work, the state q_i of the generic particle at position i is a vector $(q_{i1}, q_{i2}, q_{i3}, q_{i4}, q_{i5})$ and the first step of the evolution computation consists in building the intermediate state $\tilde{q}_i = (q_{i+2,1}, q_{i+1,2}, q_{i,3}, q_{i-1,4}, q_{i-2,5})$.

We must specify the matrix (of dimension 32×32) $U(q_1, q_2) = [u_{a_1 a_2 a_3 a_4 a_5, b_1 b_2 b_3 b_4 b_5}]$ whose elements give the probability that a state $\tilde{q} = (a_1 a_2 a_3 a_4 a_5)$ evolves in another state $q = (b_1 b_2 b_3 b_4 b_5)$.

We define the sum $S = q_{i-2} + q_{i-1} + q_{i+1} + q_{i+2}$ that is computed at time t to obtain the new state q_i at time $t + 1$ of the particle with index i, using the elements of the intermediate state, and $((x_{00}, x_{01}, x_{02}, x_{03}, x_{04}), (x_{10}, x_{11}, x_{12}, x_{13}, x_{14})) = (X_0, X_1)$ with $x_{ik} \in \{0, 1\}$ and X_i 5-elements arrays. We will often refer to the (X_0, X_1) pairs as 'rules'.

With this notation, for the generic element of U we have:

$$u_{a_1 a_2 a_3 a_4 a_5, b_1 b_2 b_3 b_4 b_5} = \begin{cases} 1 & \text{if } b_3 = X_{a_3,S} \text{ and } b_{i \neq 3} = a_i, \\ 0 & \text{otherwise} \end{cases}$$

In words, the above mechanism means that the central (or third) substrate of a particle is computed accordingly to a rule that applies to the value of the sum of substrates of the neighbouring particles, and the four 'extra' substrates are just copied from the neighbours. Instead of the above transformation, other transformations can be used, but for our aims they do not have advantages.

As we are mainly concerned with substrates and their evolution and movements, we now exemplify how the substrates are computed and interpreted in the spirit of Ref. [286]. We start at the initial time t with two particles x_1, x_2 in the following states:

$$x_1 \equiv (|0\rangle |0\rangle |0\rangle |0\rangle) \frac{1}{\sqrt{2}} (|0\rangle + |1\rangle)) = \frac{1}{\sqrt{2}} (|00000\rangle + |00001\rangle)$$

$$x_2 \equiv (|0\rangle |0\rangle |0\rangle) \left(\frac{1}{2} |0\rangle + \frac{\sqrt{3}}{2} |1\rangle \right) |0\rangle) = \frac{1}{2} |00000\rangle + \frac{\sqrt{3}}{2} |00010\rangle$$

all the other particles being in the resting state $|00000\rangle$. We use the actin automata Rule $(01100, 10011)$ [6] whose decimal form is $(12, 19)$.

The Rule (12, 19) defines the following state transition: State 3 is activated when 1 or 2 other states, coming from the neighbouring particles, are active.

We now compute the state of particle x_3 at time $t + 1$. The first step is to compute the intermediate state, that inherits substates: Substate 5 from x_1, Substate 4 from x_2, Substate 3 = Substate 2 = Substate 1 = $|0\rangle$. We get

$$\frac{1}{2\sqrt{2}}|00000\rangle + \frac{1}{2\sqrt{2}}|00001\rangle + \frac{\sqrt{3}}{2\sqrt{2}}|00010\rangle + \frac{\sqrt{3}}{2\sqrt{2}}|00011\rangle.$$

Now we apply Rule $(12, 19)$ and obtain

$$\frac{1}{2\sqrt{2}}|00000\rangle + \frac{1}{2\sqrt{2}}|00101\rangle + \frac{\sqrt{3}}{2\sqrt{2}}|00110\rangle + \frac{\sqrt{3}}{2\sqrt{2}}|00111\rangle.$$

We can now compute Substate 3 as

$$\frac{\sqrt{1}}{2\sqrt{2}}|0\rangle + \frac{\sqrt{7}}{2\sqrt{2}}|1\rangle. \tag{10.1}$$

We can *interpret* the computation as follows. Suppose that we want to assess how frequently people believe that a statement α is true OR a statement β is true. We ask a group of (say) 100 people about their belief about α and find that 50 think that it is true and 50 that it is false. We encode this fact in the state of particle x_1, reminding that as usual in quantum information the probability of a state is the square of its coefficient. Then we ask *another* group of 100 people about statement β and find that 75 think that it is true. We encode this fact in the state of particle x_2. We then form all the possible pairs of one member of the first group and one of the second, and find that 8,750 out of 10,000, that is $\frac{7}{8}$ believe that α or β is true or both, according to Eq. (10.1).

10.2 Reading automata output, concatenating automata, and interpreting superposition

We note that the proposed automaton, without other hypotheses or transformations, is not, in general, suitable for dealing with possibility distributions and the logic that can be associated with them [291, 292]. For example, for the possibility measures Π it is generally required that $\Pi(\alpha\vee\beta) = \max(\Pi(\alpha), \Pi(\beta))$, a condition that is not satisfied by our automaton. In any practical implementations, reading a state means to run the automaton and to perform a state measure, with the result necessarily being either $|0\rangle$ or $|1\rangle$. The $|0\rangle$ and $|1\rangle$ are square roots of state probabilities. When these coefficients are obtained, the interpretation in terms of $\{0, \frac{1}{2}, 1\}$ comes in. The first interpretation is obtained by dividing the interval $[0, 1]$ of probabilities of finding an excited output State $|1\rangle$ in three parts, and identifying $0 = [0, \frac{1}{3})$, $\frac{1}{2} = [\frac{1}{3}, \frac{2}{3}]$, $1 = (\frac{2}{3}, 1]$. The second interpretation is in identifying *any* superposed states with the $\frac{1}{2}$ truth value.

In scenarios of both interpretations, however, the automaton is not capable, of performing all the operations, e.g. all the gates of Łukasiewicz or Sobociński logics. We need one more assumption to deal with these constructions. We suppose to have also the possibility of interpreting any superposed states as 1. This interpretation, that is not a part of the automaton but of the system that reads and transmits the signals, is identified with the augmentation operator ∇: $\nabla 0 = 0$; $\nabla\frac{1}{2} = \nabla 1 = 1$. The practical implementation could be a little tricky, and depending on the exact characteristics of physical devices, it could be necessary to reassign this. In the following we describe a possible abstract device.

Suppose we want to apply the ∇ operator to an input qubit. As the automaton must be run a number N of times to collect statistics of the output states, we must have the possibility to repeatedly send these input qubits to the operator. The operator must measure the input to decide if it must pass a $|0\rangle$ or $|1\rangle$ state. Each measure will give either $|0\rangle$ or $|1\rangle$ and it is possible that $|1\rangle$ is read before N trials, so the operator must pass $|1\rangle$ for all the N runs. It is also possible that $|0\rangle$ is read for all the N trials, so the operator must pass $|0\rangle$ for all the runs. In both cases, the input qubits are lost in the measures, and the ∇ operator must initialise a set of N new qubits of the type that has detected reading the input. This means that the automaton must wait, or just discard the final output, until the ∇ operator starts sending qubits. In an artificial setting, it would be sufficient for the

∇ operator to send a (classical) signal to the detectors of the final output, to switch them on; in a natural setting a suitable threshold mechanism has to be found. With the aid of this operator we will be able to manage many other logic operators.

10.3 Realisations of logic operators

10.3.1 Łukasiewicz operators

Using the interpretation with three equal intervals, the automaton can perform the Łukasiewicz conjunction $\wedge_L$ and disjunction $\vee_L$, see Table 10.1(a) and 10.1(b). The Łukasiewicz implication can be obtained as $x \to_L y = \neg x \vee_L y$, see Table 10.1(c).

The automaton evolution for operators involutive NOT, $\vee_L$, and $\wedge_L$ is shown in Fig. 10.1(a–c). The Rule (00100,11011) [6], or (4,27), has been used in these computations and in the following ones. In the figures for three-valued logic, empty squares represent 0 values for the states, a black dot represents 1, and a grey one represents 1/2. Each separate group of cells represents an automaton; input and output states are highlighted with arrows and values; in some figures several automata are linked by arrows, meaning that the output of one of them is sent as input to another. When an automaton in such a schema is used to perform a specific logic operation, e.g. $\vee$, a label has been added on top of its cells.

10.3.2 The Gödel operators

From now on we identify *any* superposed states with the $\frac{1}{2}$ truth value. In this hypothesis, the automaton can perform the Gödel

Table 10.1. Operators of Łukasiewicz logic.

	(a) $\vee_L$			(b) $\wedge_L$			(c) $\to_L$		
	0	1	$\frac{1}{2}$	0	1	$\frac{1}{2}$	0	1	$\frac{1}{2}$
0	0	1	$\frac{1}{2}$	0	0	0	1	1	1
1	1	1	1	0	1	$\frac{1}{2}$	0	1	$\frac{1}{2}$
$\frac{1}{2}$	$\frac{1}{2}$	1	1	0	$\frac{1}{2}$	0	$\frac{1}{2}$	1	1

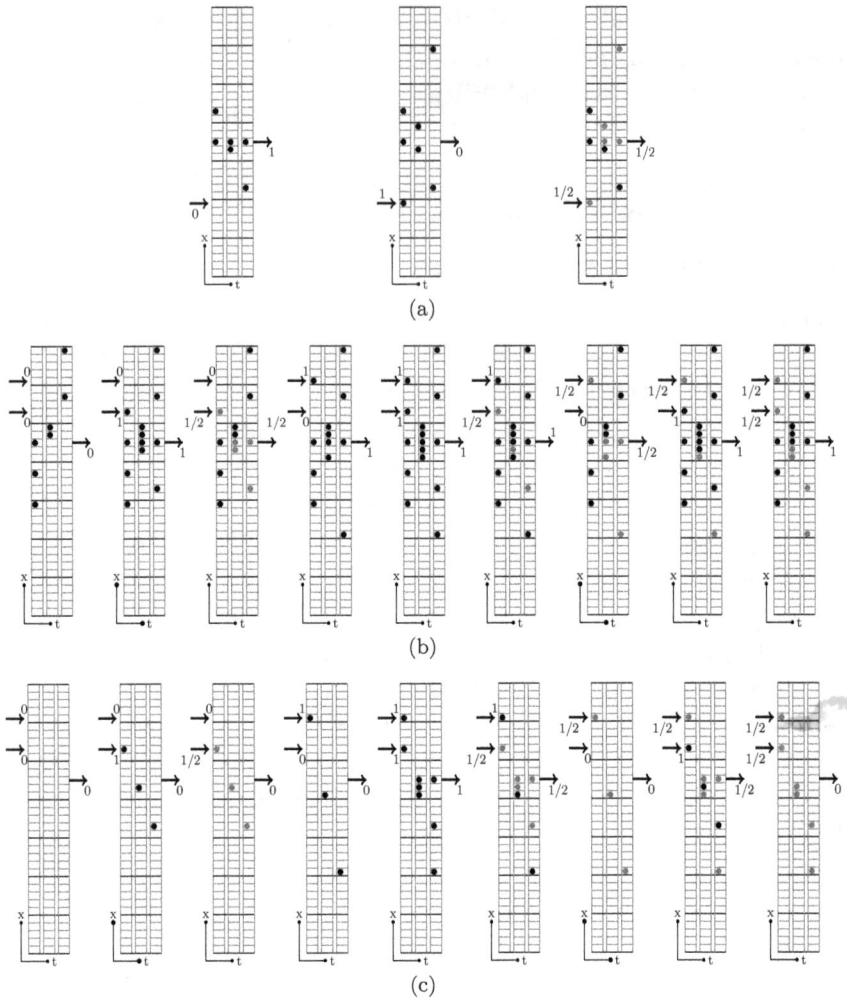

Fig. 10.1.   Automaton evolution for (a) involutive NOT, (b) $\vee_L$, (c) $\wedge_L$.

conjunction $\alpha \wedge_G \beta = \min(\alpha, \beta)$ and the disjunction $\alpha \vee_G \beta = \max(\alpha, \beta)$, see Table 10.2 and automaton simulations in Fig. 10.2. The involutive negation $\neg$ is the same as in the previous case, see Fig. 10.1(a). An implication operator can again be obtained by composing $\neg$ and $\vee_G$. Note that here operators max and min work well only because we identify $\frac{1}{2}$ with any superposed states, disregarding their actual probability.

Table 10.2. Operators max and min.

	(a) $\vee_G$			(b) $\wedge_G$		
	0	1	$\frac{1}{2}$	0	1	$\frac{1}{2}$
0	0	1	$\frac{1}{2}$	0	0	0
1	1	1	1	1	0	1	$\frac{1}{2}$
$\frac{1}{2}$	$\frac{1}{2}$	1	$\frac{1}{2}$	$\frac{1}{2}$	0	$\frac{1}{2}$	$\frac{1}{2}$

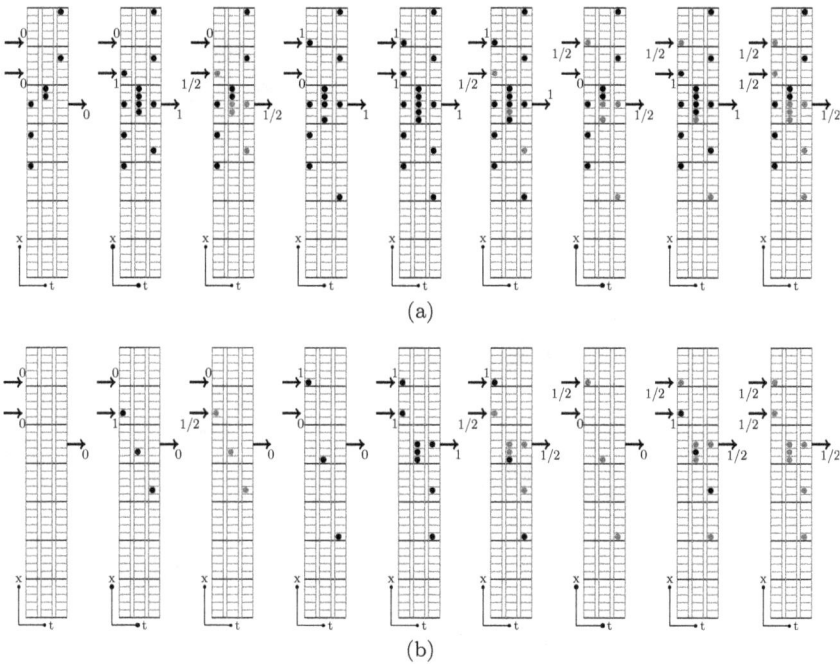

(a)

(b)

Fig. 10.2. Automaton evolution for (a) $\vee_G$ and (b) $\wedge_G$.

10.3.3 Łukasiewicz operators again

Łukasiewicz implication, Table 10.1(c), can be written as $x \to_L y = (\nabla \neg x \vee y) \wedge (\nabla y \vee \neg x)$, where $\vee = \text{Max}$ and $\wedge = \text{Min}$. We show in Fig. 10.3(a) the automata, their connections, and the evolution for the case $\frac{1}{2} \to_L \frac{1}{2}$, whose truth value is 1. It is important to stress that we cannot just send the input qubits in parallel to two

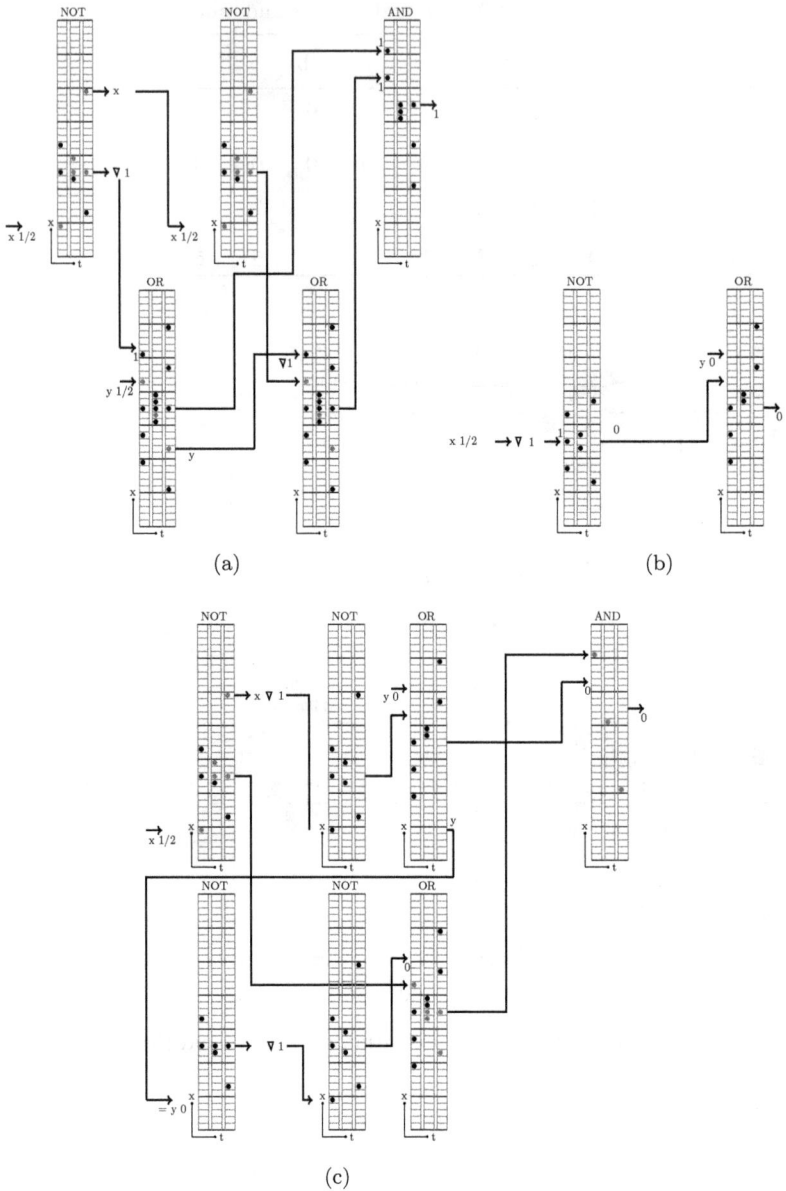

(a)

(b)

(c)

Fig. 10.3. Automaton evolution for (a) $\frac{1}{2} \to_L \frac{1}{2}$, (b) $\frac{1}{2} \to_J 0$, (c) $\frac{1}{2} \to_S 0$.

gates, as we would do if they were classical bits, because the no cloning theorem [293] affirms that quantum states cannot be copied. What actually happens, instead, is that the qubits pass through the first gate and can be used again as input for the second gate. This happens for instance for the x qubit passing the first NOT gate and for y passing the first ∨ in Fig. 10.3(a). Jaśkowski implication and Sobociński and Sette operators can be implemented in similar ways, see details in Ref. [287].

10.4 Discussion

In our research, we have demonstrated the potential of a simple partitioned quantum automaton model, inspired by actin double helix protein polymer, to perform Boolean and three-valued logical operations. However, the feasibility of realising such an automaton depends on the characteristics of the physical substrate used. Specifically, the states of the automaton must be quantum states, and they need to influence each other according to a summation rule, where a particle (node or G-actin unit) changes its state based on the activity of a fixed number of its neighbouring units. To implement this concept using actin filaments, it is necessary to determine a mechanism through which signals or data travelling along the filaments can interfere constructively to activate a state, and engage in destructive interference to deactivate the state. This mechanism would be crucial for the successful operation of the automaton. Additionally, there are other effects that must be taken into account when considering a physical system for implementation. For example, errors may occur during signal transmission or composition, which can impact the reliability of computations. Moreover, the interactions between units may fade over time, which can have both positive and negative effects on the computations. Furthermore, different sections of the system may follow different rules, adding complexity to the overall design. To build a practical computer based on the described automaton, it is essential to find efficient methods for preparing the input states and reading the output states. Moreover, routes for storing information need to be defined to enable the functioning of the computer. These considerations are important for the development of a functional and reliable computing system utilising actin filaments.

Chapter 11

Logical Gates in Actin Monomer

Andrew Adamatzky

Unconventional Computing Lab, UWE Bristol, UK

andrew.adamatzky@uwe.ac.uk

Abstract

We evaluate the computational capabilities of an individual actin molecule by investigating the distributions of logical gates achieved through the propagation of excitation patterns. Our approach involves representing the actin molecule as an excitable network of automata, known as the F-actin automaton. In this model, the state of each atom is updated based on the states of its adjacent atoms through chemical bonds (hard neighbours) and atoms in close proximity (soft neighbours). When the sum of excited hard neighbours and a weighted sum of soft neighbours falls within a specified range, a resting atom becomes excited. Through our analysis, we demonstrate that F-actin automata effectively implement logical gates, including OR, AND, XOR, and AND-NOT, by utilising interacting excitation patterns. The AND gate is the most frequently observed, while the XOR gate is the least commonly observed. Building upon these gate architectures, we further construct a one-bit half-adder and controlled-not circuits within the F-actin automata framework. Additionally, we discuss the speed and space characteristics of these molecular computers based on F-actin.

11.1 Introduction

We have explored various computational schemes in different models of actin filaments, ranging from quantum automata to lattices with Morse potential [229–232, 234]. Previous models primarily focused on coarse-grain computations using chains of F-actin units, where each monomer represented a single bit. However, to enhance the density of computing elements in actin-based systems, we aimed to incorporate multiple logical gates within a single F-actin monomer. To assess the information processing capacity of an actin molecule, we conducted calculations on the distributions of logical gates implemented by the molecule through propagating patterns of excitation. This approach was successfully tested on an automaton model of the verotoxin protein [16], and preliminary experiments have been carried out on actin molecules [18].

11.2 The model

We use a structure of F-actin molecule produced using X-ray fibre diffraction intensities obtained from well-oriented sols of rabbit skeletal muscles [190]. The structure was calculated with resolution 3.3Å in radial direction and 5.6Å along the axis (Fig. 11.1). The molecular structure is converted to a non-directed graph $\mathcal{A}$ embedded in Euclidean space: where every node represents an atom and an edge corresponds to a bond between the atoms.

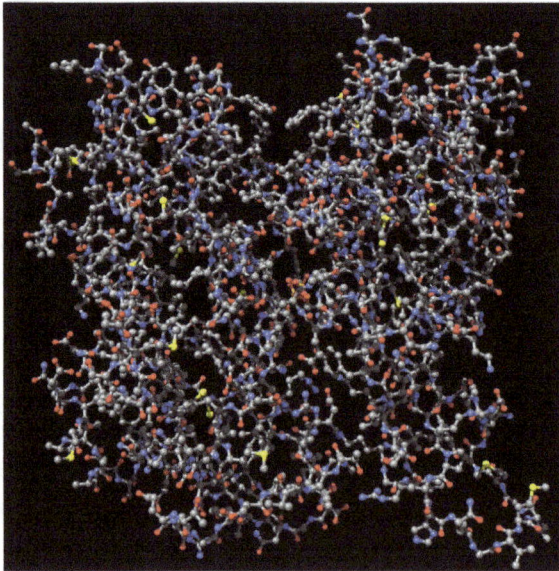

Fig. 11.1. F-actin molecule, CPK colouring, the structure of the molecule is determined in Ref. [187].

F-actin is represented by a graph $\mathcal{A} = \langle \mathbf{V}, \mathbf{E}, \mathbf{C} \rangle$, where $\mathbf{V}$ is a set of nodes, $\mathbf{E}$ is a set of edges, $\mathbf{C}$ is a set of Euclidean coordinates of nodes from $\mathbf{V}$.

The graph $\mathcal{A}$ has 2961 nodes, 3025 edges. Minimum degree is 1, maximum is 4, average is 2.044 (with standard deviation 0.8) and median degree 2. There are 883 nodes with degree 1, 1009 nodes with degree 2, 1066 nodes with degree 3, and two nodes with degree 4. The graph $\mathcal{A}$ has a diameter (longest shortest path) of 1130 nodes, and a mean distance (mean shortest path between any two nodes) 376, and a median distance 338.

Let $u(s)$ be nodes of $\mathcal{A}$ that connected with an edge with a node s, they correspond to atoms connected with a bond to atom s. We call them hard-neighbours because their neighbourhood is determined by the chemical structure of F-actin. Actin molecule is folded in 3D. Let δ be an average distance between two hard-neighbours, for F-actin $\delta = 1.43$ units. Let $w(s)$ be nodes of $\mathcal{A}$ that are at distance not exceeding ρ, in Euclidean space, from node s. We call them soft neighbours because their neighbourhood is determined by the 3D structure of F-actin. Thus, each node s has two neighbourhoods:

(1) hard neighbourhood $u(s) = \{p \in \mathbf{V} : (sp) \in \mathbf{E}\}$ (actin automata with hard neighbourhood were first proposed by us in Ref. [18]), and

(2) soft neighbourhood $w(s) = \{p \in \mathbf{V} : p \notin u(s) \text{ and } d(c_s, c_p) \leq \rho\}$, where $d(c_s, c_p)$ is a distance between nodes s and p in 3D Euclidean space and $c_s, c_p \in \mathbf{C}$.

Distribution of hard neighbourhood sizes corresponds to distribution of degrees of the nodes. Distribution of soft neighbourhood sizes is determined by distance ρ (Fig. 11.2). We have chosen $\rho = 3$ so size of $w(s)$ does not substantially exceed size of $u(s)$ (Fig 11.4(a)). Examples of neighbourhoods are shown in Fig. 11.3, a soft neighbour of a node s is not necessarily a hard neighbour of the s's neighbour. Distribution of nodes with highest numbers of soft neighbours is shown in (Fig 11.4(b)).

Each node s of $\mathcal{A}$ takes three states: resting ($\circ$), excited ($\star$), and refractory ($\bullet$). A resting node $s^t = \circ$ excites depending on a

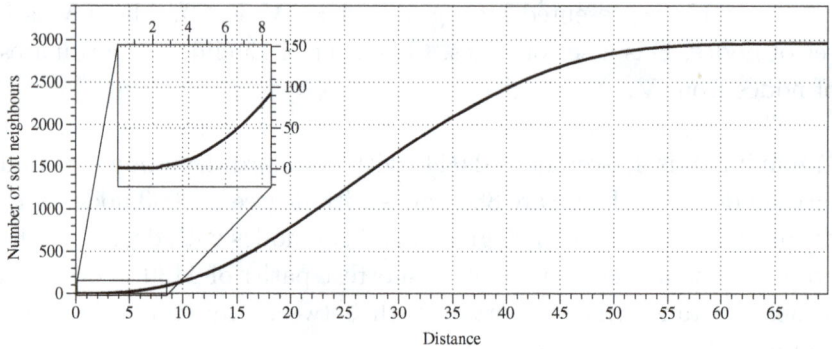

Fig. 11.2. Number of soft neighbours of a node versus distance ρ, averaged over all nodes.

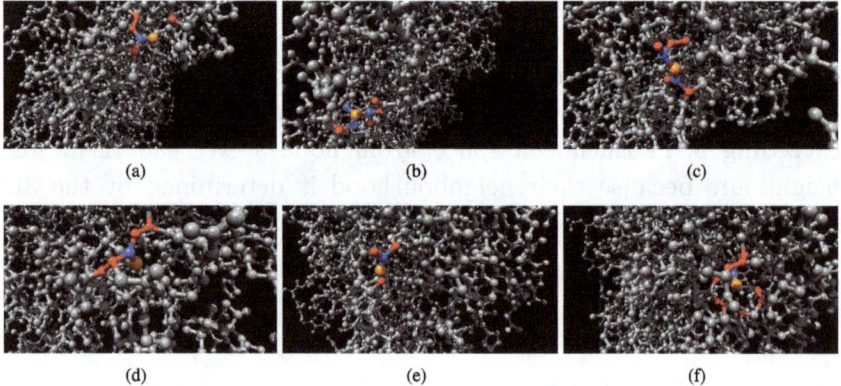

Fig. 11.3. Examples of neighbourhoods, central node is orange, hard neighbours $u(s)$ are red, soft neighbours, for $\rho = 3$, are blue $w(s)$.

number σ_s^t of its excited neighbours in neighbourhoods $u(s)$ and $w(s)$: $\sigma_s^t = \sum_{p \in u(s)} \{p^t = \star\} + \mu \sum_{p \in w(s)} \{p^t = \star\}$. We consider values of $\mu = 0.2, 0.4, 0.8$ and $\mu = \frac{\delta}{d(c_s, c_p)}$. An excited node takes refractory state. A refractory node takes resting state. We consider two types of excitations. In threshold excitation, a resting node s excites if a σ_s^t exceeds threshold θ:

$$
s^{t+1} = \begin{cases} \star, & \text{if } \sigma_s^t > \theta, \\ \bullet, & \text{if } s^t = \star, \\ \circ, & \text{otherwise.} \end{cases} \tag{11.1}
$$

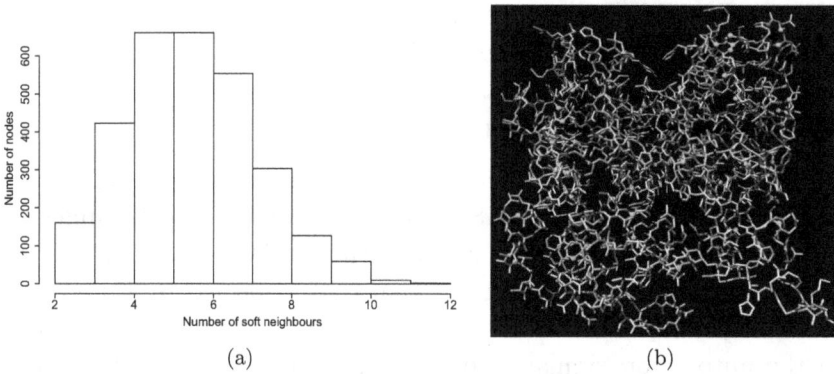

(a) (b)

Fig. 11.4. Soft neighbourhood. (a) Distribution of soft neighbourhood sizes for $\rho = 3$. (b) Nodes with maximum number of soft neighbours are coloured: Red (12 neighbours), blue (10 neighbours), green (9 neighbours).

In interval excitation, a resting node s excites if σ_s^t belongs to a discrete interval $\Theta = [\theta_1, \theta_2]$, i.e. a set of discrete numbers from θ_1 to θ_2:

$$s^{t+1} = \begin{cases} \star, & \text{if } \sigma_s^t \in \Theta, \\ \bullet, & \text{if } s^t = \star, \\ \circ, & \text{otherwise.} \end{cases} \quad (11.2)$$

We search for the Boolean logical gates by selecting a pair of nodes from $\mathbf{V}$, exciting these nodes with combinations of inputs 01, 10, 11 (where '0' is a logical False and '1' is logical True) and checking states of all other nodes [16, 18].

Let two nodes i and j be selected as inputs and one node p as an output. Boolean logical variables are x and y (inputs) and z (output). Initially, automaton $\mathcal{A}$ is in its resting state. Logical values of inputs are converted to initial states of input nodes as follows. If $x = 1$, then $s_i^0 = \star$, if $y = 1$, then $s_j^0 = \star$.

We allow the automaton to evolve till a limit cycle or a globally resting state is reached. During the automaton's evolution, we monitor the state of the output node p. If at some time step t we have $s_p^t = \star$, we assign $z = 1$. For every rule of excitation, and exact parameters, we performed random trials, by selecting $5 \cdot 10^4$ pairs of nodes at random, exciting them with strings $(\circ\star)$, $(\star\circ)$, $(\star\star)$, and recording outputs at all other nodes of $\mathcal{A}$.

11.3 Dynamics of excitation

We studied excitation dynamics of actin automata with only hard neighbourhoods, see Ref. [18], thus here we discuss dynamics of automata with hard and soft neighbourhoods. When a small number of nodes of otherwise resting automaton $\mathcal{A}$ governed by Eq. (11.1), $\theta = 0$, $\mu = 0.2$, is excited, the wave of excitation propagates in the automaton similarly to classical excitation waves in a non-homogeneous medium. Excitation travels omnidirectional, albeit with distortions determined by the geometry of the F-actin molecule (Fig. 11.5). Eventually, wave-fronts of the excitation wave collide with each other and annihilate in the result of the collision. The automaton returns to its resting state. A number of excited nodes necessary to kick start the excitation increase with increase of θ. Typically, when θ exceeds 1, no random pattern of excitation leads to a sustainable propagation excitation wave. If $\mathcal{A}$, governed by Eq. (11.1), is initially perturbed by assigning some pool of nodes either excited or in refractory states, the automaton evolves to short limit cycles, typically with a period of three time steps. Length of transient period versus number of initially stimulated nodes is shown in Fig. 11.6(a), it decreases with increasing size of an initially perturbed domains. The dependence is observed for various values of $\mu > 0$ (Fig. 11.6(b)).

A stimulated resting automaton $\mathcal{A}$ governed by Eq. (11.2), $\Theta = [1]$, $\mu = 0.2$ always ends its evolution in one of limit cycles, where excitation travels along some finite cyclic paths (Fig. 11.7). A length of an average limit cycle of $\mathcal{A}$ governed by Eq. (11.2) does not depend on a number of initially stimulated (made of excited or refractory) nodes. The length is around 50 time steps when a percentage of initially perturbed nodes is between 10% and 90%. (Fig.11.8(b)). Transient period p, expressed in time steps, shows quadratic dependence on a percentage of initially stimulated nodes ρ: $p = 23.24 + 4.9185\rho + (-0.051891)\rho^2$ (Fig. 11.8(b)).

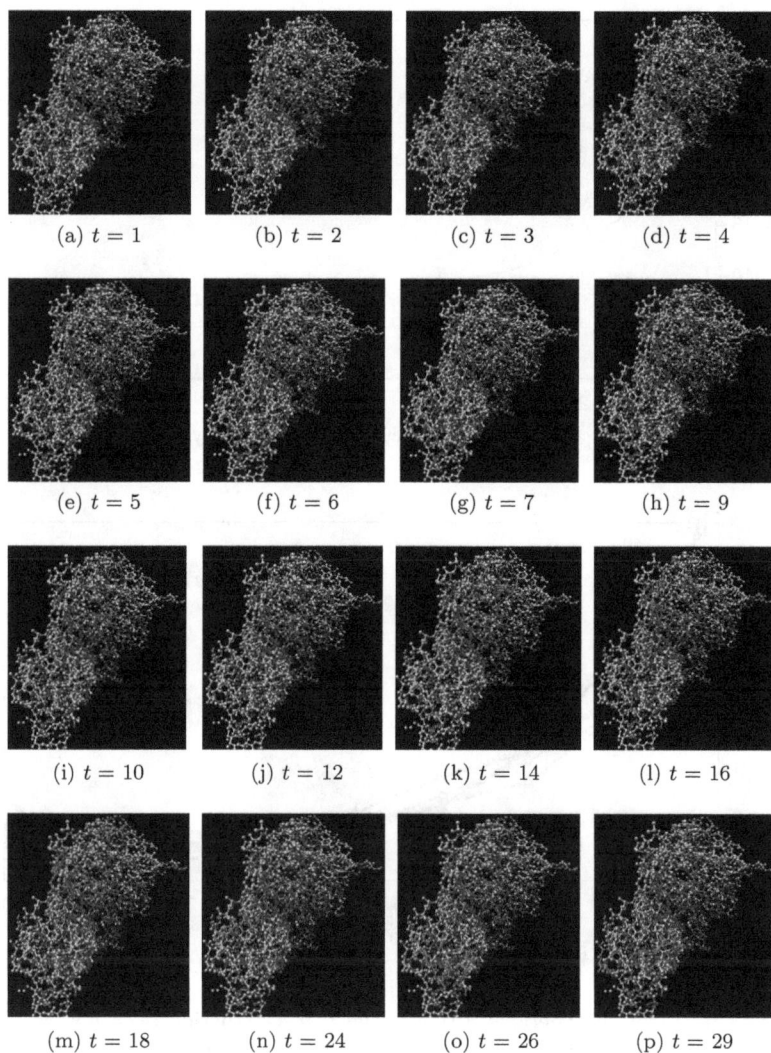

(a) $t = 1$ (b) $t = 2$ (c) $t = 3$ (d) $t = 4$

(e) $t = 5$ (f) $t = 6$ (g) $t = 7$ (h) $t = 9$

(i) $t = 10$ (j) $t = 12$ (k) $t = 14$ (l) $t = 16$

(m) $t = 18$ (n) $t = 24$ (o) $t = 26$ (p) $t = 29$

Fig. 11.5. Evolution of automaton $\mathcal{A}$ governed by Eq. (11.1), $\theta = 0$, $\mu = 0.2$. Two nodes of the resting automaton are excited initially, $t = 1$. Excited nodes are red, refractory are blue, resting are grey.

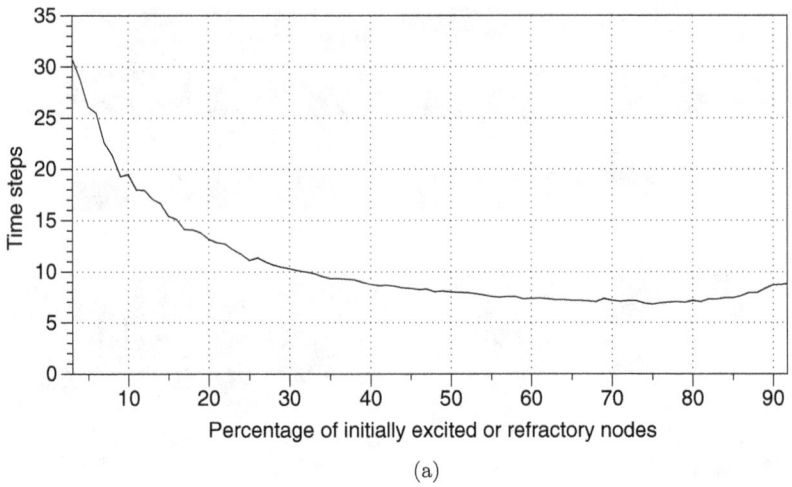

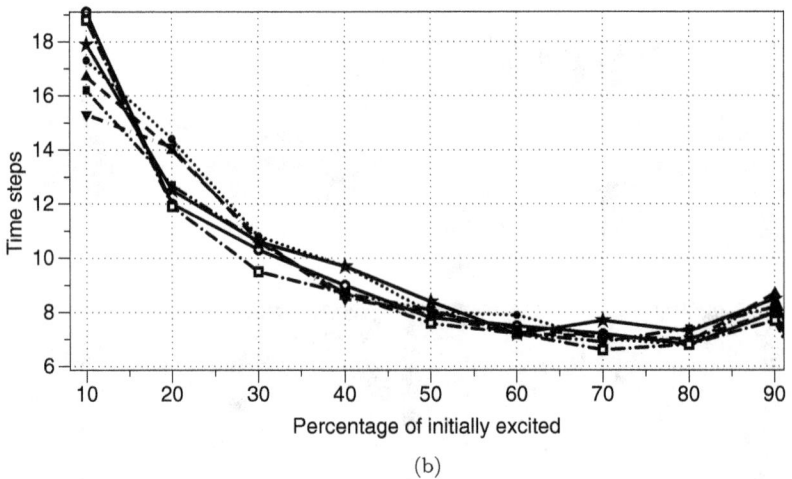

Fig. 11.6. Transient periods. (a) Average transient period of $\mathcal{A}$ governed by Eq. (11.1), $\theta = 0$, $\mu = 0.2$, depending on a number of initially excited or refractory nodes. Percentage of initially excited nodes ranges from 1 to 100, with increment 1. For each percentage we conducted 100 trials. (b) Average transient periods of $\mathcal{A}$ governed by Eq. (11.1) calculated in ten trials for several values μ, depending on a number of initially excited or refractory nodes. The values μ are as follows $\mu = 0.001$, circle; $\mu = 0.05$, solid disc; $\mu = 0.1$, triangle up; $\mu = 0.2$, triangle down; $\mu = 0.3$, empty square, $\mu = 1$, solid square, $\mu = 5$, empty star.

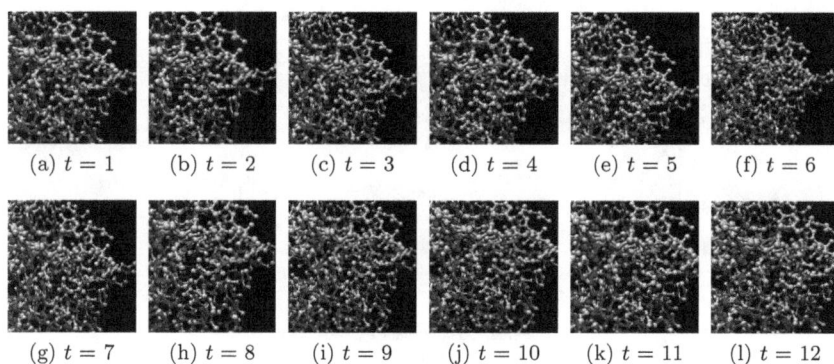

(a) $t = 1$ (b) $t = 2$ (c) $t = 3$ (d) $t = 4$ (e) $t = 5$ (f) $t = 6$

(g) $t = 7$ (h) $t = 8$ (i) $t = 9$ (j) $t = 10$ (k) $t = 11$ (l) $t = 12$

Fig. 11.7. Evolution of automaton $\mathcal{A}$ governed by Eq. (11.2), $\Theta = [1]$, $\mu = 0.2$. Two nodes of the resting automaton are excited initially, $t = 1$. Excited nodes are red, refractory are blue, resting are grey.

11.4 Boolean gates

Typically, gates discovered in $\mathcal{A}$ have two input nodes and several nodes that represent the same output.

Automaton $\mathcal{A}$ with threshold excitation Rule (11.1), $\theta = 0$, implements AND and AND-NOT gates, and also identity gates where output represents a value of one of the inputs. Almost every pair of nodes selected has some output nodes that represent these gates (Fig. 11.1). No XOR gate was found in $\mathcal{A}$ governed by Rule (11.1)

Table 11.1. Statistics of two-inputs one-output gates discovered in threshold excitation $\mathcal{A}$ without soft neighbours, Rule (11.1), $\mu = 0.2$, $\theta = 0$. The table shows average ratio m of input pairs of nodes, i.e. a number of nodes detected in all samples divided by a number of samples, and average number n of output nodes per pair of input nodes, with its standard deviation. The gates are discovered in 50K trials. Identity gates are not shown.

	$x + y$	$x\bar{y}$	$\bar{x}y$
m	0.9997	0.9965	0.9936
n	2,370	278	277
std. dev. n	152	116	118

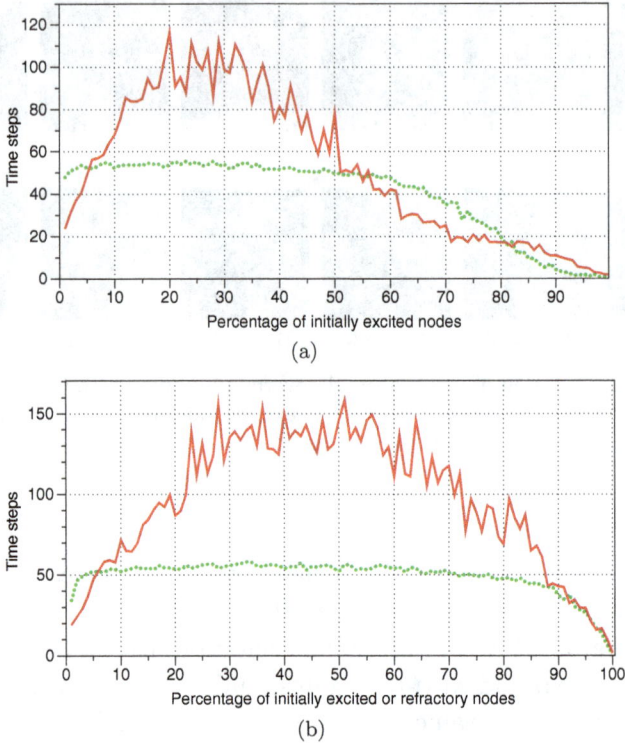

Fig. 11.8. Lengths of transient periods (solid red) and of limit cycles (dotted green) of automaton $\mathcal{A}$, governed by Eq. (11.2), $\Theta = [1]$, $\mu = 0.2$, dependent on a number of initially stimulated nodes. (a) A percentage of nodes is initially excited. (b) A percentage of nodes is initially excited or made refractory. Percentage of stimulated nodes ranges from 1 to 95, with increment 1. For each percentage, we conducted 100 trials.

with or without soft neighbours, therefore, we will not study this particular automaton further.

Automata governed by interval excitation Rule (11.2) realise input selector gates ($z = x$ and $z = y$), OR, AN, XOR, and AND-NOT ($z = \overline{x}y$ and $z = x\overline{y}$) gates (Table 11.1). Frequencies of gates discovered are characterised in Table 11.2. Frequencies of gates are similar for automata where a sum of excited soft neighbours are weighted by distance between a node and its soft neighbour $\mu = \frac{\delta}{d(c_s,c_p)}$ (Table 11.2(a)) and automata where weights of the

Table 11.2. Statistics of two-input one-output gates discovered in interval excitation $\mathcal{A}$, Rule (11.2). The table shows average ratio m of input pairs of nodes, i.e. a number of nodes detected in all samples divided by a number of samples, and average number n of output nodes per pair of input nodes, with its standard deviation. The gates are discovered in 50K trials. Identity gates are not shown. (a) $\mu = \frac{\delta}{d(c_s, c_p)}$, $\Theta = [1]$. (b) $\mu = 0.2$, $\Theta = [1]$. (c) $\mu = 0.4$, $\Theta = [1]$. (d) $\mu = 0.2$, $\Theta = [1, 2]$. (e) $\mu = 0$, $\Theta = [1]$.

	x	y	$x + y$	xy	$x \oplus y$	$x\bar{y}$	$\bar{x}y$
				(a)			
m	0.41	0.40	0.006	0.025	0.001	0.003	0.003
n	34	33	27	6	5	10	6
std. dev. n	30	33	20	3	3	8	2
				(b)			
m	0.41	0.42	0.007	0.022	0.001	0.004	0.004
n	34	34	27	6	25	14	17
std. dev n	30	30	25	7	28	18	20
				(c)			
m	0.42	0.42	0.006	0.022	0.001	0.003	0.004
n	34	34	26	6	28	15	15
std. dev n	30	30	24	7	30	19	17
				(d)			
m	0.42	0.42	0.007	0.02	0.001	0.004	0.004
n	34	33	24	6	17	14	16
std. dev n	30	30	21	5	23	17	20
				(e)			
m	0.06	0.05	0.9	0	0.5	0.9	0.9
n	2,141	2,193	2,803	0	2	40	39
std. dev n	1,260	1,236	12	0	2	9	9

soft neighbours are fixed to $\mu = 0.2$ (Fig. 11.2(b)) or $\mu = 0.4$ (Table 11.2(c)). The frequencies of gates are also robust with respect to a size of excitation interval: compare $\Theta = [1]$ (Tables 11.2(b) and 11.2(c)) and $\Theta = [1, 2]$ (Table 11.2(d)). The majority of gates discovered are input selector gates, they can be found in a half of trials. Second most common gate is AND. Gates AND-NOT and XOR are much less frequent that other gates, with XOR gate being the rarest one, are typically found just in one of thousand trials. Thus, a hierarchy of frequencies of gates is AND > OR > AND-NOT > XOR. We can see a role of soft neighbours in formation of AND gates by comparing (Table 11.2(e)) with (Table 11.2(a–d)). In the automaton without soft neighbours, $\mu = 0$, input selector gates are rare, XOR gates are discovered in half of the trials, and OR and AND-NOT gates are discovered in almost all trials.

Exemplar architectures of XOR and AND gates are shown in Fig. 11.9. In XOR gates, input nodes are located in close proximity

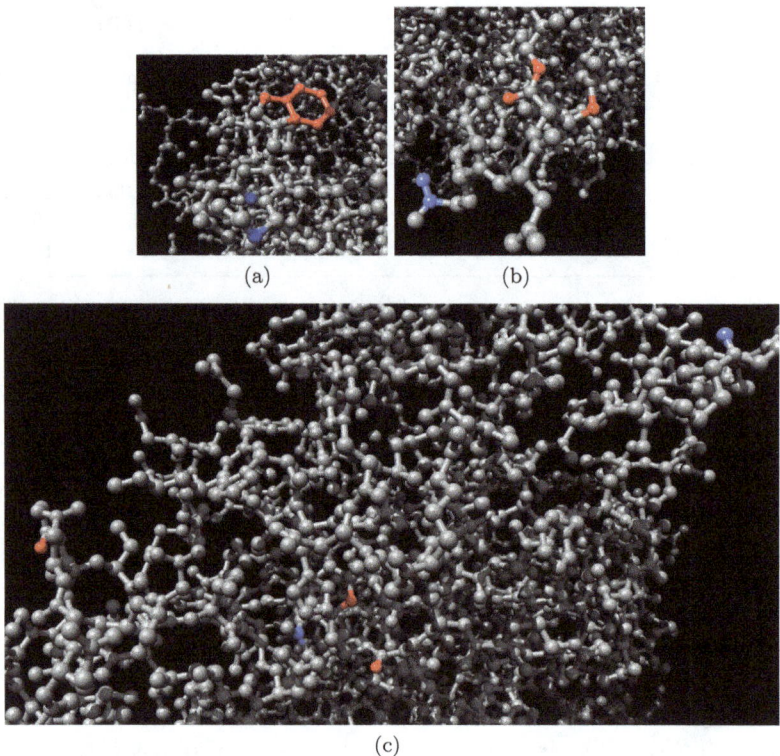

(a)   (b)

(c)

Fig. 11.9. Architectures of XOR (ab) gates, two examples, and AND (c) gate realised in $\mathcal{A}$ Rule (11.2), $\mu = 0.2$. Input nodes are blue, output nodes are red.

to each other (Fig. 11.9(a)) or even hard neighbours (Fig. 11.9(b)). The output nodes of XOR gate exhibit a similar degree of geographic proximity, e.g. output nodes in Fig. 11.9(a) are atoms of the same aromatic ring. Configuration of gate AND presents a different picture: distance between input and output nodes are large, sometimes half of the graph diameter (Fig. 11.9(c)).

11.5 One-bit half adder

A one-bit half adder is a gate with two inputs x and y and two outputs xy (sum) and $x \oplus y$ (carry). To evaluate if $\mathcal{A}$ implements a half adder we must search for two input nodes for which there are at least two output nodes: one node outputs AND and another node outputs XOR. In $5 \cdot 10^3$ samples of automaton $\mathcal{A}$ governed by Rule (11.2), $\mu = 0.2$, $\Theta = [1]$, we found just a single pair of nodes producing AND and XOR. In the same sampling size of automaton governed by Rule (11.2), $\mu = 0.2$, $\Theta = [1, 2]$, we found two pairs of input nodes that have output nodes producing XOR and AND.

Architecture of an exemplar half adder is shown in Fig. 11.10. Atoms IDs and names are as follows: x (ID=696, N atom in tyrosine), y (ID=731, CD1 atom in leucine), xy (ID=677, NE2 atom in histamine), $x \oplus y$ (ID=639, O atom in isoleucine). Dynamics of excitation in the half adder is shown for inputs $x = 0$ and $y = 1$ (Fig. 11.11), $x = 1$ and $y = 0$ (Fig. 11.12), and $x = 1$ and $y = 1$ (Fig. 11.13). Traces of excitation are shown in Fig. 11.14.

11.6 Controlled NOT gate

Controlled NOT gate first appeared in Toffoli's paper on reversible computing [250], the gate was coined 'controlled NOT' gate by Feynman [86]. Let us call the gate CNOT. The CNOT gate has two inputs x and c, and two outputs y and z: $z = c$, and $y = x$ if $c = 0$ and $y = \overline{x}$ if $c = 1$; alternative, $y = x \oplus c$. By sampling $5 \cdot 10^4$ pairs of nodes in $\mathcal{A}$ governed by Rule (11.2), $\mu = 0.2$, $\Theta = [1]$, we found 10 pairs of nodes (x, c) that have output nodes acting as control z and target y. These pairs of input nodes are shown in Fig. 11.15. Configuration of one of the CNOT gates is shown in Fig. 11.16. Output

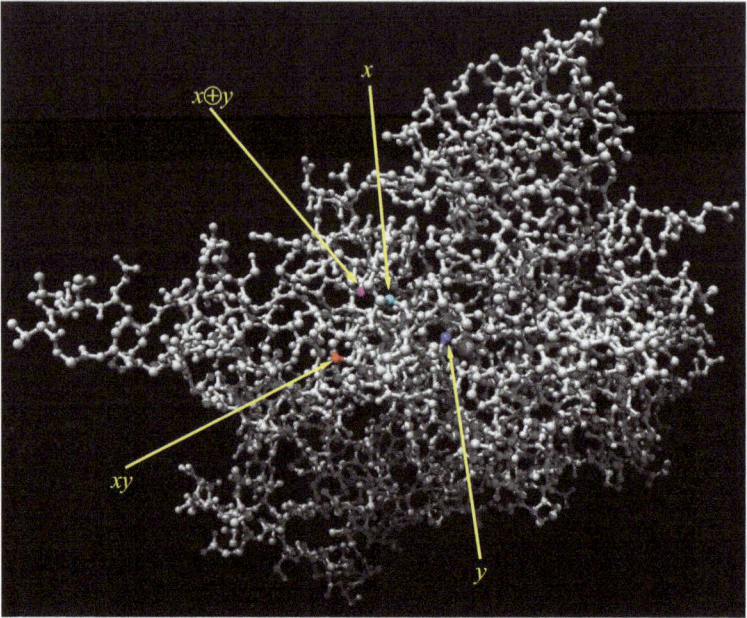

(a)

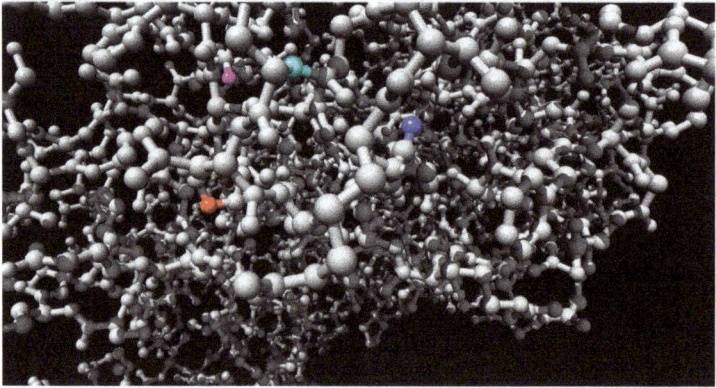

(b)

Fig. 11.10. Architecture of a one-bit half adder implemented in $\mathcal{A}$, governed by Rule (11.2), $\mu = 0.2$. Input nodes x and y are coloured cyan and blue. Output node xy is red and output node $x \oplus y$ is magenta.

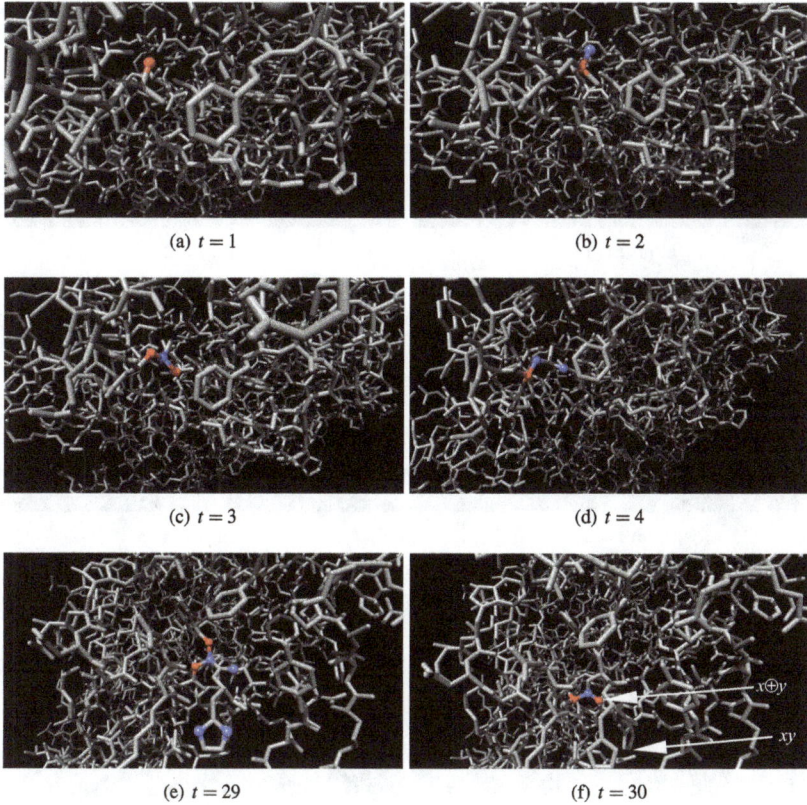

Fig. 11.11. One-bit half adder in action. Inputs are $x = 0$ and $y = 1$. Output nodes $x \oplus y$ and xy are indicated by arrows in (f). Output node xy is in resting state, output node $x \oplus y$ is excited. Excited nodes are red, refractory are blue.

nodes lie between input nodes. Nodes outputting $z = c$ are in two clusters, divided y nodes.

11.7 Discussion

In our extensive investigation, we thoroughly examined a large number of input pairs within the F-actin molecule, totalling nearly $9 \cdot 10^6$ pairs. From this comprehensive sampling, we selected a subset of $5 \cdot 10^4$ pairs for closer analysis. Our observations revealed the existence of diverse types of logical gates that can be realised through the interaction of excitation patterns within the F-actin structure.

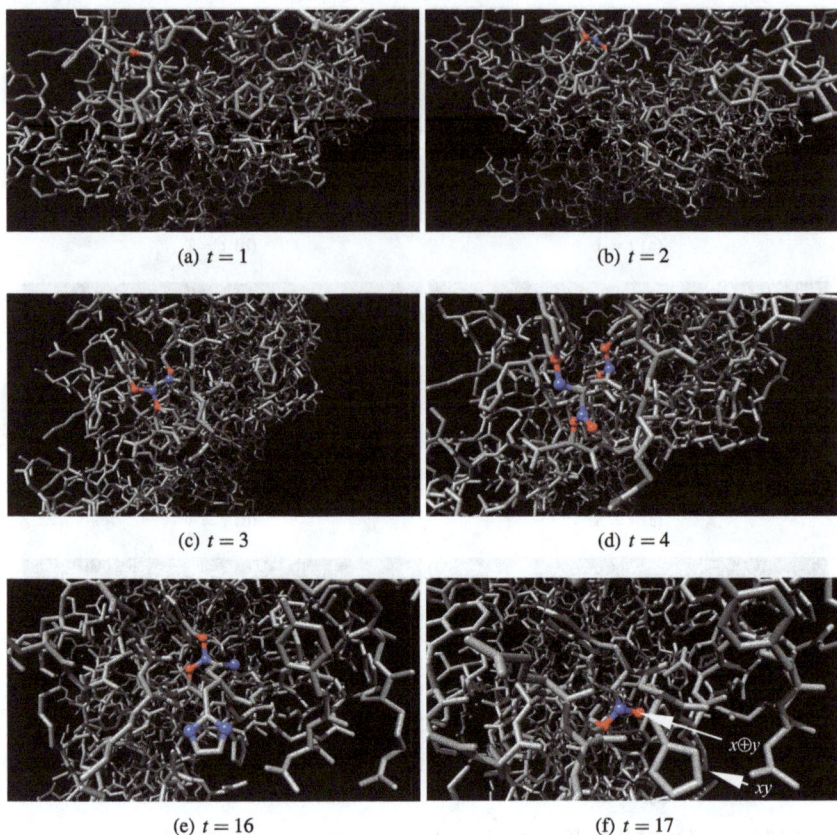

Fig. 11.12. One-bit half adder in action. Inputs are $x = 1$ and $y = 0$. Output nodes $x \oplus y$ and xy are indicated by arrows in (f). Output node xy is in resting state, output node $x \oplus y$ is excited. Excited nodes are red, refractory are blue.

We classified the excitation rules into two categories: threshold excitation and interval excitation. Among these, the threshold excitation automata were found to only implement input selectors, OR, and AND-NOT gates. Although these gates have some practical value, their potential for future computing applications is limited. As a result, we focused the remainder of our research on exploring interval excitation automata.

Within the interval excitation class, we made significant findings. The most prevalent gates observed were input selector gates, where

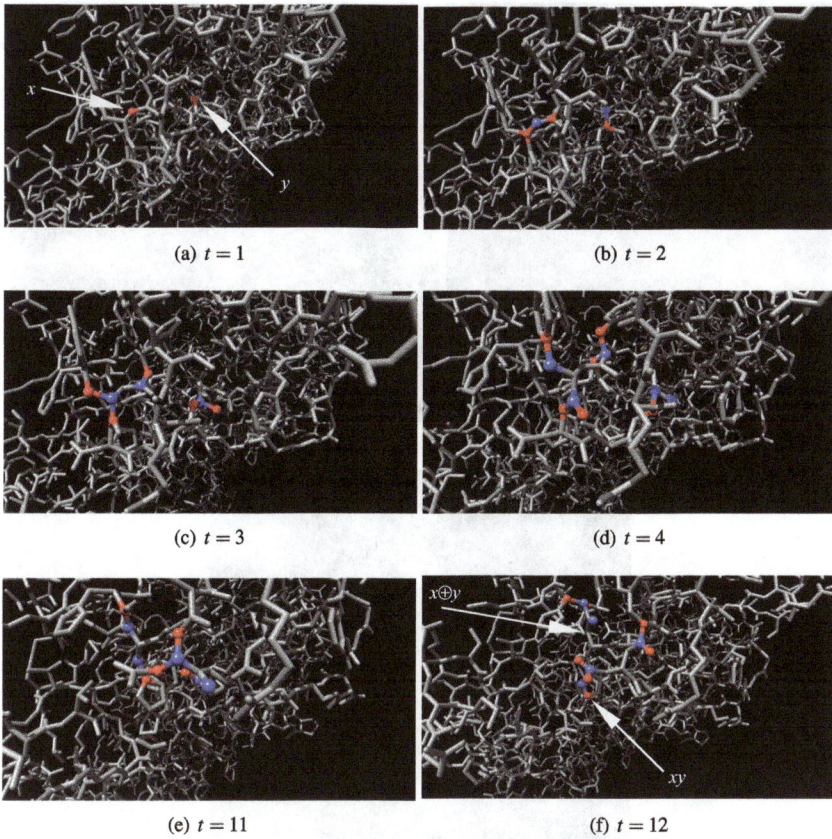

(a) $t = 1$

(b) $t = 2$

(c) $t = 3$

(d) $t = 4$

(e) $t = 11$

(f) $t = 12$

Fig. 11.13. One-bit half adder in action. Inputs are $x = 1$ and $y = 1$. Input nodes are indicated by arrows in (a), they both are excited at $t = 1$. Output nodes $x \oplus y$ and xy are indicated by arrows in (f). Output node xy is excited, output node $x \oplus y$ is in resting state. Excited nodes are red, refractory are blue.

the output corresponds to one of the input values. These gates were encountered in nearly half of the sampled cases, indicating their high frequency of occurrence. The AND gate followed as the next commonly encountered gate, with a frequency of approximately 0.02. On the other hand, the XOR gate proved to be the rarest, appearing with a frequency of 0.001. The frequencies of AND-NOT gates fell within the range of 0.003–0.004.

These findings provide valuable insights into the distribution and prevalence of specific gates within the excitation patterns of F-actin.

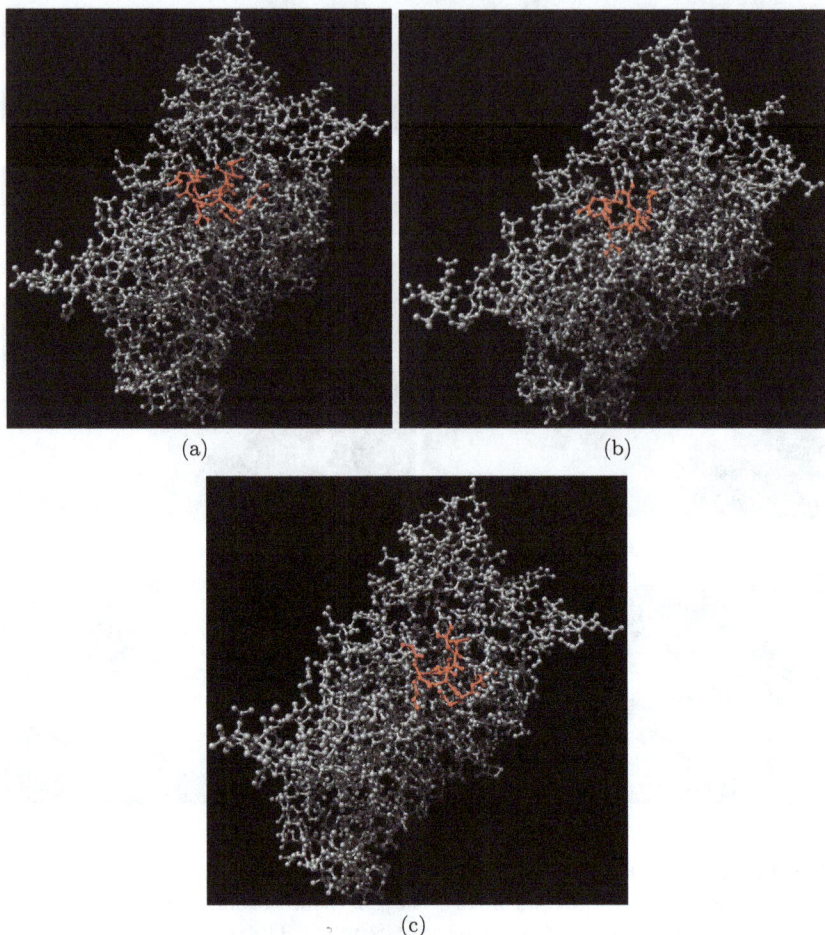

Fig. 11.14. Excitation paths in half adder for inputs (a) $x = 0$, $y = 1$, (b) $x = 1$, $y = 0$, (c) $x = 1$, $y = 1$. Nodes that have been excited at least once during the automaton evolution, for given states of input nodes, are coloured red.

This knowledge sheds light on the potential utility of these gates for computational purposes and opens up avenues for further exploration and application.

We conducted a thorough analysis of the excitation intervals, denoted as Θ, which depend on the number of hard and soft neighbours present in the F-actin molecule. Our findings revealed that the number of hard neighbours typically ranged between 2 and 3.

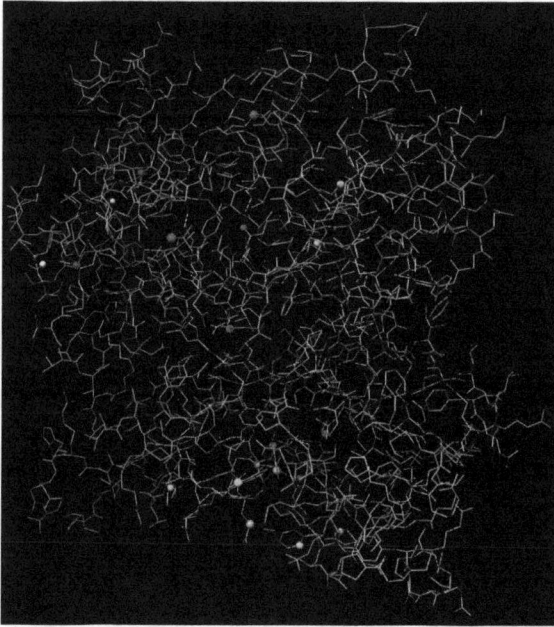

Fig. 11.15. Eleven pairs of input nodes of CNOT gates. Each pair of nodes has its unique colour.

In contrast, the number of soft neighbours exhibited variability and was dependent on the parameter ρ, as illustrated in Fig. 11.2. For our study, we specifically focused on automata with excitation intervals defined as $\Theta = [1]$ and $\Theta = [1, 2]$.

It is important to note that future research can explore other configurations of the excitation interval, such as $\Theta = [2, 3]$ or $\Theta = [1, 3, 4]$. We anticipate that the distribution of logical gates we have discovered will not be significantly impacted by the precise configuration of the excitation interval. This suggests that the specific shape of the excitation interval may not be crucial in determining the range of logical circuits observed.

Furthermore, it is important to consider the speed of operation in an experimental implementation of an F-actin computing unit in laboratory conditions. The speed is determined by the physical processes involved in the excitation and relaxation of electrons within the molecule. In this context, the excitation process occurs when an

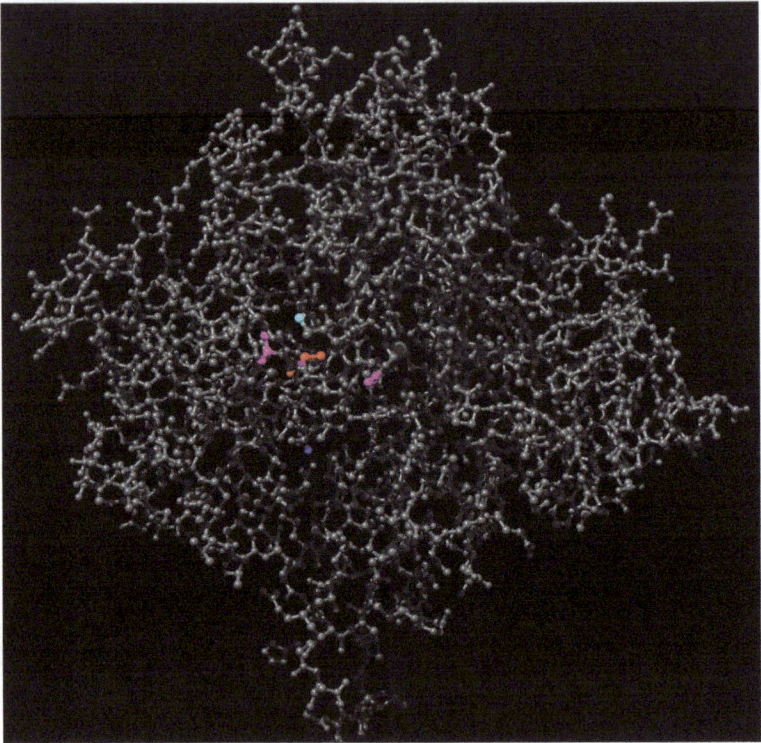

(a)

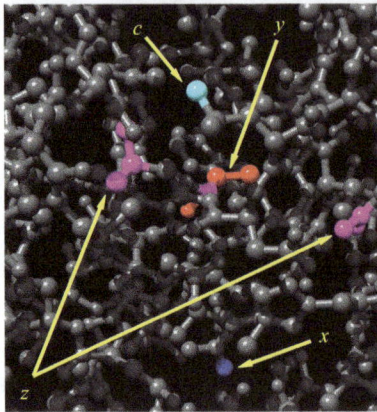

(b)

Fig. 11.16. An example of CNOT gate. (a) Location of the gate in the molecule. (b) Zoomed architecture. Cyan node is a control input c, blue node is a target input x, each of the pink nodes represents the same output z, and any of the red nodes is an output y.

electron in the ground state absorbs a photon, transitioning to a higher, yet unstable energy level. Subsequently, the electron returns to the ground state, releasing a photon in the process.

The speed at which the released photon travels is approximately $3 \cdot 10^{18}$/sec. Assuming a modeling scenario where the excitation wavefront travels a distance of $\rho = 3$ at each step of the automaton evolution, we can estimate that one step of the automaton takes approximately 10^{-18} sec, or one attosecond, of real time. For instance, the previously discussed half-adder circuit produces results in at most 30 steps of automaton evolution (Fig. 11.11), which corresponds to 30 attoseconds. Therefore, the upper boundary of the operating frequency of actin-based computing devices would be approximately 30 PHz.

To measure the outputs of the F-actin unit, controlled light waves and pulse trains can be utilised, as demonstrated in various experimental techniques such as single attosecond pulse generation [102], attosecond probing [184], and attosecond science in general [45,63]. These techniques allow for precise control and measurement of light pulses on the attosecond timescale. Alternatively, if less exotic measurement devices are used, a train of 10^3 impulses with the same data inputs can be sent to the F-actin unit, and the output can be recorded using accumulation devices such as capacitors. This would enable the extraction of the output signal through post-processing and analysis of the accumulated data.

If we are unable to excite a single node in an F-actin unit, we can explore the use of larger probes to stimulate the system. The question then arises whether the frequency distribution of realisable gates in F-actin units would be affected by the use of larger probes. To investigate this, we conducted $5 \cdot 10^4$ trials where each trial involved selecting a pair of nodes and stimulating all nodes within a sphere with a radius of 10 around the input node. It is worth noting that output nodes often form clusters that exceed 10, so the focus is primarily on the input nodes. The frequencies of the discovered logical gates are shown in Table 11.3. By comparing the frequencies of gates obtained through probe-based inputting (Table 11.3) with the frequencies of gates realisable during single node stimulation (Table 11.2(d)),

Table 11.3.   Frequencies of gates realisable in $\mathcal{A}$ governed by Rule (11.2), $\mu = 0.2$ and $\Theta = [1, 2]$. Inputs, for values '1', are excited by probes of Radius 10 Å. m is a ratio of input node pairs in the sample, i.e. a number of nodes detected in all samples divided by a number of samples, n is a number of output nodes for each pair of input nodes, and its standard deviation.

	x	y	$x+y$	xy	$x \oplus y$	$x\bar{y}$	$\bar{x}y$
m	0.622	0.640	0.050	0.033	0.004	0.074	0.070
n	52	50	15	27	19	30	39
std. dev n	44	41	19	26	35	30	35

we observe that the use of probes increases the number of gates discovered. Specifically, the frequency of the AND-NOT gate increases by approximately 18 times, the frequency of the OR gate increases by approximately 7 times, and the frequency of the XOR gate increases by approximately 5 times. This suggests that utilising larger probes for inputting in F-actin units can enhance the variety of logical gates that can be realised, potentially expanding the computational capabilities of the system.

To evaluate the density of gates in an F-actin molecule, we can start by considering the lowest frequency gate, which in this case is the XOR gate with a frequency of 0.001 (Table 11.2(d)). Assuming there are at most approximately $8 \cdot 10^6$ pairs of nodes in an F-actin molecule, this implies that there are at least $6 \cdot 10^3$ XOR gates in a single F-actin molecule. The maximum diameter of an actin filament is approximately 8 nm according to Refs. [182,241]. An actin filament is composed of overlapping units of F-actin, so the diameter of a single F-actin unit is approximately 4 nm. Based on the estimated minimum number of XOR gates per F-actin molecule (6K gates), we can calculate the gate density. Considering the projection of these gates onto a $2 \cdot 2$ nm square (approximately the size of a single F-actin unit), we can estimate that there are approximately $16 \cdot 16^{18}$ gates per square inch. This estimation provides an idea of the gate density in an F-actin molecule and highlights the potential for a high density of gates within a small spatial area.

Based on our analysis of $5 \cdot 10^4$ pairs of input nodes, we discovered one instance of a one-bit half adder. Extrapolating from this data and assuming a lower boundary of available pairs as $8 \cdot 10^6$, we can estimate that there are approximately 160 one-bit half adders in an F-actin molecule. This quantity would allow for the construction of around 80 one-bit full adders. Consequently, it is reasonable to infer that a 64-bit full adder could be assembled within a single F-actin unit. If we were to perform a 64-bit addition using this full adder, it would take approximately 1920 attoseconds, which is equivalent to roughly 2 femtoseconds.

These estimations highlight the remarkable potential of F-actin-based molecular computers to execute binary arithmetic computations at an exceptionally rapid pace. They suggest that such systems have the ability to complete a 64-bit addition in a matter of femtoseconds, demonstrating their capacity for high-speed processing in the realm of binary arithmetic.

Chapter 12

Discovering Functions in Actin Filament Automata

Andrew Adamatzky

Unconventional Computing Lab, UWE, Bristol, UK

Abstract

We employ an automaton network to simulate an actin filament, where each atom can exist in either two or three states. The state updates occur simultaneously for all atoms and depend on the ratios of their neighbouring atoms in specific states. We consider two distinct state transition rules. In the semi-totalistic Game of Life-like actin filament automaton, atoms possess binary states (0 and 1). Their state updates are determined by the ratios of neighbouring atoms in the state of 1. On the other hand, the excitable actin filament automaton incorporates three states for atoms: resting, excited, and refractory. A resting atom transitions to the excited state if the ratio of its excited neighbours falls within a specified interval. Transitions from the excited state to the refractory state and from the refractory state to the resting state occur unconditionally. To conduct our computational experiments, we employ perturbation and excitation dynamics on the actin filament automata to map an 8-bit input string to an 8-bit output string. We designate eight domains within the actin filament as input/output ports. To set a port to the value of 'True', we perturb or excite a certain percentage of nodes within the corresponding domain. After a specified time interval, we examine the outputs at the ports. A port is deemed to be in the state of 'True' if the number of excited nodes in its domain surpasses a certain threshold. Through a series of computational trials, we discover a range of eight-argument Boolean functions by mapping all possible configurations of eight-element binary strings to the excitation outputs of the input/output domains. These experiments allow us to identify a diverse set of functions that can be computed within the actin filament automaton, enabling the manipulation of information and the transformation of input configurations to output responses.

12.1 Introduction

We adopt a novel approach to compute on a protein polymer, specifically an actin filament, by utilising larger portions of the filament as input/output (I/O) regions and investigating the Boolean functions implemented through input-to-output mappings. This approach is less restrictive compared to previous implementations. What makes our approach unique is that we use a detailed model that considers multiple actin units arranged in a helical structure. This detailed model allows us to explore the computational capabilities of the actin filament in a more comprehensive manner, uncovering new insights and features that have not been previously considered.

12.2 Actin filament automata

We employed a pseudo-atomic model of an F-actin filament (Fig. 12.1) reconstructed by Galkin et al. [94] at 4.7 Å resolution using a direct electron detector, cryoelectron microscopy, and the forces imposed on actin filaments in thin films.[1] The model has 14800 atoms and is composed of six F-actin molecules. Following our previous convention [19], we represent an F-actin filament as a graph $\mathcal{F} = \langle \mathbf{V}, \mathbf{E}, \mathbf{C}, \mathbf{Q}, \mathbf{f} \rangle$, where $\mathbf{V}$ is a set of nodes, $\mathbf{E}$ is a set of edges, $\mathbf{C}$ is a set of Euclidean coordinates of nodes from $\mathbf{V}$, $\mathbf{Q}$ is

Fig. 12.1. A pseudo-atomic model of F-actin [93] in Corey–Pauling–Kolun colouring.

---

[1]PDB file can be downloaded here https://www.rcsb.org/structure/3J8I.

a set of node states, $f : \mathbf{Q} \times [0, 1] \rightarrow \mathbf{Q}$ is a node state transition function, calculating next state of a node depending on its current state and ratios of excited neighbours belonging to a sub-interval of $[0, 1]$. Each atom from a pseudo-atomic model of an F-actin filament is represented by a node from $\mathbf{V}$ with its 3D coordinates being a member of $\mathbf{C}$; atomic bonds are represented by $\mathbf{E}$. Each node $p \in \mathbf{V}$ takes states from a finite set $\mathbf{Q}$. All nodes update their states simultaneously in discrete time. A node p updates its state depending on its current state p^t and ratio $\gamma(p)^t$ of its neighbours being in some selected state $\star$. We consider two types of node neighbours. Let $u(p)$ be nodes from $\mathbf{V}$ that are connected with an edge with a node p, they correspond to atoms connected by the chemical bonds with atom p. We call them hard neighbours because their neighbourhood is determined by the chemical structure of F-actin. The ratio of nodes with one hard neighbour, is 0.298, two hard neighbours, 0.360, three hard neighbours, 0.341, and four hard neighbours, 0.001.

The actin molecule is folded in the 3D Euclidean space. Let δ be an average distance between two hard neighbours, for F-actin $\delta = 1.43\text{Å}$ units. Let $w(p)$ be the set of nodes of $\mathcal{F}$ that are at a distance not exceeding ρ, in the Euclidean space, from node p. We call them soft neighbours because their neighbourhood is determined by the 3D structure of F-actin. Thus, each node p has two neighbourhoods: hard neighbourhood $u(p) = \{s \in \mathbf{V} : (ps) \in \mathbf{E}\}$ (actin automata with hard neighbourhood were first proposed by us in Ref. [19]), and soft neighbourhood $w(p) = \{s \in \mathbf{V} : s \notin u(p) \text{ and } d(c_p, c_s) \leq \rho\}$, where $d(c_p, c_s)$ is a distance between nodes p and s in 3D Euclidean space and $c_s, c_p \in \mathbf{C}$. Interactions between a node and its hard neighbours takes place via atomic bounds and via the node and its soft neighbours via ionic currents. We have chosen $\rho = 10$ Å, which is seven times more than an average Euclidean distance 1.42 Å between two hard neighbours. Examples of neighbourhoods are shown in Fig. 12.2. The distribution of a number of soft neighbours versus a ratio of nodes with such number of soft neighbours is shown in Fig. 12.3; nearly half of the nodes (ratio 0.45) have from 133 to 185 neighbours. The ratio $\gamma(p)^t$ is calculated as $\gamma(p)^t = \frac{|s \in u(p) : s^t = \star| + \mu \cdot |s \in w(p) : s^t = \star|}{|u(p)| + |w(p)|}|$, where $|\mathbf{S}|$ is a number of elements

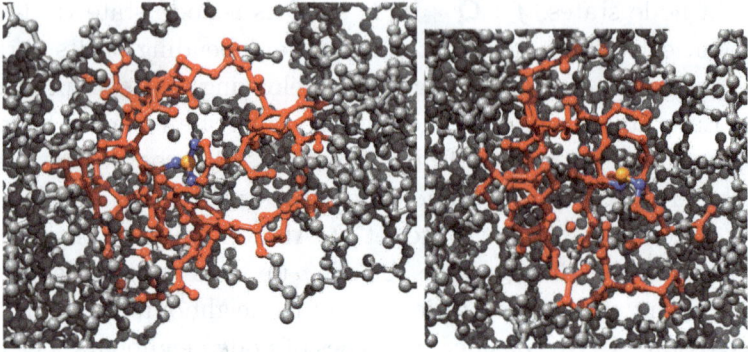

Fig. 12.2. Examples of neighbourhoods. Central nodes, 'owners' of the neighbourhoods are coloured orange, their hard neighbours are blue, and their soft neighbours are red.

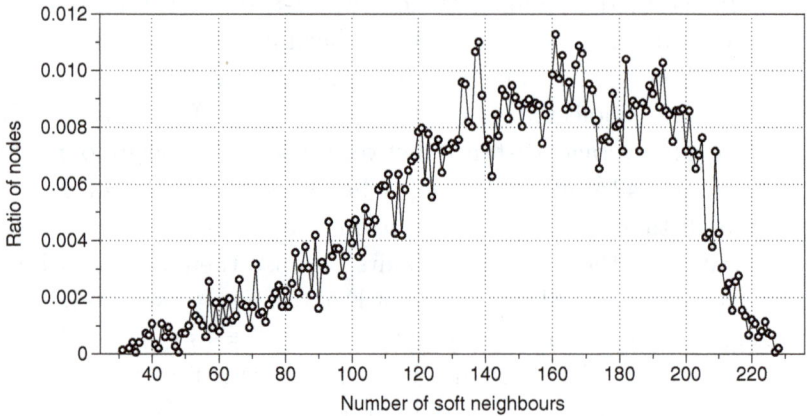

Fig. 12.3. Distribution of a ratio of nodes vs. numbers of their soft neighbours, $\rho = 10$.

in the set $\mathbf{S}$ and μ is a weight of soft neighbours; we used $\mu = 0.9$ in experiments reported.

We consider two species of family $\mathcal{F}$: semi-totalistic automaton $\mathcal{G} = \langle \mathbf{V}, \mathbf{E}, \mathbf{C}, \{\star, \circ\}, f^G \rangle$ and excitable automaton $\mathcal{E} = \langle \mathbf{V}, \mathbf{E}, \mathbf{C}, \{\star, \circ, \bullet\}, f^E \rangle$. The rules f^G and f^E are defined as follows:

$$p^{t+1} = f^G(p) = \begin{cases} \star, & \text{if } ((p^t = \circ) \wedge (\theta'_\circ \leq \gamma(p)^t \leq \theta''_\circ)) \\ & \quad \vee ((p^t = \star) \wedge (\theta'_\star \leq \gamma(p)^t \leq \theta''_\star)), \quad (12.1) \\ \circ, & \text{otherwise,} \end{cases}$$

$$p^{t+1} = f^E(p) = \begin{cases} \star, & \text{if } ((p^t = \circ) \wedge (\theta'_\circ \leq \gamma(p)^t \leq \theta''_\circ)), \\ \bullet, & \text{if } p^t = \circ, \\ \circ, & \text{otherwise.} \end{cases} \tag{12.2}$$

We have chosen intervals $[\theta'_\circ, \theta''_\circ] = [\theta'_\star, \theta''_\star] = [0.25, 0.375]$ for $\mathcal{G}$ and $[\theta'_\circ, \theta''_\circ] = [0.15, 0.25]$ for $\mathcal{E}$ because they support localised modes of excitation, i.e. a perturbation of the automata at a single site or a compact domain of several sites does not lead to an excitation spreading all over the actin chain. Localised excitations emerging at different input domains can interact with each other and the results of their interactions in the output domains will represent values of a logical function computed.

Automaton $\mathcal{G}$ is a Game of Life-like automaton [12, 67]. Speaking in the Game of Life lingo, we can say that a dead node $\circ$ becomes alive $\star$ if a ratio of live nodes in its neighbourhood lies inside the interval $[\theta'_\circ, \theta''_\circ]$; a live node $\star$ remains alive if a ratio of live nodes in its neighbourhood lies inside the interval $[\theta'_\star, \theta''_\star]$. Automaton $\mathcal{E}$ is a Greenberg–Hastings [106]-like automaton: a resting node $\circ$ excites if a ratio of excited nodes in its neighbourhood lies inside interval $[\theta'_\circ, \theta''_\circ]$; and an excited node $\star$ takes refractory state $\bullet$ in the next step of development, while a refractory $\bullet$ returns to resting state $\circ$. Rules of Conway's Game of Life could be interpreted as Eq. (12.1) having perturbation intervals $[\theta'_\circ, \theta''_\circ] = [0.375, 0.375]$ (i.e. exact value 0.375) and $[\theta'_\star, \theta''_\star] = [0.25, 0.375]$, rules of Greenberg–Hastings automata in terms of Eq. (12.2) having interval $[\theta'_\circ, \theta''_\circ] = [0.125, 1]$. The exact intervals of perturbation for the Game of Life and the Greenberg–Hastings automata are proven to be not useful for mining functions. This is because $\mathcal{G}$ with the Game of Life interval does not show any sustainable dynamics of excitation, and $\mathcal{E}$ with the Greenberg–Hastings interval exhibits 'classical' waves of excitation, where two colliding waves annihilate (Fig. 12.4).

12.3 Discovering functions

We encode Boolean values '0' (FALSE) and '1' (TRUE) in perturbations of selected domains **D** and extract a range of mappings $\{0,1\}^m \to \{0,1\}^m$, $m \in \mathbf{N}$, implementable by the actin filament

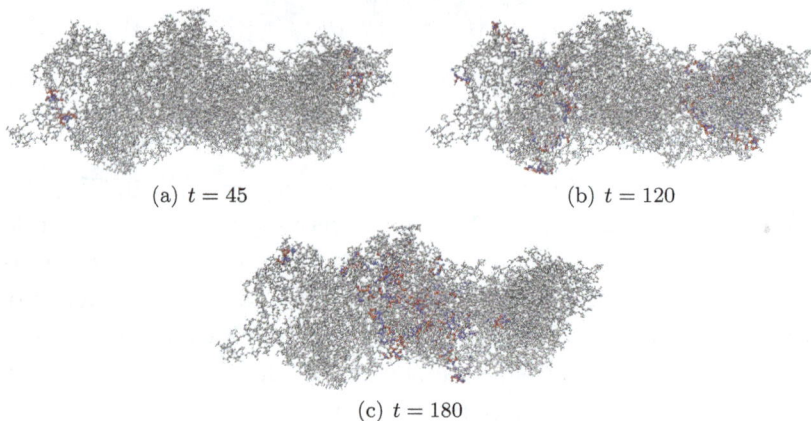

(a) $t = 45$ (b) $t = 120$

(c) $t = 180$

Fig. 12.4. Annihilation of excitation wave-fronts in $\mathcal{E}$ for $[\theta'_\circ, \theta''_\circ] = [0.125, 1]$.

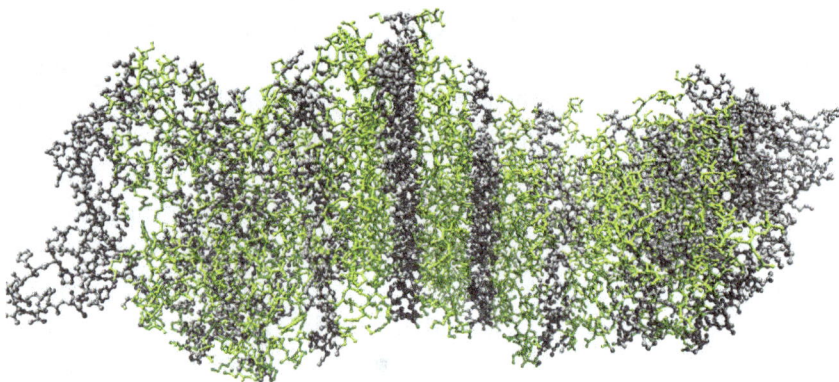

Fig. 12.5. Nodes of I/O domains $D_0, \ldots, D_7$ are shown by green colour.

automaton. Assume input and output tuples $\mathbf{I} \in \{0,1\}^m$ and $\mathbf{O} \in \{0,1\}^m$, $m = 8$, the actin automaton implements $\mathbf{I} \to \mathbf{D} \to \mathbf{O}$. We implement computation on actin filament automaton as follows. Eight cylinders across the (xy)-plane with coordinates $D_i = \{p \in \mathcal{V} : \text{abs}(p_x - k(i)) < r_s\}$, $0 \leq i < 8$, $k(i) = 15 \cdot (i + 1)$, are assigned as input–output domains (Fig. 12.5). These are mapped onto Boolean inputs $\mathbf{I} = (I_0, \ldots, I_7)$ and outputs $\mathbf{O} = (O_0, \ldots, O_7)$ as follows: $I_z = 1$ if $\sum_{p^0 \in D_z} > \kappa$, otherwise $I_z = 0$, and $O_z = 1$ if $\sum_{p^\zeta \in D_z} > \kappa$, otherwise $O_z = 0$, we have chosen $\kappa = 0$ and $\zeta = 40$.

Domains from **D** at time step $t = 0$ are excited with probability p determined by values of inputs **I**: if a node p belongs to $\mathbf{D_i}$ and $s_i = 1$, the node takes state $\star$ at the beginning of evolution, $p^0 = \star$ with probability p. We read outputs after $\zeta = 40$ steps of automaton evolution. As soon as 40 iterations occurred ($t = 41$), we measure states of nodes in the domains $\mathbf{D_i}$, $s_i \in \{0, 1\}$, and assign outputs depending on the excitation: $O_i = 1$ if $|\{p \in \mathbf{D} : \mathbf{p^t} = \star\}| > \kappa$, $\kappa = 0$. Stimulation runs for h trials (repeated simulation of automaton) with all possible configurations of **I**, $h = 100$, where frequencies of outputs are calculated as $W_i = w_i + I_i^T$, $0 \leq i < 8$ where T is a trial number, $T = 1, \ldots, h$. By the end of the experiments, we normalise **W** as $w_i = w_i/h$, h is the number of trials.

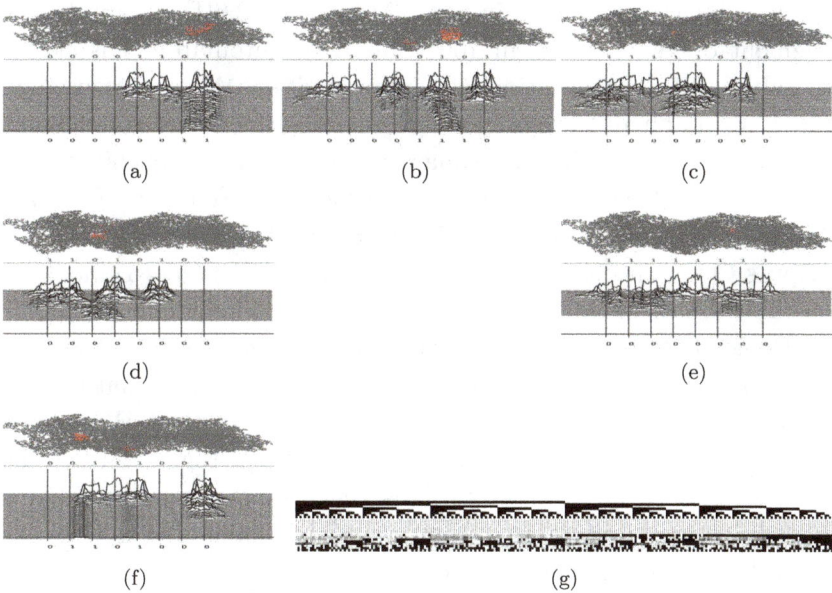

Fig. 12.6. Discovering Boolean functions in automaton $\mathcal{G}$. (a–f) Examples of excitation dynamics in automaton $\mathcal{G}$, $\theta'_\circ = \theta'_\star = 0.25$ and $\theta''_\circ = \theta''_\star = 0.375$. Projection of actin filament on z plane is shown in grey; projection of nodes being in state $\star$ by the moment of recording inputs are shown in red. Plots show values of activity, i.e. a number of nodes in state $\star$ along the x coordinate. See videos of experiments at https://doi.org/10.5281/zenodo.1312141. (g) Visualisation of register mapping implemented by automaton $\mathcal{G}$.

Table 12.1. Fragment of experimentally obtained mapping **S** to **W** for automaton $\mathcal{G}$.

$(I_0I_1I_2I_3I_4I_5I_6I_7)$	w_0	w_1	w_2	w_3	w_4	w_5	w_6	w_7
1011100	0	0	0.01	0.01	0.14	0.03	0.01	0
1011101	0.01	0.03	0.01	0.01	0.2	0.03	0.01	0.03
1011110	0	0	0	0	0.14	0.02	0.01	0.01
1011111	0	0.01	0.01	0.02	0.25	0.04	0.02	0.02
1100000	0	0.04	0.04	0	0	0	0	0
1100001	0	0.02	0.02	0	0	0	0	0.01
1100010	0	0.03	0.05	0	0	0.01	0.02	0
1100011	0	0.05	0.03	0	0	0	0	0.02
1100100	0	0.06	0.04	0	0	0.02	0	0
1100101	0.01	0.06	0.04	0	0	0.04	0.03	0.02

Examples of perturbation dynamics of automaton $\mathcal{G}$ for various input sequences are shown in Fig. 12.6(a) and 12.6(f). Example of a fragment of **W** obtained in 100 trials with automaton $\mathcal{G}$ is shown in Table 12.1. Visualisation of mapping **S** $\to$ **W** is presented in Fig. 12.6(g). There lexicograpically ordered elements of **S** are shown by black ('1') and white ('0') squares: top row from (0000000) on the right to (11111111) on the left. Corresponding elements of **W** are shown by gradations of grey $255 \cdot w_i$. From **W** we extract values of outputs **O** for various ranges of $\gamma \in [0, 1]$ as follows: $O_i = 1$ if $w_i > \gamma$, and $O_i = 0$ otherwise.

Boolean functions, in the form $O_i = f(I_0, \ldots, I_7)$, realisable by automata $\mathcal{G}$ and $\mathcal{E}$ are listed in Table 12.2. In automaton $\mathcal{G}$, a ratio ϵ of I/O transitions where at least one element of **W** exceeds γ shows quadratic decrease with increase of γ (Fig. 12.8(a)); the same applies to automaton $\mathcal{E}$. This reflects both a decrease in a number of functions realisable on output domains and a decrease of the functions complexity in terms of the arguments. A number of functions implementable in $\mathcal{F}$ polynomially decrease with increase of $\theta'_\circ$ (Fig. 12.8(b)).

12.4 Discussion

We demonstrated an implementation of logical functions on automaton models of actin filaments. The approach was inspired by the

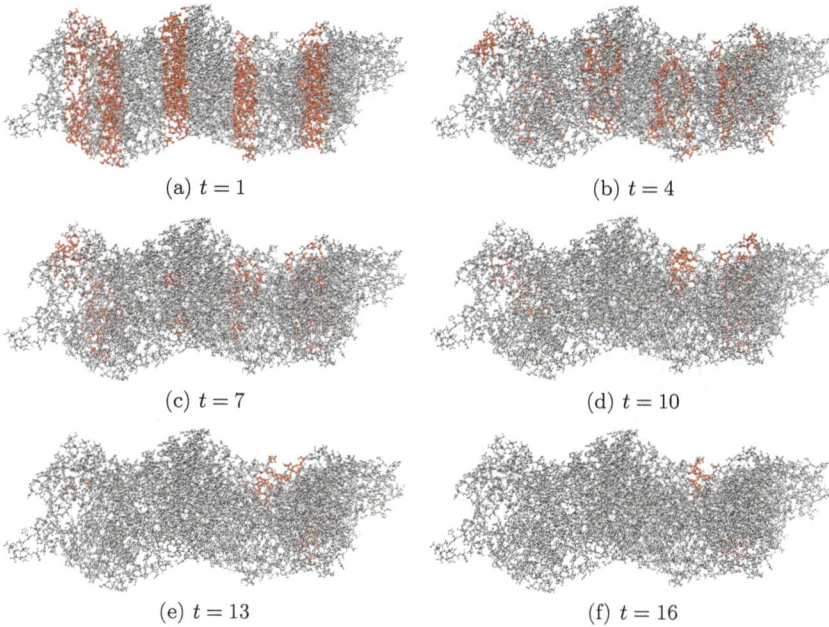

(a) $t = 1$

(b) $t = 4$

(c) $t = 7$

(d) $t = 10$

(e) $t = 13$

(f) $t = 16$

Fig. 12.7. Snapshots of excitation dynamics of automaton $\mathcal{G}$ in response to the input 11010101. See videos of experiments at https://doi.org/10.5281/zenodo.13 12141.

'evolution in materio' framework [115, 117, 181] on implementing computation without knowing the exact physical structure of a computing substrate. Propagating patterns in the automaton Game of Like-like automaton $\mathcal{G}$ can be seen as the discrete analogies of vibration excitation [72, 198, 199, 237]. The dynamics of Greenberg–Hastings excitable automaton $\mathcal{E}$ is a finite state machine analogue of the ionic waves, theoretical models of which are well studied in a context of tubulin microtubules and actin filaments [202, 215, 218, 227, 254]. How feasible is the approach? So far there are no experimental data on vibration modes of a single strand, or even a bundle of actin filaments or tubulin tubes, of a cytoskeleton polymer [150]. Outputs of the actin filament processors can be measured using controlled light waves and pulse trains [45, 63, 102, 184]. There are ways to measure the vibration of a cell membrane, as demonstrated in Ref. [137]. The vibration of the membrane might reflect vibrations

Table 12.2. Functions implemented by (a) $\mathcal{G}$ automaton, $\theta'_\circ = \theta'_\star = 0.25$ and $\theta''_\circ = \theta''_\star = 0.375$, and (b) $\mathcal{E}$ automaton, $\theta'_\circ = 0.15$ and $\theta''_\circ = 0.25$. for various values of reliability threshold γ.

γ	Functions

(a)

0.15 $O_1 = I_0 \cdot \overline{I_1} \cdot I_2 \cdot \overline{I_3} \cdot I_4 \cdot I_5 \cdot I_6 \cdot I_7;$
$\quad O_2 = \overline{I_0} \cdot \overline{I_1} \cdot I_2 \cdot I_3 \cdot I_7 \cdot (I_4 \cdot I_5 \cdot \overline{I_6} + I_4 \cdot \overline{I_5} \cdot I_6 + \overline{I_4} \cdot I_5 \cdot I_6);$
$\quad O_4 = \overline{I_0} \cdot \overline{I_1} \cdot \overline{I_2} \cdot I_3 \cdot I_4 \cdot (\overline{I_5} \cdot \overline{I_7} + I_6 \cdot I_7 + I_5 \cdot I_6 \cdot \overline{I_7})$

0.2 $O_4 = I_3 \cdot I_4 \cdot (\overline{I_0} \cdot I_1 \cdot \overline{I_2} \cdot I_5 \cdot I_7 + \overline{I_0} \cdot \overline{I_2} \cdot \overline{I_5} \cdot I_6 \cdot \overline{I_7} + I_0 \cdot \overline{I_1} \cdot \overline{I_2} \cdot \overline{I_5} \cdot I_6 + \overline{I_0} \cdot I_1 \cdot I_2 \cdot \overline{I_5} \cdot I_6 \cdot \overline{I_7} + I_0 \cdot I_1 \cdot I_2 \cdot \overline{I_5} \cdot I_6 \cdot I_7 + I_0 \cdot \overline{I_1} \cdot \overline{I_2} \cdot \overline{I_5} \cdot \overline{I_6} \cdot I_7 + \overline{I_0} \cdot \overline{I_1} \cdot \overline{I_2} \cdot \overline{I_5} \cdot I_6 \cdot \overline{I_7} + \overline{I_0} \cdot \overline{I_1} \cdot \overline{I_2} \cdot \overline{I_5} \cdot I_6 \cdot I_7)$

0.22 $O_4 = \overline{I_2} \cdot I_3 \cdot I_4 \cdot (\overline{I_0} \cdot I_1 \cdot I_5 \cdot I_6 \cdot I_7 + \overline{I_0} \cdot \overline{I_1} \cdot I_5 \cdot I_6 \cdot \overline{I_7} + I_0 \cdot \overline{I_1} \cdot I_5 \cdot I_6 \cdot I_7 + \overline{I_0} \cdot \overline{I_1} \cdot I_5 \cdot I_6 \cdot \overline{I_7} + \overline{I_0} \cdot \overline{I_1} \cdot I_5 \cdot I_6 \cdot I_7)$

0.23 $O_4 = \overline{I_0} \cdot \overline{I_2} \cdot I_3 \cdot I_4 \cdot (\overline{I_1} \cdot I_5 \cdot I_6 \cdot \overline{I_7} + \overline{I_1} \cdot \overline{I_5} \cdot I_6 \cdot I_7 + I_1 \cdot I_5 \cdot I_6 \cdot I_7 + \overline{I_1} \cdot \overline{I_5} \cdot \overline{I_6} \cdot \overline{I_7})$

0.24 $O_4 = \overline{I_0} \cdot \overline{I_2} \cdot I_3 \cdot I_4 \cdot I_6 \cdot I_7 \cdot (\overline{I_1} \cdot \overline{I_5} + I_1 \cdot I_5)$

0.25 $O_4 = \overline{I_0} \cdot I_1 \cdot \overline{I_2} \cdot I_3 \cdot I_4 \cdot I_5 \cdot I_6 \cdot I_7$

(b)

0.7 $O_0 = \overline{I_0} \cdot I_1 \cdot \overline{I_2} \cdot I_7 \cdot (\overline{I_3} \cdot \overline{I_4} \cdot I_5 \cdot \overline{I_6} + I_3 \cdot I_4 \cdot \overline{I_5} \cdot I_6)$
$\quad O_1 = I_0 \cdot \overline{I_1} \cdot I_2 \cdot \overline{I_3} \cdot I_4 + \overline{I_0} \cdot I_1 \cdot \overline{I_2} \cdot I_3 \cdot I_4 \cdot \overline{I_5} \cdot I_7 + \overline{I_0} \cdot I_1 \cdot \overline{I_2} \cdot I_3 \cdot \overline{I_4} \cdot I_5 \cdot \overline{I_6} + \overline{I_0} \cdot I_1 \cdot \overline{I_2} \cdot I_3 \cdot \overline{I_4} \cdot I_6 \cdot \overline{I_7} + \overline{I_0} \cdot I_1 \cdot I_2 \cdot \overline{I_3} \cdot I_4 \cdot \overline{I_5} \cdot \overline{I_6} \cdot I_7 + \overline{I_0} \cdot I_1 \cdot \overline{I_2} \cdot I_3 \cdot I_4 \cdot I_5 \cdot I_6 \cdot I_7 + I_0 \cdot \overline{I_1} \cdot I_2 \cdot I_3 \cdot \overline{I_4} \cdot I_5 \cdot \overline{I_6} \cdot I_7 + \overline{I_0} \cdot I_1 \cdot \overline{I_2} \cdot I_3 \cdot \overline{I_4} \cdot I_5 \cdot I_6 \cdot I_7$
$\quad O_2 = I_0 \cdot \overline{I_1} \cdot I_2 \cdot \overline{I_3} \cdot I_4 + \overline{I_0} \cdot I_1 \cdot I_2 \cdot \overline{I_3} \cdot I_4 \cdot \overline{I_5} + I_0 \cdot \overline{I_1} \cdot I_2 \cdot I_3 \cdot \overline{I_4} \cdot I_5 \cdot \overline{I_7} + \overline{I_0} \cdot I_1 \cdot \overline{I_2} \cdot I_3 \cdot \overline{I_4} \cdot I_6 \cdot \overline{I_7} + \overline{I_0} \cdot I_1 \cdot \overline{I_2} \cdot I_3 \cdot \overline{I_4} \cdot I_6 \cdot I_7 + \overline{I_0} \cdot I_1 \cdot I_3 \cdot \overline{I_4} \cdot I_5 \cdot \overline{I_6} \cdot I_7 + \overline{I_0} \cdot I_1 \cdot \overline{I_2} \cdot I_3 \cdot I_4 \cdot \overline{I_5} \cdot \overline{I_6} \cdot I_7 + I_0 \cdot I_1 \cdot I_2 \cdot I_3 \cdot \overline{I_4} \cdot I_5 \cdot \overline{I_6} \cdot I_7 + \overline{I_0} \cdot I_1 \cdot \overline{I_2} \cdot I_3 \cdot \overline{I_4} \cdot I_5 \cdot I_6 \cdot I_7 + I_0 \cdot \overline{I_1} \cdot I_2 \cdot I_3 \cdot \overline{I_4} \cdot I_5 \cdot I_6 \cdot I_7$
$\quad O_3 = I_2 \cdot \overline{I_3} \cdot I_4 \cdot \overline{I_5} \cdot \overline{I_6} \cdot I_7 + I_0 \cdot I_3 \cdot \overline{I_4} \cdot I_5 \cdot \overline{I_6} \cdot I_7 + \overline{I_0} \cdot I_1 \cdot I_2 \cdot I_3 \cdot \overline{I_4} \cdot I_5 \cdot \overline{I_7} + \overline{I_0} \cdot I_1 \cdot I_3 \cdot \overline{I_4} \cdot I_5 \cdot \overline{I_6} \cdot I_7 + \overline{I_1} \cdot I_2 \cdot I_3 \cdot \overline{I_4} \cdot I_5 \cdot I_6 \cdot \overline{I_7} + I_0 \cdot \overline{I_1} \cdot I_2 \cdot I_3 \cdot \overline{I_4} \cdot I_5 \cdot \overline{I_7} + \overline{I_0} \cdot I_1 \cdot I_2 \cdot I_3 \cdot \overline{I_4} \cdot I_5 \cdot I_6 + I_0 \cdot \overline{I_1} \cdot I_2 \cdot I_3 \cdot \overline{I_4} \cdot I_5 \cdot I_6 \cdot I_7 + \overline{I_0} \cdot I_1 \cdot I_2 \cdot I_3 \cdot \overline{I_4} \cdot I_5 \cdot \overline{I_6} \cdot I_7 + I_0 \cdot I_1 \cdot I_2 \cdot I_3 \cdot \overline{I_4} \cdot I_5 \cdot I_6 \cdot \overline{I_7} + I_0 \cdot \overline{I_1} \cdot I_2 \cdot \overline{I_3} \cdot I_4 \cdot \overline{I_5} \cdot I_6 \cdot \overline{I_7} + \overline{I_0} \cdot I_1 \cdot I_2 \cdot \overline{I_3} \cdot I_4 \cdot \overline{I_5} \cdot I_6 \cdot I_7 + \overline{I_0} \cdot I_1 \cdot I_2 \cdot \overline{I_3} \cdot I_4 \cdot \overline{I_5} \cdot \overline{I_6} \cdot I_7$
Functions realises on outputs O_4 to O_7 are not shown.

0.8 $O_2 = I_3 \cdot \overline{I_4} \cdot I_5 \cdot I_7 \cdot (\overline{I_0} \cdot I_1 \cdot \overline{I_2} + I_0 \cdot \overline{I_1} \cdot I_2 \cdot \overline{I_6})$
$\quad O_3 = \overline{I_6} \cdot (I_0 \cdot \overline{I_1} \cdot I_3 \cdot \overline{I_4} \cdot I_5 \cdot I_7 + \overline{I_0} \cdot I_1 \cdot I_3 \cdot \overline{I_4} \cdot I_5 \cdot I_7 + I_0 \cdot \overline{I_1} \cdot I_2 \cdot I_3 \cdot \overline{I_4} \cdot I_5 \cdot \overline{I_7} + I_0 \cdot \overline{I_1} \cdot I_2 \cdot \overline{I_3} \cdot I_4 \cdot \overline{I_5} \cdot I_7 + I_0 \cdot \overline{I_1} \cdot \overline{I_2} \cdot I_3 \cdot \overline{I_4} \cdot I_5 \cdot I_7)$
$\quad O_4 = \overline{I_6} \cdot (I_2 \cdot I_3 \cdot \overline{I_4} \cdot I_5 \cdot I_7 + \overline{I_1} \cdot I_2 \cdot \overline{I_3} \cdot I_4 \cdot \overline{I_5} \cdot I_7 + I_0 \cdot I_1 \cdot \overline{I_2} \cdot I_3 \cdot \overline{I_4} \cdot I_5 + I_0 \cdot \overline{I_1} \cdot \overline{I_2} \cdot I_3 \cdot \overline{I_4} \cdot I_5 \cdot I_7)$
$\quad O_5 = \overline{I_2} \cdot I_3 \cdot \overline{I_4} \cdot I_5 \cdot \overline{I_6} \cdot I_7 \cdot (I_0 + \overline{I_1})$

0.9 $O_5 = \overline{I_1} \cdot I_3 \cdot \overline{I_4} \cdot I_5 \cdot \overline{I_6} \cdot I_7 \cdot (I_0 + \overline{I_2})$

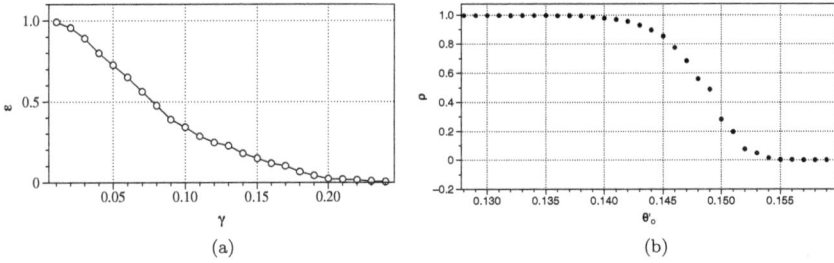

Fig. 12.8. (a) Ratio ϵ of transitions where at least one entry in **W** exceeds γ. (b) Dependence of the ratio ρ of outputs in state 1 to an overall number of outputs of the lower threshold of excitation θ'_o; upper threshold $\theta''_o = 0.25$ was kept constant.

of cytoskeleton networks attached to the membrane [64]; however, it shows a cumulative effect of vibration of a cytoskeleton network. Due to the polarity of actin units, vibration modes are manifested in electromagnetic perturbation, which can be measured when existing experimental techniques will be perfected [198, 199].

Chapter 13

Computing on Actin Bundles Network

Andrew Adamatzky[*,§], Florian Huber[†], and Jörg Schnauß[‡,¶]

*Unconventional Computing Lab, UWE Bristol, UK
†Netherlands eScience Center, Science Park 140,
1098 XG Amsterdam, The Netherlands
‡Soft Matter Physics Division,
Peter Debye Institute for Soft Matter Physics,
Faculty of Physics and Earth Science,
Leipzig University, Germany and Fraunhofer Institute
for Cell Therapy and Immunology (IZI),
DNA Nanodevices Group, Leipzig, Germany
§andrew.adamatzky@uwe.ac.uk
¶joerg.schnauss@uni-leipzig.de

Abstract

The implementation of computing circuits using actin filaments is promising due to their unique characteristics such as ionic currents, mechanical solitons, and voltage solitons. However, controlling the propagation of these localised phenomena on a single actin filament poses experimental challenges. To address this limitation, we focus on studying the propagation of excitation waves on bundles of actin filaments. In our computational experiments, we investigate a two-dimensional slice of an actin bundle network. Rather than controlling individual filaments, our approach explores the collective behaviour of the bundle to implement circuits. By arranging electrodes in various configurations, we demonstrate the feasibility of implementing circuits with two inputs and one output within the actin bundle network. This approach allows us to leverage the computational potential of actin filaments, while circumventing the difficulties associated with controlling excitations on single filaments.

13.1 Introduction

The concept of using collisions of signals travelling along one-dimensional nonlinear structures for computation has a rich history. It can be traced back to the mid-1960s when Atrubin developed a chain of finite state machines capable of executing multiplication [44]. Fisher designed cellular automata for generating prime numbers [88], and Waksman proposed an eight-state solution for the firing squad synchronisation problem [267]. In 1986, Park, Steiglitz, and Thurston designed a parity filter in cellular automata with soliton-like dynamics of localisations [192]. Their design led to the development of a one-dimensional particle machine that performs computation by colliding particles in cellular automata, effectively embedding computation within a bulk media [243].

Inspired by the theoretical concepts of collision-based computing [3, 10], our focus is on translating these ideas into nano-computing at a sub-cellular level. We consider actin networks as ideal candidates for serving as a computing substrate. The notion of sub-cellular computing on cytoskeleton networks was initially proposed by Hameroff and Rasmussen in the 1980s, specifically in the context of microtubule automata [112, 114, 209]. Priel, Tuszynski, and Cantiello further analysed how information processing could be realised in actin-tubulin networks found in neuron dendrites [206]. For the purpose of this chapter, our focus is solely on actin networks.

Current theoretical models in the field of actin computing primarily focus on processing information at the level of individual actin units or small chains of units. While these models are theoretically promising for computing purposes, their practical implementation in experimental laboratory conditions presents significant challenges. In this study, we propose an alternative approach to computing on actin networks.

Instead of considering a single actin filament, our approach investigates the propagation of excitation waves on bundles of actin filaments. By examining the collective behaviour and overall density of the conductive material formed by the arrangement of actin bundles, we take into account factors such as crowding effects. Importantly, our approach does not require the involvement of additional accessory proteins, simplifying the experimental setup [221, 222].

In the following sections, we present our preliminary results based on this novel approach, highlighting the potential of utilising actin bundles for computational purposes.

13.2 Model

The FitzHugh–Nagumo (FHN) equations [89, 185, 194] are a qualitative approximation of the Hodgkin–Huxley model [48] of electrical activity of living cells:

$$\frac{\partial v}{\partial t} = c_1 u(u - a)(1 - u) - c_2 uv + I + D_u \nabla^2, \qquad (13.1)$$

$$\frac{\partial v}{\partial t} = b(u - v), \qquad (13.2)$$

where u is a value of a trans-membrane potential, v, a variable accountable for a total slow ionic current, or a recovery variable responsible for a slow negative feedback, I is a value of an external stimulation current. The current through intra-cellular spaces is approximated by $D_u \nabla^2$, where D_u is a conductance. Detailed explanations of the 'mechanics' of the model are provided in Ref. [212], here we shortly repeat some insights. The term $D_u \nabla^2 u$ governs a passive spread of the current. The terms $c_2 u(u-a)(1-u)$ and $b(u-v)$ describe the ionic currents. The term $u(u - a)(1 - u)$ has two stable fixed points $u = 0$ and $u = 1$ and one unstable point $u = a$, where a is a threshold of an excitation.

We integrated the system using the Euler method with the five-node Laplace operator, a time step $\Delta t = 0.015$, and a grid point spacing $\Delta x = 2$, while other parameters were $D_u = 1$, $a = 0.13$, $b = 0.013$, $c_1 = 0.26$. We controlled excitability of the medium by varying c_2 from 0.09 (fully excitable) to 0.013 (non-excitable). Boundaries are considered to be impermeable: $\partial u / \partial \mathbf{n} = 0$, where $\mathbf{n}$ is a vector normal to the boundary.

In our study, we utilised still images of an actin network obtained from laboratory experiments focused on the formation of regularly spaced bundle networks from homogeneous filament solutions [123]. We specifically chose this network as it is the result of an experimental protocol that reliably produces regularly spaced aster-based networks without the involvement of molecular motor-driven processes or other accessory proteins [123].

These structures exhibit remarkable stability and longevity, forming well-defined three-dimensional networks. To visualise and analyse these networks, we employed a confocal laser scanning microscope, which allowed us to capture two-dimensional representations of the three-dimensional structures. These actin networks serve as a potential hardware platform for future cytoskeleton-based computers [28].

The actin network (Fig. 13.1(a)) was projected onto a 1024×1024 nodes grid. The original image $M = (m_{ij})_{1 \leq i,j \leq n}$, $m_{ij} \in \{r_{ij}, g_{ij}, b_{ij}\}$, where $n = 1024$ and $1 \leq r, g, b \leq 255$ (Fig. 13.1(a)), was converted to a conductive matrix $C = (m_{ij})_{1 \leq i,j \leq n}$ (Fig. 13.1(b)) derived from the image as follows: $m_{ij} = 1$ if $r_{ij} > 40$, $(g_{ij} > 19)$ and $b_{ij} > 19$.

The parameter c_2 determines excitability of the medium and thus determines a range of the network coverage by excitation wavesfronts. This is illustrated in Fig. 13.2.

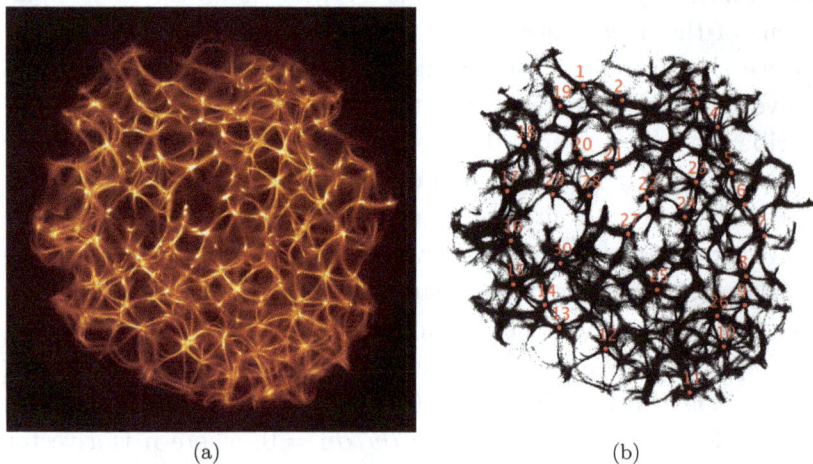

(a)   (b)

Fig. 13.1. (a) Original image, which was published in [123]. (b) The 'conductive' matrix selected from (a) Locations of the electrodes $E_1, \ldots, E_{30}$ are shown by their indexes.

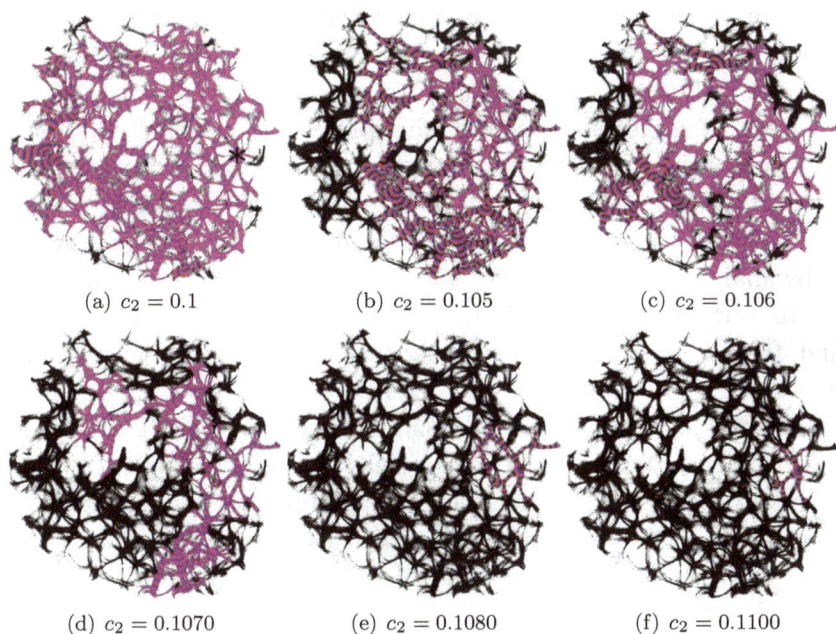

Fig. 13.2. Time lapse images of excitation wave-fronts propagating on the network displayed in Fig. 13.1(a) for selected values of c_2. In each trial, excitation was initiated at the site labelled '7' in Fig. 13.1(b) and labelled as star in (a). Excitation wave-fronts are shown in red, conductive sites, in black.

To show dynamics of both u and v, we calculated a potential p_x^t at an electrode location x as $p_x = \sum_{y:|x-y|<2}(u_x - v_x)$. Locations of the electrodes $E_1, \ldots, E_{30}$ are shown in Fig. 13.1(b).

Time-lapse snapshots were captured at regular intervals of every 150th time step. For visualisation purposes, we displayed sites with a value of u greater than 0.04. Additionally, videos and figures were created by saving simulation frames at every 100th step of the numerical integration and compiling these saved frames into a video with a playback rate of 30 frames per second (fps). You can access the videos at the following link: https://zenodo.org/record/2561273.

13.3 Results

To encode input Boolean values in our actin network, we designate two specific sites, denoted as x and y, as dedicated inputs. A logical value of 'True' or '1' is represented by an excitation at the corresponding input site. If $x = 1$, the site associated with x is excited, and if $x = 0$, the site remains unexcited.

Regarding the output values, we explore different representations. One approach is to observe the presence or absence of an excitation wave-front at a designated output site. Alternatively, output values can be represented by patterns of spike activity within the network or by analysing the frequencies of activity in specific output domains.

In our study, we present four prototype logical gates. The first type is structural gates, where the exact timing of collisions between excitation wave-fronts determines the output. Frequency-based gates encode Boolean values into the frequencies of excitations. Integral activity gates represent Boolean values based on the overall activity of the entire network. Lastly, spiking gates utilise spikes or combinations of spikes to represent logical values, and the search for these gates is facilitated by multiple output electrodes distributed throughout the network.

13.3.1 Structural gates

An example of an interaction gate is shown in Fig. 13.3. The gate is a junction of seven actin bundles, we call them 'channels' (Fig. 13.3(a)). We earmark two channels as inputs x and y, and five other channels as outputs $z_1, \ldots, z_5$. Such allocation is done for illustrative purposes. In principle, the mapping $\{0, 1\}^7 \rightarrow \{0, 1\}^7$ can be considered. To represent $x = 1$, we excite channel x, to represent $y = 1$, we excite channel y. When only channel y is stimulated, the excitation wave-fronts propagate into channels z_2 and z_3 (Fig. 13.3(b)). When only channel x is stimulated, the excitation is recorded in channels z_1, z_2, z_3 (Fig. 13.3(c)). When both channels are excited, $x = 1$ and $y = 1$, the excitation propagates into all channels (Fig. 13.3(d)). Thus, the following functions are implemented on the output channels $z_1 = x$, $z_2 = z_3 = x + y$, $z_4 = z_5 = xy$. The channel z_1 is a selector function. The channels z_2 and z_3 realise disjunction. The channels z_4 and z_5 implement conjunction. An advantage of the interaction gate is that it is cascadable, i.e. many gates can be linked together

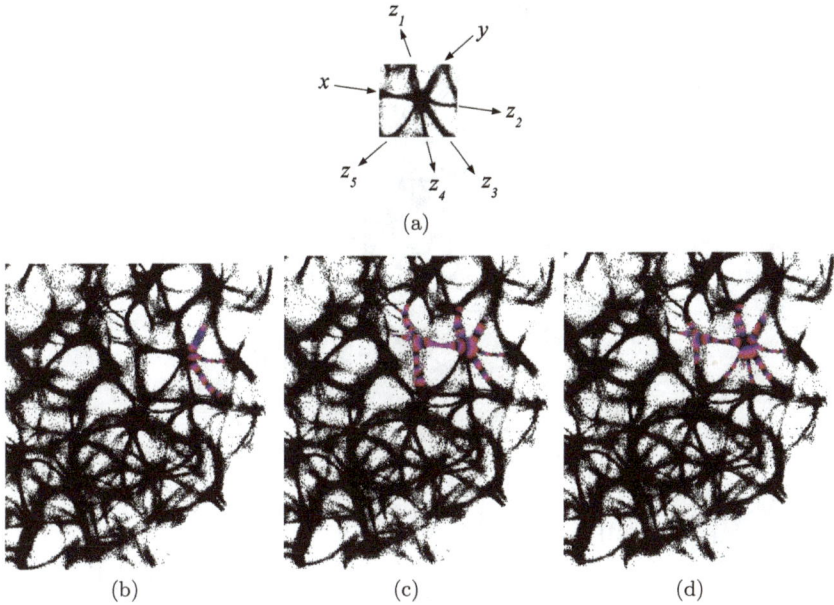

Fig. 13.3. Interaction gate. (a) A scheme of input and output channels. (b, c, d) Time lapse images of wave-fronts propagating in the gate for inputs (b) $x = 0$ and $y = 1$, (c) $x = 1$ and $y = 0$, (d) $x = 1$ and $y = 1$. Excitability of the medium is $c_2 = 0.108$.

without decoders or couplers. A disadvantage is that functioning of the gate is determined by the exact geometrical structure of the actin bundle network, which might be difficult to control precisely.

13.3.2 Frequency-based gates

For each pair of inputs (xy) $h \in \{01, 10, 11\}$, we calculated a frequency matrix $\Omega_h = (\omega_{hs})$, $s \in \mathbf{L}$, where each entity with coordinates s shows how often a node s of $\mathbf{L}$ was excited. At every iteration t of the simulation, the frequency at every node s is updated as follows: $\omega_s^t = \omega_s^t + 1$ if $u_s^t > 0.1$. When the simulation ends, the frequencies in all nodes are normalised as $\omega_s = \omega_s / \max\{\omega_z | z \in \mathbf{L}\}$. For each Ω_h, we selected domains of higher frequency as having entities $\omega_s > 0.72$. These domains are shown in Fig. 13.4. This unique mapping allows to implement any two-inputs-one-output logical gate

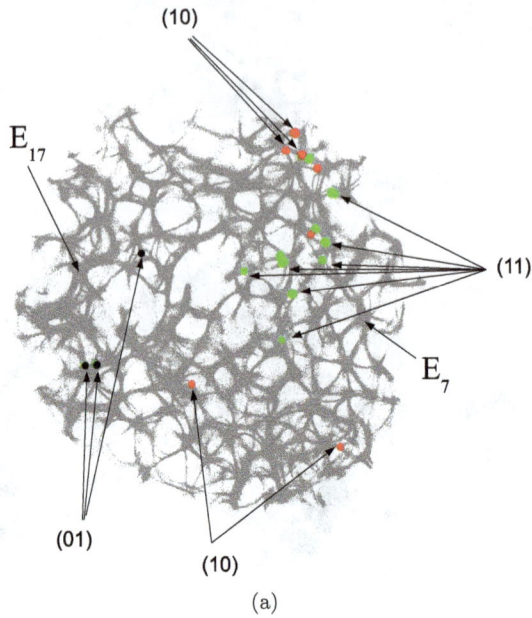

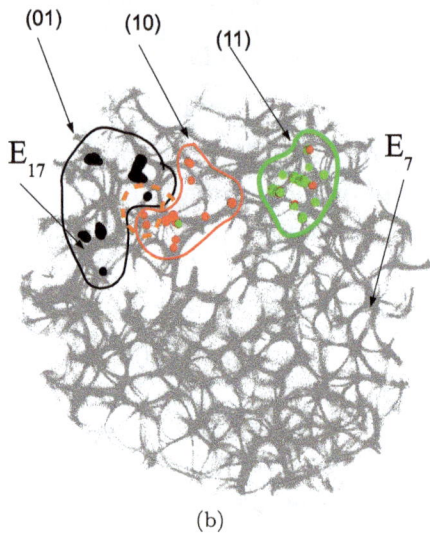

Fig. 13.4. Domains with the highest frequency of excitation represent spatially separated outputs. They are shown by black discs for inputs $x = 0$ and $y = 1$, red discs for inputs $x = 1$ and $y = 0$, and green discs for inputs $x = 1$ and $y = 1$. Inputs x and y are sites E_7 and E_{17} shown by arrows. (a) $c_2 = 0.1$. (b) $c_2 = 0.107$.

by placing electrodes in the required unique domains. For example, by placing electrodes in the domains that represent outputs for both input pair (01) and input pair (10) (black and red discs in Fig. 13.4), we can realise a XOR gate.

While in excitable mode, $c_2 = 0.1$, domains corresponding to different input tuples are somewhat dispersed in the network (Fig. 13.4(a)), the sub-excitable medium, $c_2 = 0.107$, shows compact and well spatially separated domains (Fig. 13.4(b)). Moreover, for $c_2 = 0.107$, we even observe a localised domain (shown by orange-dashed contour in Fig. 13.4(b)), where input tuples (01) and (10) are displayed and thus a XOR gate is realised.

13.3.3 Overall level of activity

At every iteration t, we measured the activity of the network as a number of conductive nodes x with $u_x^t > 0.1$. A stimulation of the resting network evokes travelling wave-fronts, which collide with each other and may annihilate or form new wave-fronts in the result of the collisions. The wave-fronts also travel along cycling pathways in the network. Typically, e.g. after $8 \cdot 10^4$ iterations for $c_2 = 0.1$ and 10^5 iterations for $c_2 = 0.107$, the system falls in one of the limit cycles of the overall level of activity with a range of superimposed oscillations (Fig. 13.5). We found no evidence that shapes of the superimposed spikes in activity reflect the exact combination of inputs; however, an average level of activity definitely does. A correspondence between input tuples and average level of activity A as percentage of the total number of conductive nodes is the following:

		(xy)	
c_2	(01)	(10)	(11)
0.1	0.068	0.05	0.08
0.107	0.006	0.02	0.02

By selecting an interval of A' as a logical TRUE we can implement a range of gates. Consider the scenario $c_2 = 0.1$: xy for $A' = [0.075, 0.085]$, $x\overline{y}$ for $A' = [0.045, 0.055]$, $\overline{x}y$ for $A' = [0.063, 0.073]$, $x \oplus y$ $A' = [0.045, 0.073]$. In the scenario $c_2 = 0.107$, a range of gates,

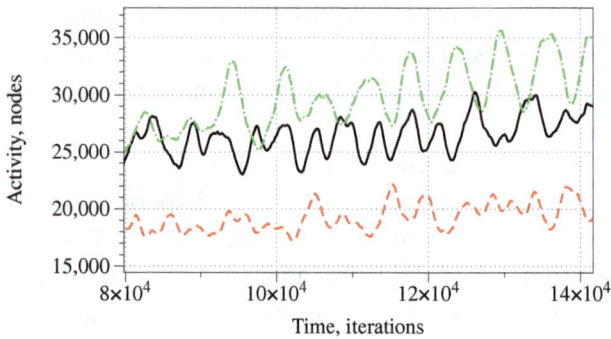

(a)

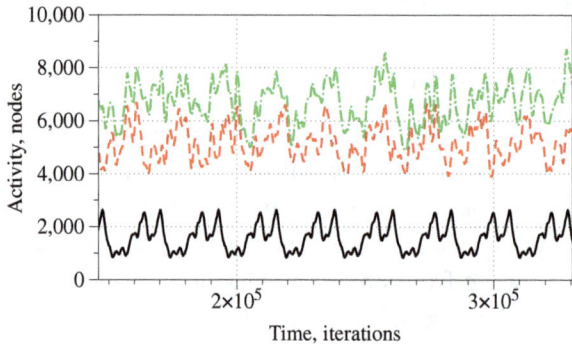

(b)

Fig. 13.5. Activity for input pairs (xy) (stimulation sites are E_7 and E_{17} in (Fig. 13.1(b))): (01) — black solid, (10) — red dashed, (11) — dashed dotted. (a) $c_2 = 0.1$. (b) $c_2 = 0.107$.

implementable by assigning an activity interval to TRUE, is limited to $\overline{x}y$ for $A' = [0.005, 0.007]$ and x for $A' = [0.015, 0.025]$.

13.3.4 Spiking gates

A spiking activity of the network shown in Fig. 13.1(a), with $c_2 = 0.1$ in a response to stimulation via electrodes E_7 and E_{17} recorded from electrodes $E_1, \ldots, E_{30}$ is shown in Fig. 13.6. We here assume that each spike represents logical TRUE and that spikes occurring within less than $2 \cdot 10^2$ iterations happen simultaneously. Then a representation of gates by spikes and their combinations will be as shown in Table 13.1.

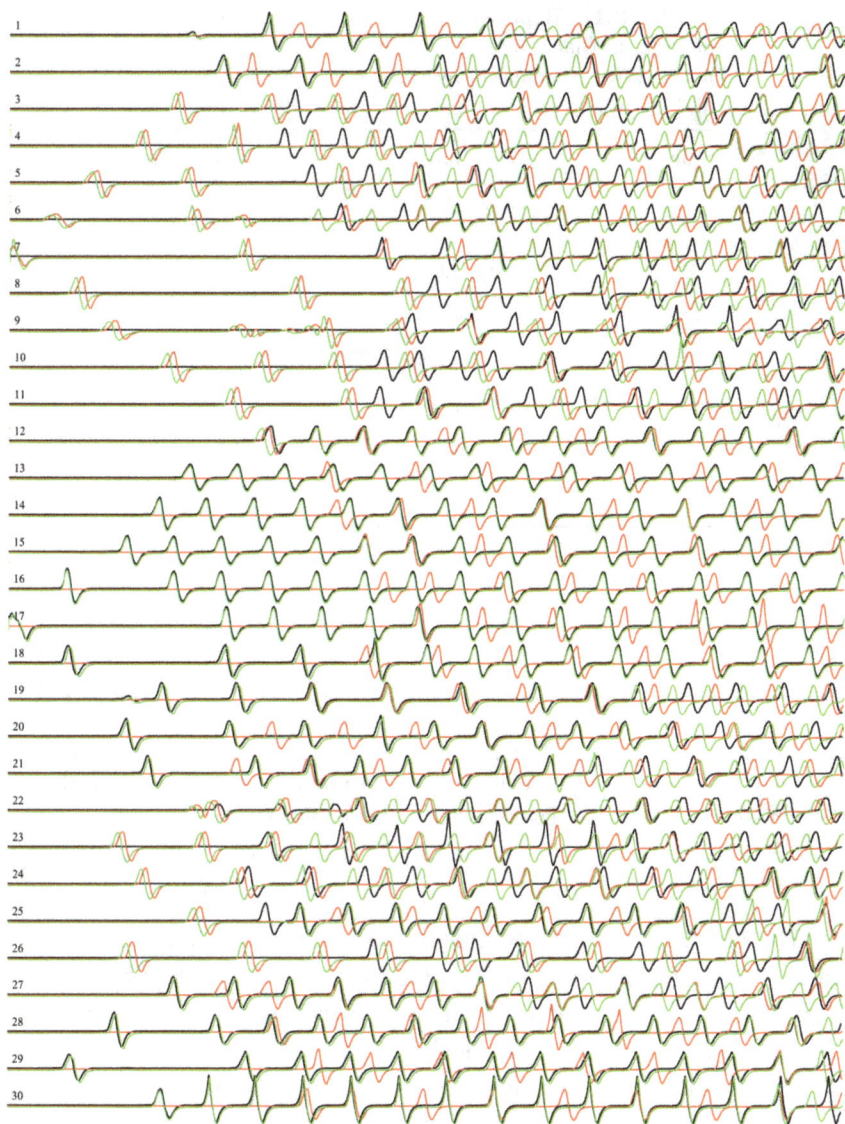

Fig. 13.6. Potential recorded at 30 electrodes (Fig. 13.1(b)) during c. $14.2 \cdot 10^4$ iterations. Indexes of electrodes are shown on the left. Black lines show potential when the network was stimulated by input pair (01), red by (10) and green by (11). Excitability of the medium is $c_2 = 0.1$.

Table 13.1. Representation of gates by combinations of spikes. Black lines show the potential when the network was stimulated by input pair (01), red by (10) and green by (11).

Spikes	Gate	Notations
	OR	$x + y$
	SELECT	y
	XOR	$x \oplus y$
	SELECT	x
	NOT-AND	$\overline{x}y$
	AND-NOT	$x\overline{y}$
	AND	xy

Fig. 13.7. Spikes recorded at the electrode E_7. The moment of initial stimulation is shown by '*'. Black lines show the potential when the network was stimulated by input pair (01), red by (10) and green by (11).

By selecting specific intervals of recordings, we can realise several gates in a single site of recording. In this particular case, we assumed that spikes are separated if their occurrences lie more than 10^3 iterations apart. An example is shown in Fig. 13.7.

To estimate the logical richness of the network, we calculated frequencies of logical gates' discoveries. For each of the recording sites we calculated a number of gates realised during $14.2 \cdot 10^4$ iterations (Table 13.2). In terms of 'frequency' of appearance of gates during the simulation, the gates can be arranged in the following hierarchy, from the most frequently found gate to the least frequent gate: SELECT $\triangleright$ {AND-NOT, NOT-AND} $\triangleright$ AND $\triangleright$ OR $\triangleright$ XOR.

Table 13.2. Number of gates discovered for each recording site. Excitability of the medium is $c_2 = 0.1$.

$i(E_i)$	$x+y$	y	$x \oplus y$	x	$\bar{x}y$	$x\bar{y}$	xy	Total
1	0	7	1	3	2	5	3	21
2	0	8	1	3	2	5	3	22
3	2	2	0	6	5	3	4	22
4	2	3	0	6	5	3	3	22
5	2	2	0	6	4	3	4	21
6	1	4	1	5	4	2	5	22
7	1	4	1	4	2	2	2	16
8	0	3	1	6	3	1	2	16
9	1	2	0	7	3	1	0	14
10	1	2	0	8	4	0	0	15
11	1	3	0	4	3	3	3	17
12	2	7	0	3	0	3	0	15
13	3	10	0	1	0	3	0	17
14	2	11	0	1	0	4	0	18
15	0	11	0	4	0	3	0	18
16	0	11	0	3	1	4	0	19
17	2	9	0	2	0	2	0	15
18	3	8	0	2	0	2	0	15
19	1	5	0	5	2	2	2	17
20	2	8	1	1	1	5	3	21
21	2	7	0	3	1	5	3	21
22	2	4	0	7	2	0	3	18
23	3	5	1	6	2	1	4	22
24	1	5	1	7	2	1	4	21
25	2	7	0	3	2	3	1	18
26	2	2	0	6	5	2	1	18
27	2	6	1	5	2	2	3	21
28	1	10	0	3	0	5	0	19
29	0	9	0	4	0	4	0	17
30	2	9	0	3	0	3	1	18
Average	1.43	6.13	0.30	4.23	1.90	2.73	1.80	18.53
Std. dev	0.94	3.06	0.47	1.96	1.67	1.46	1.65	2.57
Median	2.00	6.50	0.00	4.00	2.00	3.00	2.00	18.00

The model can realise two-inputs-two-outputs logical gates when we consider values of two recording electrodes at the same specified interval. For example, for a one-bit half adder: one output is AND and another output is XOR, and for a Toffoli gate: one output is SELECT and another XOR.

13.4 Discussion

In our numerical experiments, we have successfully demonstrated the implementation of logical gates within actin bundle networks. To achieve this, we explored different methods of mapping the network's excitation dynamics to the output space. Specifically, we provided detailed constructions of structural gates, frequency-based gates, and overall activity-based gates. Additionally, we conducted an extensive analysis of spiking gates and established a hierarchy of gates: SELECT▷AND-NOT, NOT-AND▷OR, AND▷XOR. This gate hierarchy, with some variations, aligns with gate hierarchies observed in living slime mould, plants, and the Belousov–Zhabotinsky chemical medium.

The dominant gate, SELECT, signifies the reachability of the recording site from the stimulation site. In this case, excitation from one electrode can reach the recording site, while excitation from another electrode cannot. The AND-NOT and NOT-AND gates represent situations where the wave-front from one input site obstructs the wave-front from another input site, such as through its refractory tail. The AND gate symbolises the requirement for wave-fronts originating from both input sites to intersect at some point during their propagation, often through regions with lower excitability, such as narrow channels suddenly expanding. If wave-fronts from both input sites can reach the recording site without cancelling each other out, an OR gate is implemented. The XOR gate represents the cancellation of wave-fronts originating from different input sites.

These modelling results are highly promising as they demonstrate that computation can be achieved within actin bundle networks by observing the excitation dynamics at selected domains. Our experiments were conducted using a two-dimensional projection of a slice from a three-dimensional actin bundle network. Future studies will consider complete three-dimensional networks. However, it is important to note that the distribution of logical gates may be more influenced by the connectivity of the graph rather than its dimensionality. Based on our previous experiments with other disorganised substrates, we anticipate that the geometry of gate distribution will remain consistent [21, 23, 116, 251].

In our position paper [28], we discuss various potential approaches for the input/output (I/O) interface in the experimental implementation of actin bundle networks, considering the challenges they pose for future studies. One approach we propose is the use of multi-electrode arrays (MEA) technology [92, 240], which has been successfully tested with disorganised ensembles of carbon nanotubules [147]. MEA allows for electrical impulses to generate inputs, following a conventional approach. Another option is the utilisation of nanofiber light-addressable potentiometric sensors [228]. These sensors provide an alternative means of generating inputs.

For recording the outputs of the actin bundle computer, MEA can be employed, or existing systems designed for imaging voltage in neurons can be adapted [195]. Additionally, single-molecule fluorescence methods, such as Forster resonance energy transfer [31, 69, 201], offer another possibility for recording outputs.

It is important to note that these are potential realisations of the I/O interface for actin bundle networks, and further research and experimentation will be necessary to explore their feasibility and optimise their implementation. The development and refinement of these interface approaches will contribute to the advancement of actin bundle computing and enable practical applications in the future.

In our experimental system, we intentionally excluded accessory proteins and a constant energy supply to study the emergent behaviour of actin networks based solely on the minimisation of free energy [123]. Despite the absence of these factors, the resulting network structures exhibited remarkable stability over extended periods, with self-repairing capabilities observed in some cases.

While our static actin networks demonstrated computational capabilities, the inherent dynamism of actin networks can be harnessed for additional advantages. For example, by introducing caged ATP and using UV light as an energy source, we can induce dynamic reconfigurations within the network. This dynamic behaviour allows for a broader range of logical functions compared to static networks, as the network's geometry determines the structure of the function. Furthermore, this dynamic nature enables the application of evolutionary computing and machine learning techniques, similar to those successfully demonstrated in highly dynamic systems such as thin layer Belousov–Zhabotinsky media [53–55]. We envision that similar approaches can be adapted to actin bundle networks in future studies.

In our recent work building upon our previous experiments [123], we have made advancements in biasing the architectural design of actin networks. By employing artificial constructs based on DNA, we can emulate the functions of natural accessory proteins, such as cross-linkers, and programmatically alter the properties of actin structures [156]. These DNA-based constructs offer a means to manipulate and control the behaviour and characteristics of actin networks, opening up new possibilities for tailored architectural designs and functional properties.

By combining the inherent dynamism of actin networks with programmable DNA-based constructs, we can further enhance the computational capabilities and versatility of actin bundle networks. These advancements pave the way for future studies that explore the interplay between network dynamics, architectural design, and functional properties in actin-based computing systems.

The geometrical design of DNA-based constructs offers great flexibility in controlling the properties of actin networks. By adjusting the number of binding domains and modifying the underlying DNA template, we can precisely influence binding angles and the formation of bundles at each junction. These DNA constructs serve as molecular precursors that guide the architecture of the actin system without directly interfering with bundle formation or information transport. This level of control over junction properties allows us to manipulate the computational potential of actin networks. Furthermore, it is possible to combine different types of structural proteins, enabling parallel processing of information.

In addition to using DNA-based constructs, electrical fields present another approach for programming the geometry of actin bundle networks. Through the application of electrical fields, it is possible to shape and control the arrangement of actin bundles. This method provides an additional means of influencing the network's structure and behaviour, further enhancing our ability to program and manipulate actin-based computing systems.

By integrating these approaches, involving DNA-based constructs and electrical fields, we can achieve even greater control over the architecture and functionality of actin bundle networks. These advancements enable us to explore new avenues in the design and implementation of actin-based computing systems, opening up possibilities for advanced computation and information processing.

Chapter 14

Actin Droplet Machine

Andrew Adamatzky*,§, Florian Huber†, and Jörg Schnauß‡,¶

*Unconventional Computing Laboratory, UWE Bristol, UK
†Netherlands eScience Center, Science Park 140, 1098 XG Amsterdam,
The Netherlands
‡Soft Matter Physics Division, Peter Debye Institute for Soft Matter
Physics, Faculty of Physics and Earth Science, Leipzig University,
Germany and Fraunhofer Institute for Cell Therapy and Immunology
(IZI), DNA Nanodevices Group, Leipzig, Germany
§andrew.adamatzky@uwe.ac.uk
¶joerg.schnauss@uni-leipzig.de

Abstract
The actin droplet machine is a computational model that emulates a
three-dimensional network of actin bundles within a droplet containing physiological solution. This model demonstrates the capability of
implementing mappings for sets of binary strings. The actin bundle
network allows for the propagation of travelling excitations or impulses,
which can be harnessed for computational purposes. The machine is
designed to interface with a specific set of k electrodes. These electrodes
serve as input/output channels for the machine. Stimuli, represented as
binary strings of length k, are applied to the machine through impulses
generated on the electrodes. The machine records the responses to
these stimuli in the form of impulses and subsequently converts them
back into binary strings. The state of the machine is represented by a
binary string of length k. Each position in the string corresponds to
an electrode, where a '1' indicates the presence of a recorded impulse
on that electrode, and '0' indicates its absence. The design of the actin
droplet machine involves the establishment of state transition graphs,
which depict the possible transitions between different states of the
machine in response to specific input stimuli. These graphs provide
insights into the behaviour and functionality of the machine. The actin
droplet machine holds potential as an elementary processor in future

massively parallel computers composed of biopolymers. It offers a unique approach to harnessing the computational capabilities of actin networks and demonstrates the feasibility of using biologically inspired systems for information processing tasks.

14.1 Introduction

Our previous work, as presented in Ref. [25], explored the concept of computing with excitation waves propagating on a conductive material, specifically networks of actin bundles arranged by crowding effects without the need for additional accessory proteins [221, 222]. In that study, we demonstrated how logical gates can be discovered on a two-dimensional slice of the actin bundle network by representing Boolean inputs and outputs as spikes in the network activity. By employing numerical integration of the Fitzhugh–Nagumo model, we showed that a two-dimensional actin network can realise k-ary Boolean functions $G : 0, 1^k \to 0, 1$ when utilising k input electrodes and one output electrode.

We present a novel concept and computer modelling implementation of the actin network machine that realises a mapping $F : 0, 1^k \to 0, 1^k$, where k represents the number of electrodes. In this mapping, the presence of an impulse on an electrode is denoted by 1', while the absence is denoted by 0'. The machine operates as a finite state machine at a higher level, while the structure of the mapping F is determined by the interactions of impulses propagating on the three-dimensional network of actin bundles at a lower level.

In addition to the numerical integration method used in Ref. [25], we propose an alternative approach utilising an automaton model of a three-dimensional actin network. Extensive evidence supports the suitability of automaton models as discrete tools for accurately representing the dynamics of spatially extended nonlinear excitable media [94, 161, 271], propagation phenomena [152], action potentials [43, 278], and electrical pulses in the heart [81, 219, 237]. One significant advantage of automaton models is their efficiency in terms of computational resources compared to typical numerical integration approaches.

The results presented in this chapter provide a perspective from a "computer engineering" standpoint regarding the implementation of computation using travelling localisation on actin bundle networks. Our focus is primarily on the technical aspects and functionality of this approach, and we refrain from speculating about potential biological implications of the phenomena described. Exploring the biological significance of these findings may be a subject for future studies and investigations.

14.2 Methods

The overall approach is the following: we simulate the actin bundle network using three-dimensional arrays of finite state machines, cellular automata. We select several domains of the network and assign them as inputs and outputs. We represent Boolean logic values with spikes of electrical activity, which are schematically represented as a virtual experiment in Fig. 14.1. We stimulate the network with

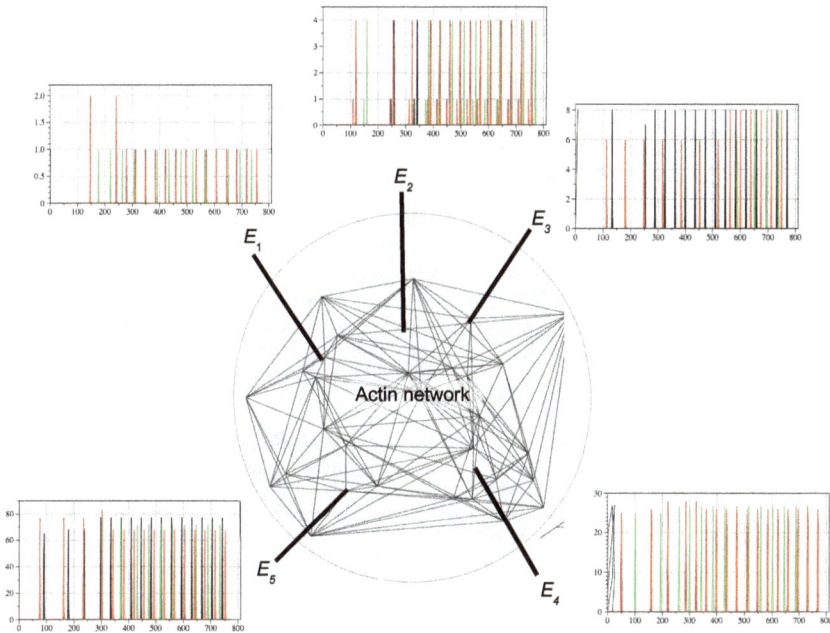

Fig. 14.1. A schematic representation of a virtual experiment. The actin bundle network is shown as a three-dimensional Delaunay triangulation. Electrodes are shown by thick lines and labelled E_1 to E_5. Exemplary trains of spikes are shown near the electrodes.

all possible configurations of input strings and record spikes on the outputs. Based on the mapping of configurations of input spikes to output spikes, we reconstruct logical functions implemented by the network. In our design of the actin droplet machine, we consider outputs recorded on all electrodes at a given time step as a binary string and then represent the actin droplet machine as a finite state machine whose states are binary strings of a given length.

14.2.1 Three-dimensional actin network

As a template for our actin droplet machine, we used an actual three-dimensional actin bundle network produced in laboratory experiments with purified proteins (Fig. 14.2). The underlying experimental method was shown to reliably produce regularly spaced bundle networks from homogeneous filament solutions inside small isolated droplets in the absence of molecular motor-driven processes or other accessory proteins [123]. These structures effectively form very stable and long-living three-dimensional networks, which can be readily imaged with confocal microscopy resulting in stacks of optical two-dimensional slices (Fig. 14.2). Dimensions of the network are the following: size along x coordinate is 225 μm (width), along y coordinate is 222 μm (height), along z coordinate is 112 μm (depth), voxel width is 0.22 μm, height, 0.22 μm, and depth, 4 μm.

Original image: $A_z = (a_{ijz})_{1 \leq i,j \leq n, 1 \leq z \leq m}$, $a_{ijz} \in \{r_{ijz}, g_{ijz}, b_{ijz}\}$, where $n = 1024$, $m = 30$, $r_{ijz}, g_{ijz}, b_{ijz}$ are RGB values of the element at ijz, $1 \leq r_{ijz}, g_{ijz}, b_{ijz} \leq 255$ was converted to a conductive matrix $\mathbf{C} = (c_{ijz})_{1 \leq i,j \leq n, 1 \leq z \leq m}$ as follows: $c_{ijz} = 1$ if $r_{ijz} > 40$, $g_{ijz} > 19$, and $b_{ijz} > 19$. The conductive matrices are shown in Fig. 14.3. The three-dimensional conductive matrix is compressed along z-axis to reduce consumption of computational resources, scenario of the non-compressed matrix will be considered in future papers.

14.2.2 Automaton model

To model activity of an actin bundle network we represent it as an automaton $\mathcal{A} = \langle \mathbf{C}, \mathbf{Q}, r, h, \theta, \delta \rangle$. $\mathbf{C} \subset \mathbf{Z}^3$ is a set of voxels, or a

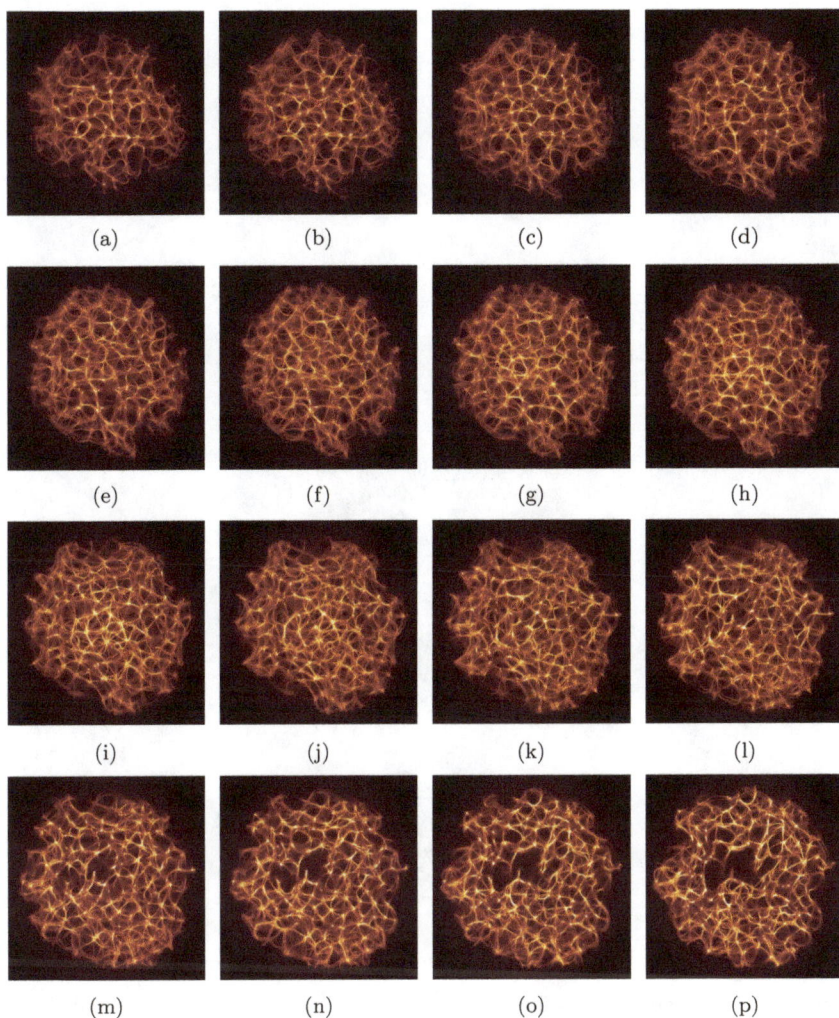

Fig. 14.2. Exemplary z-slices of a three-dimensional actin bundle network reconstructed as described in Ref. [123].

conductive matrix $\mathbf{C}$ defined in Section 14.2.1. Each voxel $p \in \mathbf{C}$ takes states from the set $\mathbf{Q} = \{\star, \bullet, \circ\}$, excited ($\star$), refractory ($\bullet$), resting ($\circ$) and is complemented by a counter h_p to handle the temporal decay of the refractory state. Following discrete time steps, each voxel p updates its state depending on its current state and

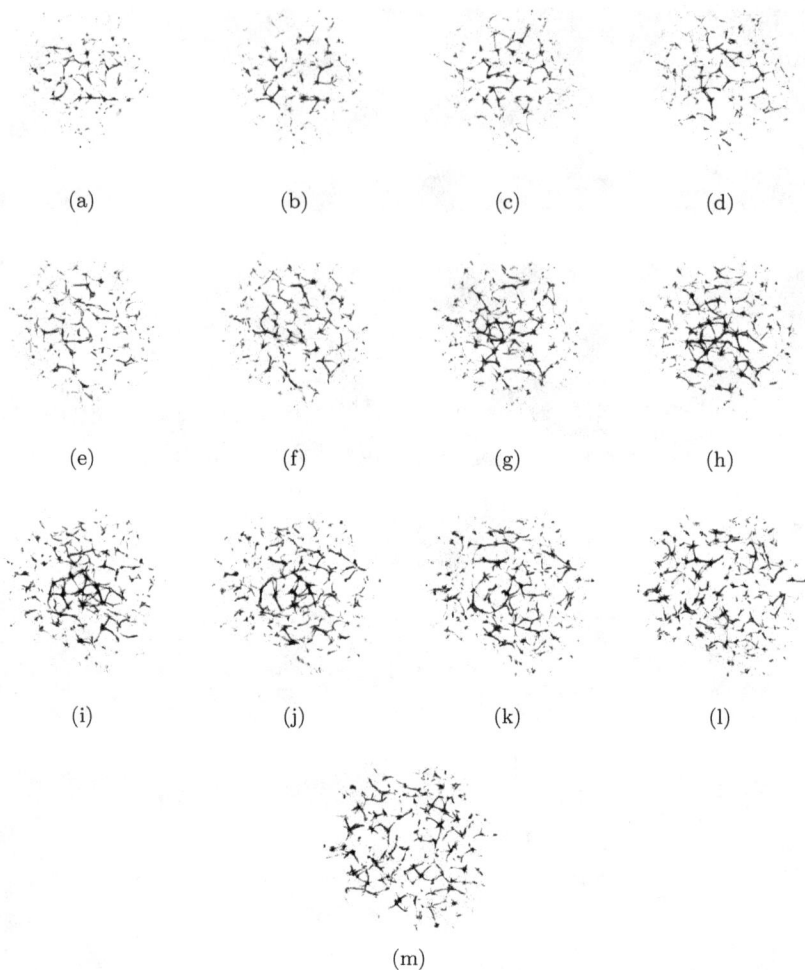

(a)   (b)   (c)   (d)

(e)   (f)   (g)   (h)

(i)   (j)   (k)   (l)

(m)

Fig. 14.3. Exemplary z-slices of 'conductive' geometries C selected from the three-dimensional actin bundle network shown in Fig. 14.2, which were reconstructed as described in Ref. [123].

the states of its neighbourhood $u(p) = \{q \in \mathbf{C} : d(p, q) \leq r\}$, where $d(p, q)$ is an Euclidean distance between voxels p and q; $r \in \mathbf{N}$ is a neighbourhood radius. $\theta \in \mathbf{N}$ is an excitation threshold, and $\delta \in \mathbf{N}$ is refractory delay. All voxels update their states in parallel and by the same rule:

$$p^{t+1} \begin{cases} \star, & \text{if } (p^t = \circ) \text{ and } (\sigma(p)^t > \theta), \\ \bullet, & \text{if } (p^t = \star) \text{ or } ((p^t = \bullet) \text{ and } (h_p^t > 0)), \\ \circ, & \text{otherwise,} \end{cases}$$

$$h_p^{t+1} = \begin{cases} \delta, & \text{if } (p^{t+1} = \bullet) \text{ and } (p^t = \star), \\ h_p^t - 1, & \text{if } (p^{t+1} = \bullet) \text{ and } (h_p^t > 0), \\ 0, & \text{otherwise.} \end{cases}$$

Every resting ($\circ$) voxel of **C** excites ($\star$) at the moment $t + 1$ if a number of its excited neighbours at the moment t, $\sigma(p)^t = |\{q \in u(p) : q^t = \star\}|$, exceeds a threshold θ. An excited voxel $p^t = \star$ takes the refractory state $\bullet$ at the next time step $t + 1$ and at the same moment a counter of refractory state h_p is set to the refractory delay δ. The counter is decremented, $h_p^{t+1} = h_p^t - 1$ at each iteration until it becomes 0. When the counter h_p becomes zero, the voxel p returns to the resting state $\circ$. For all results shown in this manuscript, the neighbourhood radius was set to $r = 3$. Choices of θ and δ are considered in Section 14.2.4.

14.2.3 Interfacing with the network

To stimulate the network and to record activity of the network, we assigned several domains of **C** as electrodes. We calculated a potential p_x^t at an electrode location $c \in$ **C** as $p_c = |z : d(c, z) < r_e$ and $z^t = \star|$, where $d(c, z)$ is an Euclidean distance between sites x and z in three-dimensional space. We have chosen an electrode radius of $r_e = 4$ voxels and conducted two families of experiments with two configurations of electrodes.

In the first family of experiments $\mathcal{E}_1$, we studied frequencies of two-input-one-output Boolean functions implementable in the network. We used 10 electrodes, their coordinates are listed in Table 14.1 and a configuration is shown in Figure 14.4(a). Electrodes E_0 representing input x and E_9 representing input y are

A. Adamatzky et al.

Table 14.1. Coordinates of electrodes in experiments family $\mathcal{E}_1$.

e	i	j	z
1	369	567	6
2	509	580	10
3	631	590	10
4	382	322	12
5	533	331	23
6	626	463	7
7	358	676	22
8	369	424	7
9	572	691	17
10	705	394	17

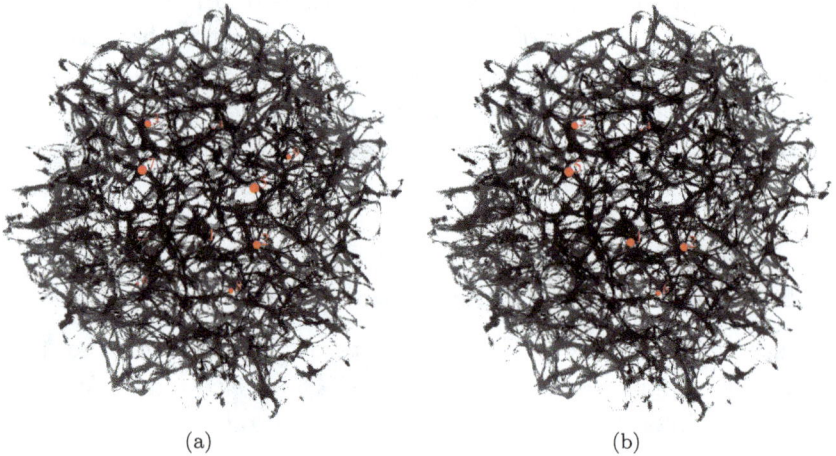

(a)   (b)

Fig. 14.4. Configurations of electrodes in the three-dimensional network of actin bundles used in (a) $\mathcal{E}_1$ and (b) $\mathcal{E}_2$. Depth of the network is shown by level of grey. Sizes of the electrodes are shown in perspective.

the input electrodes, all others are output electrodes representing outputs $z_1, \ldots, z_8$. Results are presented in Section 14.2.4. In the second family of experiments $\mathcal{E}_2$, we used six electrodes (Table 14.2 and Fig. 14.4(b)). All electrodes were considered as inputs during stimulation and outputs during recording of the network activity.

Table 14.2. Coordinates of electrodes in experiments family $\mathcal{E}_2$.

e	i	j	z
1	369	567	6
2	509	580	10
3	631	590	10
4	382	322	12
5	533	331	23
6	369	424	7
7	572	691	17
8	705	394	17

Exemplary snapshots of excitation dynamics on the network are shown in Fig. 14.5. Domains corresponding to the two electrodes e_0 and e_9 (Table 14.1 and Fig. 14.4(a)) have been excited (Fig. 14.5(a)). The excitation wave fronts propagate away from e_0 and e_9 (Fig. 14.5(b)). The fronts traverse the whole breadth of the network (Fig. 14.5(c)). Due to the presence of circular conductive paths in the network, repetitive patterns of activity emerge (Fig. 14.5(d)). Recordings of potential and videos of experiments are available within the Zenodo repository [24].

14.2.4 Selecting excitation threshold and refractory delay to maximise a number of logical gates

To design an actin droplet machine with complex behaviour, we need to find, values of refractory delay and excitation threshold for which the actin bundles' network executes a maximum of Boolean gates. To map dynamics of the network onto sets of gates, we undertook the following trials of stimulation:

(1) Fixed refractory delay $\delta = 20$ and excitation threshold $\theta = 4, 5, \ldots, 12$,
(2) Fixed excitation threshold $\theta = 7$, and refractory delay $\delta = 10, 15, 17, \ldots, 24, 30$.

An example of the network spiking activity as a response to stimulation is shown in Fig. 14.1. We stimulated the network with

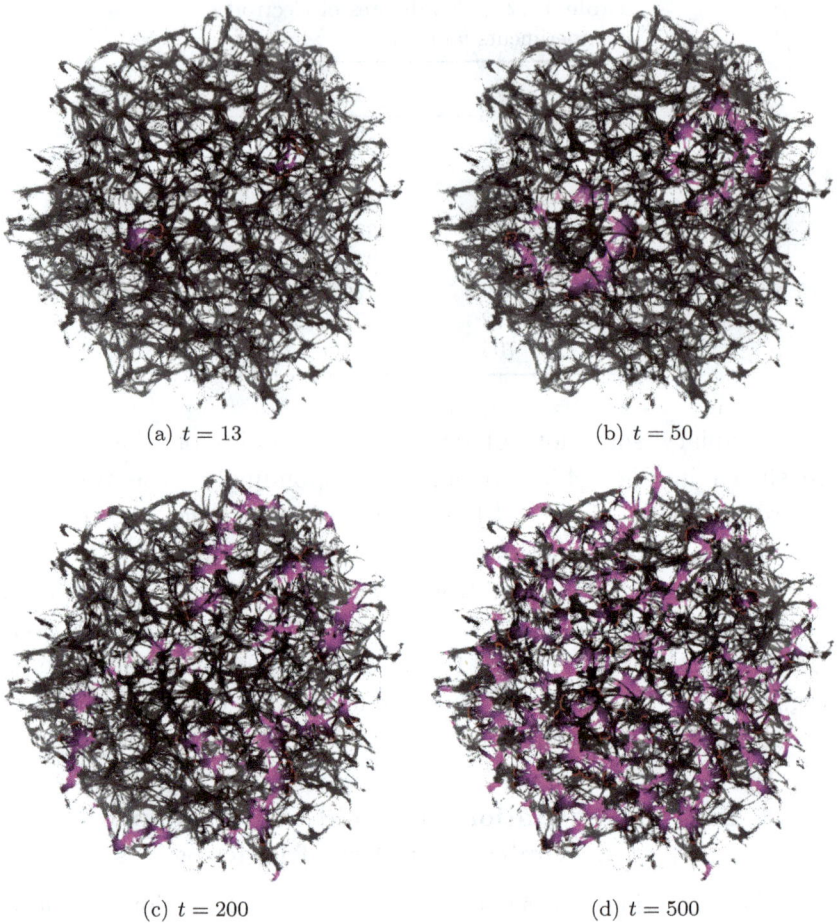

(a) $t = 13$

(b) $t = 50$

(c) $t = 200$

(d) $t = 500$

Fig. 14.5. Snapshots of excitation dynamics on the network. The excitation wave front is red and the refractory tail is magenta. The excitation threshold is $\theta = 7$ and the refractory delay is $\delta = 20$.

all possible configurations of inputs, recorded the network's electrical dynamics, and then extracted logical gates as follows. For each possible combination (i, j, k), $1 \leq i, j, k \leq 6$, $i \neq j$, $i \neq k$, $j \neq k$, we considered electrodes $\mathcal{E}_i$ and $\mathcal{E}_j$ to be inputs, representing Boolean variables x and y, respectively, and electrode $\mathcal{E}_k$ as output electrode, representing the result of a Boolean function. To input $x =$TRUE we applied a current to electrode $\mathcal{E}_i$, to input $y =$TRUE to electrode $\mathcal{E}_j$.

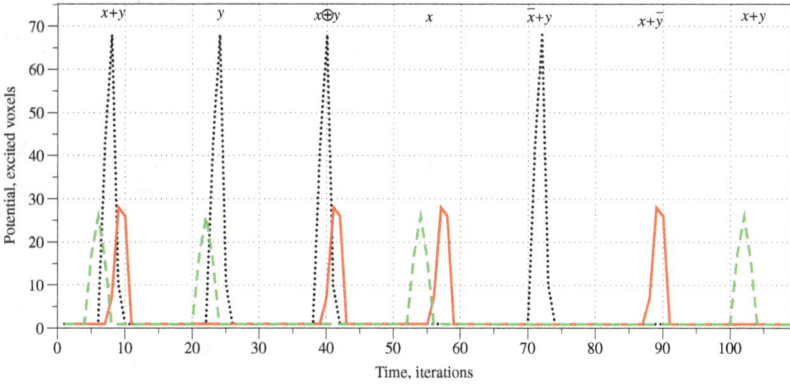

Fig. 14.6. Representation of two-inputs-one-output Boolean gates by combinations of spikes. Black dotted line shows the potential at an output electrode when the network was stimulated by input pair (x, y) =(FALSE, TRUE), red solid by (TRUE, FALSE), and green dashed by (x, y) =(TRUE, TRUE).

Then we recorded the potential at electrode $\mathcal{E}_k$. Two-input-one-output logical functions were extracted from the spiking events as follows. Assume each spike represents logical TRUE and that spikes being less than six iterations closer to each other happen at the same moment. Then a representation of gates by spikes and their combination will be as shown in Fig. 14.6.

For each combination (ρ, θ), we counted the numbers of gates OR $(x + y)$, AND (xy), XOR $(x\oplus)$, NOT-AND $(\overline{x}y)$, AND-NOT $(x\overline{y})$, and SELECT (x, y). We found that overall the total number of gates $\nu(\theta)$ realised by the network decreases with increase of θ (Fig. 14.7(a)). The function $\nu(\theta)$ is nonlinear and could be adequately described by a five degree polynomial. The function reaches its maximal value at $\theta = 7$ (Fig. 14.7(a)). OR gates are most commonly realised at $\theta = 11$, AND gates at $\theta = 6$, and *xor* gates at $\theta = 5$ as well as $\theta = 7$ (Fig. 14.7(b)). The number of AND-NOT gates implemented by the network reaches its highest value at $\theta = 6$, then drops sharply after θ_8 (Fig. 14.7(c)). NOT-AND gates are more common at $\theta = 5, 7, 9, 11$, while SELECT(x) has its peak at $\theta = 7$ and SELECT(y) at $\theta = 8, 9$ (Fig. 14.7(c)). The total number of gates realised in the network with the excitability threshold fixed to $\theta = 7$ decreases with the increase of δ. Oscillations of $\nu(\delta)$ are visible at $15 \leq \delta \leq 25$ (Fig. 14.7(d)).

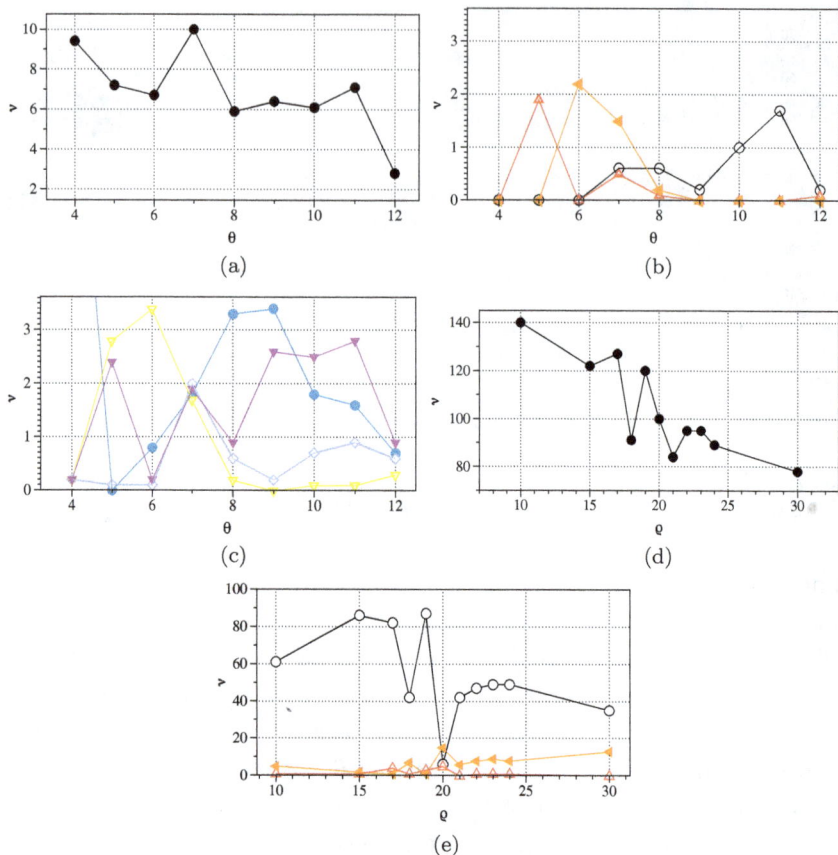

Fig. 14.7. The average number ν of gates realisable on each of the electrodes $e_1, \ldots, e_8$ depends on the threshold θ of excitation when the refractory delay δ is fixed to 20 (abc) and on refractory delays δ when the threshold θ is fixed to 7 (def). (a) Number of gates ν vs. threshold θ, $\delta = 20$. (b) Number of OR (black circle), AND (orange solid triangle) and XOR (red blank triangle) gates, $\delta = 20$. (c) Number of NOT-AND (yellow blank triangle), AND-NOT (magenta solid triangle), SELECT(x) (cyan blank rhombus), SELECT(y) (light blue disc), $\delta = 20$. (d) Number of gates ν vs. delay δ, $\theta = 7$. (e) Number of OR (black circle), AND (orange solid triangle), and XOR (red blank triangle) gates, $\theta = 7$.

The three highest values of $\nu(\delta)$ are achieved at $\delta = 10, 17$, and 20. Let us look now at the dependence of the numbers of OR, AND, and XOR gates of the refractory delay δ in Fig. 14.7(e). The number of OR gates increases with δ increasing from 10 to 15, but then drops

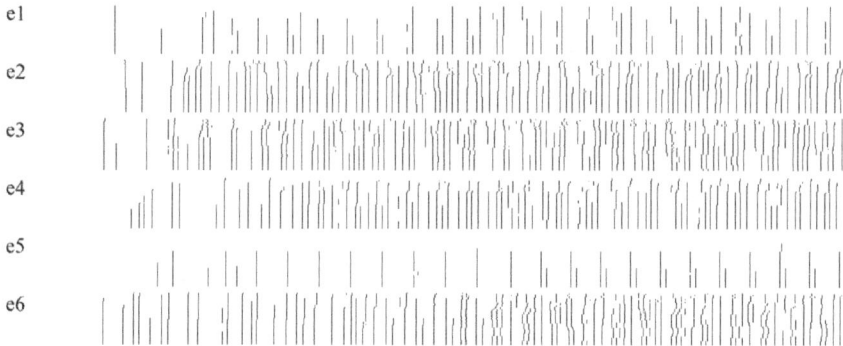

Fig. 14.8. All spikes recorded at each electrode for input binary strings from 1 to 63. The representation is implemented as follows. We stimulate the $\mathcal{M}$ with strings from $\{0,1\}^6$ and represent a spike detected at time t by a black pixel at position t along the horizontal axis. A plot of each electrode e_i represents a binary matrix $\mathbf{S} = (s_{zt})$, where $1 \leq z \leq 63$ and $1 \leq t \leq 1000$: $s_{zt} = 1$ if the input configuration was z and a spike was detected at moment t, and $s_{zt} = 1$ otherwise.

substantially at $\delta = 18$ to reach its maximum at $\delta = 19$. Numbers of gates AND and XOR behave similarly to each other. They both have a pronounced peak at $\delta = 20$ (Fig. 14.7(e)).

The gate frequency analysis presented in this section allows us to choose $\theta = 7$ and $\delta = 20$ for an actin droplet machine constructed in the next section.

14.3 Actin droplet machine

An actin droplet machine is defined as a tuple $\mathcal{M} = \langle \mathcal{A}, k, \mathbf{E}, \mathbf{S}, F \rangle$, where $\mathcal{A}$ is an actin network automaton, defined in Section 14.2.2, k is a number of electrodes, $\mathbf{E}$ is a configuration of electrodes, $\mathbf{S} = \{0,1\}^k$, F is a state-transition function $F : \mathbf{S} \rightarrow \mathbf{S}$ that implements a mapping between sets of all possible configurations of binary strings of length k. In the experiments reported here, $k = 6$.

In our experiments we have chosen six electrodes, their locations are shown in Fig. 14.4(b) and exact coordinates in Table 14.2. Thus, $F : \{0,1\}^6 \rightarrow \{0,1\}^6$ and the machine $\mathcal{M}$ has 64 states. We represent the inputs and the machine states in decimal encoding.

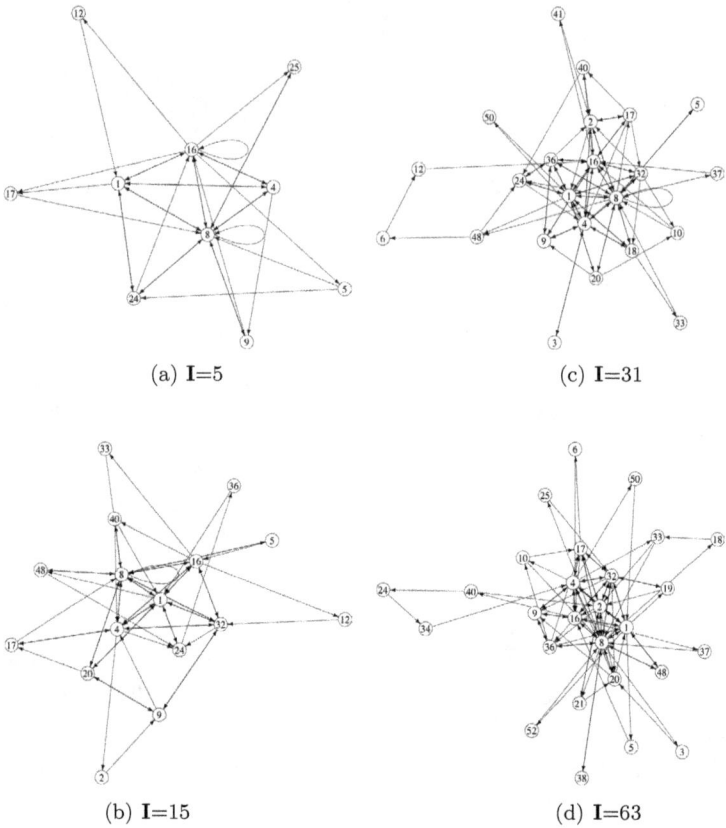

(a) **I**=5

(c) **I**=31

(b) **I**=15

(d) **I**=63

Fig. 14.9. State transitions of machine $\mathcal{M}$ for selected inputs I. A node is a decimal encoding of the $\mathcal{M}$ state $(e_0^t \ldots e_5^t)$.

Spikes detected in response to every input from $\{0,1\}^6$ are shown in Fig. 14.8.

Global transition graphs of $\mathcal{M}$ for selected inputs are shown in Fig. 14.9. Nodes of the graphs are states of $\mathcal{M}$, edges show transitions between the states. These directed graphs are defined as follows. There is an edge from node a to node b if there is such $1 \leq t \leq 1000$ that $\mathcal{M}^t = a$ and $\mathcal{M}^{t+1} = b$.

Let us now define a weighted global transition graph $\mathcal{G} = \langle \mathbf{Q}, \mathbf{E}, w \rangle$, where $\mathbf{Q}$ is a set of nodes (isomorphic to the $\{0,1\}^6$), and $\mathbf{E}$ is a set of edges, and weighting function $w : \mathbf{E} \to [0,1]$

(a)

(b)

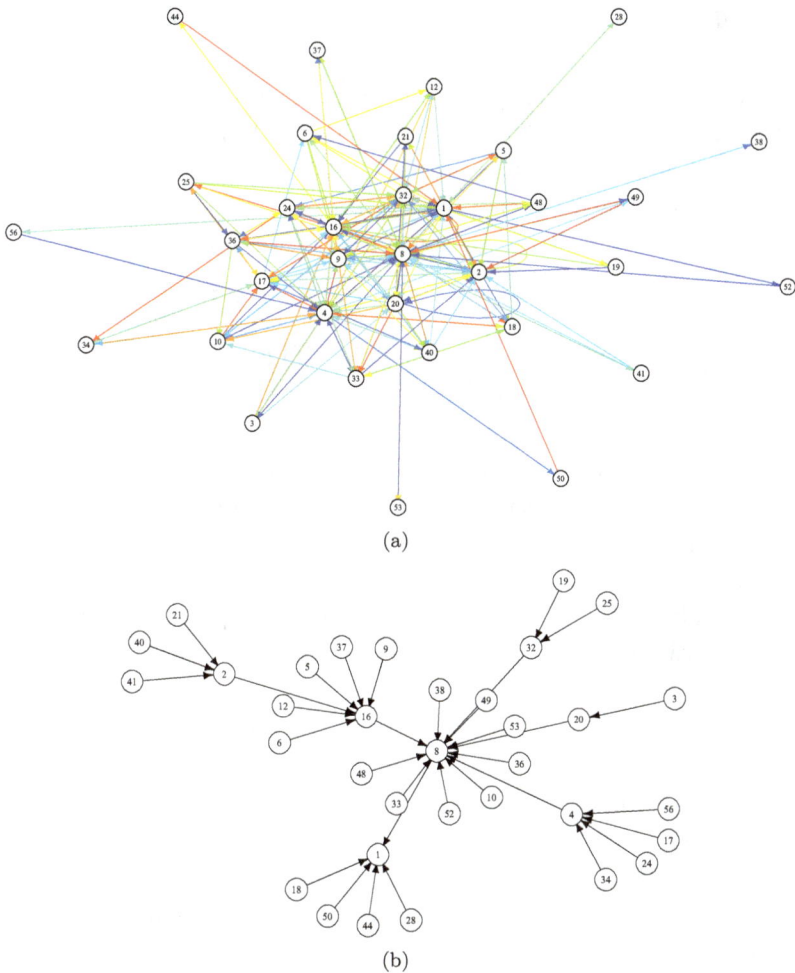

Fig. 14.10. (a) Global graph of $\mathcal{M}$ state transitions. Edge weights are visualised by colours: from lowest weight in orange to highest weight in blue. (b) Pruned global graph of $\mathcal{M}$: only transitions with maximum weight for any given predecessors are shown, each node/state has at most one outgoing edge.

assigning a number of a unit interval to each edge. Let $a, b \in \mathbf{Q}$ and $e(a, b) \in \mathbf{E}$, then a normalised weight is calculated as $w(e(a, b)) = \dfrac{\sum_{i \in \mathbf{Q}, t \in \mathbf{T}} \chi(s^t = a \text{ and } s^{t+1} = b)}{\sum_{d \in \mathbf{Q}, t \in \mathbf{T}} \sum_{\mathbf{Q}, t \in \mathbf{T}} \chi(s^t = a \text{ and } s^{t+1} = d)}$, with χ taking the value '1' when the conditions are true and '0' otherwise. In words, $w(e(a, b))$ is a number of transitions from a to b observed in the evolution of $\mathcal{M}$ for all possible inputs from Q during time interval $\mathbf{T}$ normalised by the total number of transitions from a to all other nodes. The graph G is visualised in Fig. 14.10(a). Nodes that have predecessors are 1–6, 8–10, 12, 16–21, 24, 25, 28, 32–34, 36–38, 40, 41, 44, 48–50, 52, 53, 56. Nodes without predecessors are 7, 11, 13–15, 22, 23, 26, 27, 29–31, 35, 39, 42, 43, 45–47, 51, 54, 55, 57–63.

Let us convert G to an acyclic non-weighted graph of more likely transitions $G^*\langle \mathbf{Q}, \mathbf{E}^* \rangle$, where $e(a, b) \in \mathbf{E}^*$ if $w(e(a, b)) = max\{w(e(a, c)) | e(a, c) \in \mathbf{E}\}$. That is for each node we select an outgoing edge with maximum weight. The graph is a tree, see Fig. 14.10(b). Most states apart from 1, 2, 4, 8, 16, 20, 32 are Garden-of-Eden configurations, which have no predecessors. Indegrees $\nu()$ of not-Garden-of-Eden nodes are $\nu(20) = 1, \nu(32) = 2, \nu(2) = 3, \nu(4) = 4, \nu(1) = 5, \nu(16) = 6, \nu(8) = 12$. There is one fixed point, the state 1, corresponding to the situation when a spike is recorded only on electrode e_5; it has no successors.

By analysing G we can characterise the richness of $\mathcal{M}$'s responses to input stimuli. We define a richness as a number of different states over all inputs, as shown in Table 14.3, and distribution in Fig. 14.12(a). The number of states produced increases from under five for beginning of $\mathcal{M}$ evolution and then reaches circa seven states in average. Oscillations around this value are seen in (Fig. 14.12(a)). Figure 14.12(b) shows the number of different nodes, generated in evolution of $\mathcal{M}$, stimulated by a given input. Less than fifteen different states are found in the evolution in responses to inputs 1–21 (21 corresponds to binary input string 010101); then a number of different nodes stay around 25. The diagram Fig. 14.12(c) shows how many inputs might lead to a given state/node of $\mathcal{M}$. Some of the states/nodes are seen to be Garden-of-Eden configurations $\mathbf{E}$ (nodes without predecessors) and thus could not be generated by stimulating $\mathcal{M}$ by sequences from $\mathbf{Q} - \mathbf{E}$.

Assume $\mathbf{T}$ is a set of temporal moments when the machine responded at least to one input string with a non-zero state.

Table 14.3. Fifty four state transitions of $\mathcal{M}$ over all possible inputs: t is a transition step, $\mu(t)$ is a number of different states appearing over all possible inputs, $\mathbf{P}(t)$ is a set of nodes appeared at t.

t	$\mu(t)$	$\mathbf{P}(t)$
1	3	8, 9, 1,
2	3	16, 32, 8,
3	3	1, 16, 32,
4	3	8, 1, 16,
5	3	1, 8, 16,
6	3	16, 8, 1,
7	4	8, 1, 16, 4,
8	4	1, 16, 8, 5,
9	5	16, 1, 8, 4, 5,
10	4	16, 1, 8, 4,
11	5	8, 1, 16, 20, 4,
12	4	1, 16, 8, 20,
13	6	16, 8, 1, 17, 4, 20,
14	8	8, 16, 17, 4, 20, 1, 32, 2,
15	8	1, 16, 8, 4, 2, 10, 20, 32,
16	6	16, 4, 8, 1, 10, 32,
17	5	16, 1, 4, 8, 9,
18	7	8, 16, 4, 1, 17, 10, 9,
19	6	1, 8, 16, 17, 4, 10,
20	8	16, 1, 8, 17, 4, 24, 10, 2,
21	9	8, 16, 1, 17, 32, 24, 9, 4, 10,
22	6	16, 1, 8, 32, 9, 4,
23	7	8, 1, 16, 4, 32, 9, 17,
24	6	1, 16, 17, 4, 32, 8,
25	7	16, 1, 8, 4, 17, 32, 9,
26	6	8, 16, 4, 12, 1, 17,
27	6	1, 8, 16, 4, 17, 32,
28	6	16, 8, 4, 1, 24, 32,
29	7	8, 1, 4, 16, 12, 24, 32,
30	7	16, 1, 8, 4, 17, 2, 32,
31	9	8, 1, 24, 16, 12, 4, 2, 17, 32,
32	7	1, 16, 8, 24, 17, 2, 40,
33	9	16, 8, 1, 4, 40, 17, 24, 32, 2,
34	7	8, 1, 16, 24, 40, 4, 32,
35	6	1, 16, 8, 4, 24, 2,
36	6	16, 8, 1, 17, 4, 32,
37	7	8, 16, 17, 4, 1, 40, 2,

(*Continued*)

Table 14.3. (*Continued*)

t	$\mu(t)$	$\mathbf{P}(t)$
38	7	1, 8, 16, 17, 4, 24, 2,
39	7	16, 1, 8, 17, 9, 4, 2,
40	7	8, 16, 4, 1, 24, 40, 2,
41	10	1, 8, 16, 9, 17, 4, 18, 24, 40, 2,
42	8	16, 1, 8, 4, 18, 33, 40, 24,
43	9	8, 1, 16, 4, 24, 33, 18, 32, 34,
44	9	1, 16, 8, 4, 17, 33, 24, 32, 40,
45	7	16, 8, 4, 1, 12, 24, 34,
46	7	8, 1, 16, 4, 24, 18, 34,
47	5	1, 16, 8, 4, 33,
48	5	16, 8, 1, 4, 17,
49	8	8, 1, 16, 4, 20, 32, 24, 19,
50	6	1, 16, 8, 4, 17, 32,
51	8	16, 8, 1, 4, 17, 32, 41, 19,
52	9	8, 16, 4, 1, 32, 33, 41, 2, 19,
53	10	1, 8, 16, 4, 20, 10, 2, 41, 32, 19,
54	9	16, 1, 8, 5, 17, 4, 2, 32, 19,

Configurations at each transition t can be considered as outputs representing the function $g : 0,1^6 \to 0,1^6$. As we can see in Table 14.3, transitions at $t = 41$ and $t = 53$ correspond to the highest number of different binary strings $(e_1, \ldots, e_6)$. The graph corresponding to $g(41)$ at $t = 41$ is shown in Fig. 14.11 and is not connected. The small component consists of fixed point 40 (string '101000') with two leaves 39 ('100111') and 38 ('100110'). The largest component has a tree structure at large, with cycle 2 ('000010') – 1 ('000001') as a root. Other nodes with most predecessors are 8 ('001000'), 16 ('010000'), and 18 ('010010').

From the transitions $g(41)$ we can reconstruct Boolean functions realised at each of the six electrodes (the functions are minimised and represented in a disjunctive normal form):

e_0 : $f_0(x_0, \ldots, x_5) = x_0 \cdot \overline{x_1} \cdot x_2 \cdot x_3 + \overline{x_0} \cdot x_1 \cdot \overline{x_3} \cdot \overline{x_4} \cdot \overline{x_5} + x_0 \cdot \overline{x_1} \cdot x_2 \cdot$
$\overline{x_3} \cdot x_4 + \overline{x_0} \cdot \overline{x_2} \cdot x_3 \cdot x_4 \cdot x_5 + \overline{x_0} \cdot x_1 \cdot \overline{x_2} \cdot x_3 \cdot \overline{x_4} + \overline{x_1} \cdot x_2 \cdot \overline{x_3} \cdot \overline{x_4} \cdot$
$x_5 + \overline{x_0} \cdot \overline{x_1} \cdot \overline{x_2} \cdot \overline{x_3} \cdot \overline{x_4} \cdot x_5 + \overline{x_0} \cdot x_1 \cdot \overline{x_2} \cdot x_3 \cdot x_4 \cdot \overline{x_5}$

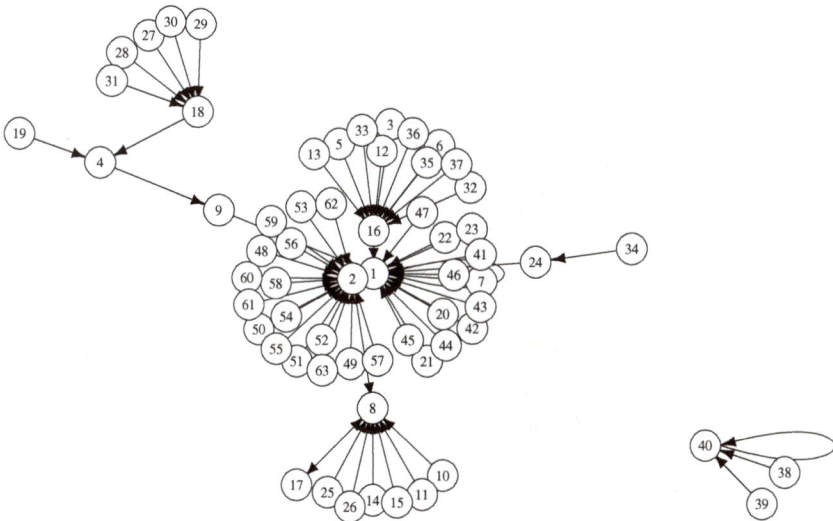

Fig. 14.11. Graph of g at $t = 41$.

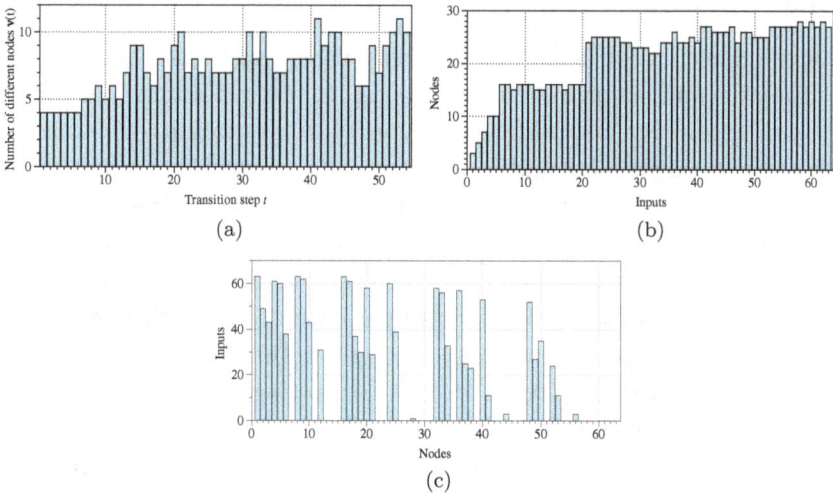

(a)

(b)

(c)

Fig. 14.12. Distributions characterising richness of $\mathcal{M}$'s responses. (a) Different states per transition over all inputs. Horizontal axis shows steps of $\mathcal{M}$ transitions. Vertical axis is a number of different states. (b) Nodes per input. Horizontal axis shows decimal values of input strings. Horizontal axis shows a number of different states/nodes are generated in the evolution of $\mathcal{M}$. (c) Inputs per node.

$$e_1 : f_1(x_0, \ldots, x_5) = x_0 \cdot \overline{x_1} \cdot x_2 \cdot x_3 + \overline{x_0} \cdot x_1 \cdot \overline{x_3} \cdot \overline{x_4} \cdot \overline{x_5} + x_0 \cdot \overline{x_1} \cdot x_2 \cdot$$
$$\overline{x_3} \cdot x_4 + \overline{x_0} \cdot \overline{x_2} \cdot x_3 \cdot x_4 \cdot x_5 + \overline{x_0} \cdot x_1 \cdot \overline{x_2} \cdot x_3 \cdot \overline{x_4} + \overline{x_1} \cdot x_2 \cdot \overline{x_3} \cdot \overline{x_4} \cdot$$
$$x_5 + \overline{x_0} \cdot \overline{x_1} \cdot \overline{x_2} \cdot \overline{x_3} \cdot \overline{x_4} \cdot x_5 + \overline{x_0} \cdot x_1 \cdot \overline{x_2} \cdot x_3 \cdot x_4 \cdot \overline{x_5}$$

$$e_2 : f_2(x_0, \ldots, x_5) = x_0 \cdot \overline{x_1} \cdot x_2 \cdot x_3 + \overline{x_0} \cdot x_1 \cdot \overline{x_3} \cdot \overline{x_4} \cdot \overline{x_5} + x_0 \cdot \overline{x_1} \cdot x_2 \cdot$$
$$\overline{x_3} \cdot x_4 + \overline{x_0} \cdot \overline{x_2} \cdot x_3 \cdot x_4 \cdot x_5 + \overline{x_0} \cdot x_1 \cdot \overline{x_2} \cdot x_3 \cdot \overline{x_4} + \overline{x_1} \cdot x_2 \cdot \overline{x_3} \cdot \overline{x_4} \cdot$$
$$x_5 + \overline{x_0} \cdot \overline{x_1} \cdot \overline{x_2} \cdot \overline{x_3} \cdot \overline{x_4} \cdot x_5 + \overline{x_0} \cdot x_1 \cdot \overline{x_2} \cdot x_3 \cdot x_4 \cdot \overline{x_5}$$

$$e_3 : f_3(x_0, \ldots, x_5) = x_0 \cdot \overline{x_1} \cdot x_2 \cdot x_3 + \overline{x_0} \cdot x_1 \cdot \overline{x_3} \cdot \overline{x_4} \cdot \overline{x_5} + x_0 \cdot \overline{x_1} \cdot x_2 \cdot$$
$$\overline{x_3} \cdot x_4 + \overline{x_0} \cdot \overline{x_2} \cdot x_3 \cdot x_4 \cdot x_5 + \overline{x_0} \cdot x_1 \cdot \overline{x_2} \cdot x_3 \cdot \overline{x_4} + \overline{x_1} \cdot x_2 \cdot \overline{x_3} \cdot \overline{x_4} \cdot$$
$$x_5 + \overline{x_0} \cdot \overline{x_1} \cdot \overline{x_2} \cdot \overline{x_3} \cdot \overline{x_4} \cdot x_5 + \overline{x_0} \cdot x_1 \cdot \overline{x_2} \cdot x_3 \cdot x_4 \cdot \overline{x_5}$$

$$e_4 : f_4(x_0, \ldots, x_5) = x_0 \cdot \overline{x_1} \cdot x_2 \cdot x_3 + \overline{x_0} \cdot x_1 \cdot \overline{x_3} \cdot \overline{x_4} \cdot \overline{x_5} + x_0 \cdot \overline{x_1} \cdot x_2 \cdot$$
$$\overline{x_3} \cdot x_4 + \overline{x_0} \cdot \overline{x_2} \cdot x_3 \cdot x_4 \cdot x_5 + \overline{x_0} \cdot x_1 \cdot \overline{x_2} \cdot x_3 \cdot \overline{x_4} + \overline{x_1} \cdot x_2 \cdot \overline{x_3} \cdot \overline{x_4} \cdot$$
$$x_5 + \overline{x_0} \cdot \overline{x_1} \cdot \overline{x_2} \cdot \overline{x_3} \cdot \overline{x_4} \cdot x_5 + \overline{x_0} \cdot x_1 \cdot \overline{x_2} \cdot x_3 \cdot x_4 \cdot \overline{x_5}$$

$$e_5 : f_5(x_0, \ldots, x_5) = x_0 \cdot \overline{x_1} \cdot x_2 \cdot x_3 + \overline{x_0} \cdot x_1 \cdot \overline{x_3} \cdot \overline{x_4} \cdot \overline{x_5} + x_0 \cdot \overline{x_1} \cdot x_2 \cdot$$
$$\overline{x_3} \cdot x_4 + \overline{x_0} \cdot \overline{x_2} \cdot x_3 \cdot x_4 \cdot x_5 + \overline{x_0} \cdot x_1 \cdot \overline{x_2} \cdot x_3 \cdot \overline{x_4} + \overline{x_1} \cdot x_2 \cdot \overline{x_3} \cdot \overline{x_4} \cdot$$
$$x_5 + \overline{x_0} \cdot \overline{x_1} \cdot \overline{x_2} \cdot \overline{x_3} \cdot \overline{x_4} \cdot x_5 + \overline{x_0} \cdot x_1 \cdot \overline{x_2} \cdot x_3 \cdot x_4 \cdot \overline{x_5}$$

14.4 Discussion

Early concepts of sub-cellular computing, such as microtubule automata and information processing in actin-tubulin networks, lacked specificity regarding the type of computation or information processing that could be performed by cytoskeleton networks and how it was achieved [112, 114, 206, 209]. To address this, we conducted several concrete implementations of logical gates and functions on individual actin filaments and their intersections using soliton collisions [231, 235]. We also employed a reservoir computing approach to discover functions on single actin units and filaments [17, 20]. However, we realised that expecting the initiation and recording of travelling localisations (solitons, impulses) on a single actin filament was impractical. To overcome this, we developed a numerical model of spikes propagating on a network of actin filament bundles and demonstrated the implementation of Boolean gates [25].

In our recent research, we have conducted a thorough reevaluation of the concept of information processing on actin networks and introduced a novel model called the actin droplet machine. This machine represents a three-dimensional network based on an experimental actin network created within a droplet. Its primary function is to execute a mapping function F that takes binary strings of length k and maps them onto themselves.

Operating as a finite state machine, the actin droplet machine allows for the propagation of localisations or impulses along the network, where they interact with each other at a low level. By focusing on a specific electrode location, which represents a single element of the binary string, we can reconstruct k functions with k arguments. The exact structure of each k-ary function is determined by the mapping function F, which relies on the precise architecture of the three-dimensional actin network and the configuration of electrodes.

Moving forward, future research endeavours could concentrate on conducting detailed analyses of various actin network architectures in laboratory experiments. Exploring different configurations of electrodes would enable us to understand how they influence the structure of the mapping function F and the resulting distribution of implementable functions by the actin droplet machine. The ultimate goal is to successfully implement actin droplet machines in laboratory settings and investigate the potential of cascading multiple machines to create a multi-processor computing architecture.

One notable advantage of actin network computers is their dynamic reconfigurability compared to conventional static hardware. Actin networks undergo continuous reconfiguration, with filaments depolymerising and new filaments polymerising. However, this dynamic nature does not pose a disadvantage for actin network computers for several reasons: (1) These computers operate at speeds several orders of magnitude faster than actin treadmilling rates. (2) Actin networks can be stabilised, allowing for the preservation of desired structures and functionalities. (3) The dynamic reconfigurability of actin networks can be leveraged in computation, providing additional flexibility and adaptability in the implementation of various computational tasks.

In actin bundle networks, computations are implemented through the propagation of mechanical or electrical signals. The speed of signal propagation can be estimated to be around 10^6 μm/s for sound solitons or 10^5–10^8 μm/s for action potentials [220]. Let's consider the lower estimate of 10^5 μm/s. Assuming the maximum linear size of an actin droplet machine is approximately 250 μm, the machine can process approximately 400 parallel inputs per second, operating at a frequency of 0.4 kHz. The speed of actin polymerisation is typically estimated to be $4 \cdot 10^{-1}$ μm/s [200]. An actin bundle consists of up to 500 actin filaments, which are unlikely to fail simultaneously. Moreover, in our experiments, we observed that the networks remained stable for hours without significant rearrangements. It's important to note that we did not use any actin accessory proteins or provide ATP as an energy source. The structures self-assembled solely based on thermodynamic principles into a stable and frozen state. Therefore, even if we assume an active network with high treadmilling rates, we can consider the network to be fixed for at least 10 seconds, allowing us to execute up to 4×10^3 cycles of computation.

Introducing accessory proteins into actin networks can have a significant impact on their properties and lifespan. Synthetic actin crosslinkers made from DNA and peptides, as well as integrin and drebrin peptides, have been shown to enhance network stability and increase longevity [156, 175, 180]. Additional strategies involve strengthening filaments with α-actinin and stabilising them using synthetic mini-nebuline to prolong network lifespan [62, 190, 277]. By incorporating proteins such as gelsolin, cofilin, formin, and myosins, dynamic reconfigurations can be facilitated, enabling the construction of a dynamic computing system [121].

Leveraging the capability of dynamical reconfiguration in actin network computers opens up a wide range of applications. It can accelerate Boolean satisfiability solvers, serve as the foundation for reconfigurable data flow machines in atomic functional programming languages, explore dynamical genetic programming on evolvable Boolean networks, and even find applications in cryptography [51, 52, 97, 211, 284]. The dynamic nature of actin networks provides opportunities for adaptive and flexible computation, paving the way for various computational tasks.

References

1. S Aaronson and J Watrous. Closed timelike curves make quantum and classical computing equivalent. *Proceedings of the Royal Society A: Mathematical, Physical and Engineering Science*, 465(2102): 631–647, 2009.
2. S Adak, S Mukherjee, and S Das. Do there exist non-linear maximal length cellular automata? A study. In *Cellular Automata: 13th International Conference on Cellular Automata for Research and Industry, ACRI 2018, Como, Italy, September 17–21, 2018, Proceedings*, pp. 289–297. Springer, 2018.
3. A Adamatzky. Collision-based computing in biopolymers and their automata models. *International Journal of Modern Physics C*, 11: 1321–1346, 2000.
4. A Adamatzky. Topics in reaction-diffusion computers. *Journal of Computational and Theoretical Nanoscience*, 8: 295–303, 2011.
5. A Adamatzky. Slime mould processors, logic gates and sensors. *Philosophical Transactions of the Royal Society A*, 373(2046): 20140216, 2015.
6. A Adamatzky and R. Mayne. Actin automata: Phenomenology and localizations. *arxiv*, (1408.3676), 2014.
7. A Adamatzky. Collision-based computing in biopolymers and their automata models. *International Journal of Modern Physics C*, 11(7): 1321–1346, 2000.
8. A Adamatzky. *Computing in Nonlinear Media and Automata Collectives*. CRC Press, 2001.
9. A Adamatzky (ed.). *Collision-based Computing*. Springer, 2002.
10. A Adamatzky. Collision-based computing in Belousov–Zhabotinsky medium. *Chaos, Solitons & Fractals*, 21(5): 1259–1264, 2004.
11. A Adamatzky (ed.). *Game of Life Cellular Automata*, Vol. 1. Springer, 2010.
12. A Adamatzky. Topics in reaction-diffusion computers. *Journal of Computational and Theoretical Nanoscience*, 8(3): 295–303, 2011.
13. A Adamatzky. On diversity of configurations generated by excitable cellular automata with dynamical excitation intervals. *International Journal of Modern Physics C*, 23(12): 1250085, 2012.

14. A Adamatzky. Patterns of conductivity in excitable automata with updatable intervals of excitations. *Physical Review E*, 86(5): 056105, 2012.
15. A Adamatzky. Computing on verotoxin. *PhysChemPhyc*, 2017.
16. A Adamatzky. Logical gates in actin monomer. *Scientific Reports*, 7(1): 11755, 2017.
17. A Adamatzky. On dynamics of excitation and information processing in F-actin: Automaton model. *Complex Systems*, 2017.
18. A Adamatzky. On dynamics of excitation and information processing in F-actin: Automaton model. *Complex Systems*, 2(4): 295–317, 2017.
19. A Adamatzky. On discovering functions in actin filament automata. *arXiv preprint arXiv:1807.06352*, 2018.
20. A Adamatzky. On interplay between excitability and geometry. *arXiv preprint arXiv:1904.06526*, 2019.
21. A Adamatzky and LO Chua. Phenomenology of retained refractoriness: On semi-memristive discrete media. *International Journal of Bifurcation and Chaos*, 22(11): 1230036, 2012.
22. A Adamatzky, S Harding, V Erokhin, R Mayne, N Gizzie, F Baluška, S Mancuso, and GC Sirakoulis. Computers from plants we never made: Speculations. In *Inspired by Nature*, pp. 357–387. Springer, 2018.
23. A Adamatzky, F Huber, and J Schnauss. Supplementary materials for the paper "actin droplet machine". 2019. http://doi.org/10.5281/zenodo.2649293.
24. A Adamatzky, F Huber, and J Schnauß. Computing on actin filament bundles. *arXiv preprint arXiv:1903.10186*, 2019.
25. A Adamatzky and R Mayne. Actin automata: Phenomenology and localizations. *arXiv preprint arXiv:1408.3676*, 2014.
26. A Adamatzky and R Mayne. Actin automata: Phenomenology and localizations. *International Journal of Bifurcation and Chaos*, 25(2): 1550030, 2015.
27. A Adamatzky, J Tuszynski, J Pieper, DV Nicolau, R Rinalndi, G Sirakoulis, V Erokhin, J Schnauss, and DM Smith. Towards cytoskeleton computers. A proposal. In A Adamatzky, S Akl, and G Sirakoulis (eds.), *From Parallel to Emergent Computing*. CRC Group/Taylor & Francis, 2019.
28. A Adamatzky, A Wuensche, and B De Lacy Costello. Glider-based computing in reaction-diffusion hexagonal cellular automata. *Chaos, Solitons & Fractals*, 27(2): 287–295, 2006.
29. AI Adamatzky. Universal computation in excitable media: The 2+ medium. *Advanced Materials for Optics and Electronics*, 7(5): 263–272, 1997.
30. A Aloisi, AD Torre, A De Benedetto, and R Rinaldi. Bio-recognition in spectroscopy-based biosensors for* heavy metals-water and waterborne contamination analysis. *Biosensors*, 9(3): 96, 2019.
31. R Alonso-Sanz. *Discrete Systems with Memory*, Vol. 75. World Scientific, 2011.
32. R Alonso-Sanz and L Bull. One-dimensional coupled cellular automata with memory: Initial investigations. *Journal of Cellular Automata*, 5(1–2): 29–49, 2003.

33. R Alonso-Sanz and M Martin. Elementary cellular automata with elementary memory rules in cells: The case of linear rules. *Journal of Cellular Automata*, 1(1): 71–87, 2006.

34. Ramón Alonso-Sanz. Memory versus spatial disorder in the support of cooperation. *Biosystems*, 97(2): 90–102, 2009.

35. S Amoroso and G Cooper. The garden-of-eden theorem for finite configurations. *Proceedings of the American Mathematical Society*, 26(1): 158–164, 1970.

36. C Anastassiou, M Segev, K Steiglitz, JA Giordmaine, M Mitchell, M-F Shih, S Lan, and J Martin. Energy-exchange interactions between colliding vector solitons. *Physical Review Letters*, 83(12): 2332, 1999.

37. B Aoun and M Tarifi. Introduction to quantum cellular automata. *arxiv*, quant-ph(0401123), 2004.

38. I Aprodu, A Ionescu, I Banu, and C Banu. Actin monomer-monomer interaction — A molecular mechanics study. *The Annals of the University Dunarea de Jos of Galati*, 31: 51–57, 2008.

39. E Archer, B Maigret, C Escrieut, L Pradayrol, and D Fourmy. Rhodopsin crystal: New template yielding realistic models of G-protein-coupled receptors? *Trends in Pharmacological Sciences*, 24(1): 36–40, 2003.

40. S Asakura and F Oosawa. On interaction between two bodies immersed in a solution of macromolecules. *The Journal of Chemical Physics*, 22(7): 1255–1256, 1954.

41. S Asakura and F Oosawa. Interaction between particles suspended in solutions of macromolecules. *Journal of Polymer Science*, 33(126): 183–192, 1958.

42. FA Atienza, JR Carrión, AG Alberola, JLR Álvarez, JJS Muñoz, JM Sánchez, and MV Chávarri. A probabilistic model of cardiac electrical activity based on a cellular automata system. *Revista Española de Cardiología (English Edition)*, 58(1): 41–47, 2005.

43. AJ Atrubin. A one-dimensional real-time iterative multiplier. *IEEE Transactions on Electronic Computers*, 3: 394–399, 1965.

44. A Baltuška, T Udem, M Uiberacker, M Hentschel, E Goulielmakis, C Gohle, R Holzwarth, VS Yakovlev, A Scrinzi, TW Hänsch, *et al.* Attosecond control of electronic processes by intense light fields. *Nature*, 421(6923): 611–615, 2003.

45. S Bandini, A Bonomi, and G Vizzari. An analysis of different types and effects of asynchronicity in cellular automata update schemes. *Natural Computing*, 11: 277–287, 2012.

46. MF Barnsley, RL Devaney, BB Mandelbrot, H-O Peitgen, D Saupe, RF Voss, Y Fisher, and M McGuire. *The Science of Fractal Images*, Vol. 1. Springer, 1988.

47. GW Beeler and H Reuter. Reconstruction of the action potential of ventricular myocardial fibres. *The Journal of Physiology*, 268(1): 177–210, 1977.

48. AS Belmont, S Dietzel, AC Nye, YG Strukov, and T Tumbar. Large-scale chromatin structure and function. *Current Opinion in Cell Biology*, 11(3): 307–311, 1999.

49. A Bianco, K Kostarelos, and M Prato. Applications of carbon nanotubes in drug delivery. *Current Opinion in Chemical Biology*, 9(6): 674–679, 2005.

50. L Bull. On dynamical genetic programming: Simple Boolean networks in learning classifier systems. *International Journal of Parallel, Emergent and Distributed Systems*, 24(5): 421–442, 2009.

51. L Bull. On the evolution of Boolean networks for computation: A guide RNA mechanism. *International Journal of Parallel, Emergent and Distributed Systems*, 31(2): 101–113, 2016.

52. L Bull, A Budd, C Stone, I Uroukov, BDL Costello, and A Adamatzky. Towards unconventional computing through simulated evolution: Control of nonlinear media by a learning classifier system. *Artificial Life*, 14(2): 203–222, 2008.

53. L Bull, R Toth, C Stone, BDL Costello, and A Adamatzky. Light-sensitive Belousov–Zhabotinsky computing through simulated evolution. In *Advances in Unconventional Computing*, pp. 199–212. Springer, 2017.

54. L Bull, R Toth, C Stone, BDL Costello, and A Adamatzky. Chemical computing through simulated evolution. In *Inspired by Nature*, pp. 269–286. Springer, 2018.

55. HF Cantiello, C Patenaude, and K Zaner. Osmotically induced electrical signals from actin filaments. *Biophysical Journal*, 59(6): 1284–1289, 1991.

56. HF Cantiello. Role of actin filament organization in cell volume and ion channel regulation. *Journal of Experimental Zoology*, 279(5): 425–435, 1997.

57. Riccardo Caponetto, Luigi Fortuna, Stefano Fazzino, and Maria Gabriella Xibilia. Chaotic sequences to improve the performance of evolutionary algorithms. *IEEE Transactions on Evolutionary Computation*, 7(3): 289–304, 2003.

58. B Chen, F Chen, and GJ Martínez. Glider collisions in hybrid cellular automaton with memory rule (4 3, 7 4). *International Journal of Bifurcation and Chaos*, 27(06): 1750082, 2017.

59. G Chen and X Dong. *From Chaos to Order: Methodologies, Perspectives and Applications*, Vol. 24. World Scientific, 1998.

60. AP Chetverikov, W Ebeling, and MG Velarde. Dissipative solitons and complex currents in active lattices. *International Journal of Bifurcation and Chaos*, 16(6): 1613–1632, 2006.

61. M Chu, CC Gregorio, and CT Pappas. Nebulin, a multi-functional giant. *Journal of Experimental Biology*, 219(2): 146–152, 2016.

62. MF Ciappina, J Perez-Hernandez, A Landsman, W Okell, S Zherebtsov, B Förg, J Schötz, L Seiffert, T Fennel, T Shaaran, *et al.* Attosecond physics at the nanoscale. *Reports on Progress in Physics*, 2017.

63. M Cifra, J Vanis, O Kucera, J Hasek, I Frydlova, F Jelinek, J Saroch, and J Pokorny. Electrical vibrations of yeast cell membrane. *PIERS Online*, 3(8): 1190–1194, 2007.

64. LA Cingolani and Y Goda. Actin in action: the interplay between the actin cytoskeleton and synaptic efficacy. *Nature Reviews Neuroscience*, 9(5): 344–356, 2008.

65. D. Ciucci and D. Dubois. Three-valued logics for incomplete information and epistemic logic. In *Logics in Artificial Intelligence*, pages 147–159. Springer, 2012.

66. D. Ciucci and D. Dubois. From paraconsistent three-valued logics to multiple-source epistemic logic. In *8th Conf European Society for Fuzzy Logic and Technology (EUSFLAT-13)*. Atlantis Press, 2013.

67. M Conrad. Cross-scale information processing in evolution, development and intelligence. *BioSystems*, 38(2): 97–109, 1996.

68. J Conway. The game of life. *Scientific American*, 223(4): 4, 1970.

69. TJA Craddock, JA Tuszynski, and S Hameroff. Cytoskeletal signaling: Is memory encoded in microtubule lattices by camkii phosphorylation? *PLoS Computational Biology*, 8(3): e1002421, 2012.

70. AH Crevenna, N Naredi-Rainer, DC Lamb, R Wedlich-Söldner, and J Dzubiella. Effects of hofmeister ions on the α-helical structure of proteins. *BioPhysical Journal*, 102(4): 907–915, 2012.

71. AS Davydov. Solitons and energy transfer along protein molecules. *Journal of Theoretical Biology*, 66(2): 379–387, 1977.

72. AS Davydov. Solitons in quasi-one-dimensional molecular structures. *Soviet Physics Uspekhi*, 25(12): 898, 1982.

73. AS Davydov. Solitons, bioenergetics, and the mechanism of muscle contraction. *International Journal of Quantum Chemistry*, 16(1): 5–17, 1979.

74. NG De Bruijn. A combinatorial problem. *Proceedings of the Section of Sciences of the Koninklijke Nederlandse Akademie van Wetenschappen te Amsterdam*, 49(7): 758–764, 1946.

75. D Debanne. Information processing in the axon. *Nature Reviews Neuroscience*, 5(4): 304–316, 2004.

76. L Dehmelt and Shelley Halpain. Actin and microtubules in neurite initiation: Are maps the missing link? *Journal of Neurobiology*, 58(1): 18–33, 2004.

77. P Delgado. Quantum cellular automata: Theory and applications. PhD Thesis, University of Waterloo, 2007.

78. E Derivery and A Gautreau. Generation of branched actin networks: Assembly and regulation of the *n*-wasp and wave molecular machines. *Bioessays*, 32(2): 119–131, 2010.

79. D Désérable. A family of cayley graphs on the hexavalent grid. *Discrete Applied Mathematics*, 93(2-3): 169–189, 1999.

80. D Désérable, P Dupont, M Hellou, and S Kamali-Bernard. Cellular automata in complex matter. *Complex Systems*, 20(1): 67, 2011.

81. C Dillon and Y Goda. The actin cytoskeleton: Integrating form and function at the synapse. *Annual Review of Neuroscience*, 28: 25–55, 2005.

82. M Dowle, RM Mantel, and D Barkley. Fast simulations of waves in three-dimensional excitable media. *International Journal of Bifurcation and Chaos*, 7(11): 2529–2545, 1997.

83. Milos Drutarovsky and Pavol Galajda. A robust chaos-based true random number generator embedded in reconfigurable switched-capacitor hardware. In *2007 17th International Conference Radioelektronika*, pages 1–6. IEEE, 2007.

84. C Durr, H LêThanh, and M Santha. A decision procedure for well-formed linear quantum cellular automata. *arXiv preprint cs/9906024*, 1999.

85. D Eppstein. Growth and decay in life-like cellular automata. *Game of Life cellular automata*, pages 71–97, 2010.

86. R Feynman. Quantum mechanical computers. *Foundations of Physics*, 16(6): 986, 1985.

87. JC Fiala, J Spacek, and KM Harris. Dendritic spine pathology: Cause or consequence of neurological disorders? *Brain Research Reviews*, 39(1): 29–54, 2002.

88. E Fifková and RJ Delay. Cytoplasmic actin in neuronal processes as a possible mediator of synaptic plasticity. *The Journal of Cell Biology*, 95(1): 345–350, 1982.

89. PC Fischer. Generation of primes by a one-dimensional real-time iterative array. *Journal of the ACM (JACM)*, 12(3): 388–394, 1965.

90. R FitzHugh. Impulses and physiological states in theoretical models of nerve membrane. *Biophysical Journal*, 1(6): 445–466, 1961.

91. S Flach and CR Willis. Discrete breathers. *Physics Reports*, 295(5): 181–264, 1998.

92. AS Fokas, EP Papadopoulou, and YG Saridakis. Soliton cellular automata. *Physica D: Nonlinear Phenomena*, 41(3): 297–321, 1990.

93. F Franke, D Jäckel, J Dragas, J Müller, M Radivojevic, D Bakkum, and A Hierlemann. High-density microelectrode array recordings and real-time spike sorting for closed-loop experiments: An emerging technology to study neural plasticity. *Frontiers in Neural Circuits*, 6: 105, 2012.

94. VE Galkin, A Orlova, MR Vos, GF Schröder, and EH Egelman. Near-atomic resolution for one state of F-actin. *Structure*, 23(1): 173–182, 2015.

95. M Gerhardt, H Schuster, and JJ Tyson. A cellular automation model of excitable media including curvature and dispersion. *Science*, 247(4950): 1563–1566, 1990.

96. G Giannone, BJ Dubin-Thaler, O Rossier, Y Cai, O Chaga, G Jiang, W Beaver, H-G Döbereiner, Y Freund, G Borisy, *et al.* Lamellipodial actin mechanically links myosin activity with adhesion-site formation. *Cell*, 128(3): 561–575, 2007.

97. A Gierer and H Meinhardt. A theory of biological pattern formation. *Kybernetik*, 12: 30–39, 1972.

98. C Giraud-Carrier. A reconfigurable data flow machine for implementing functional programming languages. *Sigplan Notices*, 29(9): 22–28, 1994.

99. M Glaser, J Schnauß, T Tschirner, BUS Schmidt, M Moebius-Winkler, JA Käs, and DM Smith. Self-assembly of hierarchically ordered structures in dna nanotube systems. *New Journal of Physics*, 18(5): 055001, 2016.

100. SW Golomb and LR Welch. Grm, and haw, shift register sequences, 1967.

101. Georg A Gottwald and Ian Melbourne. On the implementation of the 0–1 test for chaos. *SIAM Journal on Applied Dynamical Systems*, 8(1): 129–145, 2009.

102. AP Goucher. Universal computation and construction in Gol cellular automata. *Game of Life Cellular Automata*, 505–517, 2010.

103. E Goulielmakis, M Schultze, M Hofstetter, VS Yakovlev, J Gagnon, M Uiberacker, AL Aquila, EM Gullikson, DT Attwood, R Kienberger, et al. Single-cycle nonlinear optics. *Science*, 320(5883): 1614–1617, 2008.

104. P Grassberger. Long-range effects in an elementary cellular automaton. *Journal of Statistical Physics*, 45: 27–39, 1986.

105. J Gravner and D Griffeath. The one-dimensional exactly 1 cellular automaton: replication, periodicity, and chaos from finite seeds. *Journal of Statistical Physics*, 142(1): 168–200, 2011.

106. JM Greenberg and SP Hastings. Spatial patterns for discrete models of diffusion in excitable media. *SIAM Journal on Applied Mathematics*, 34(3): 515–523, 1978.

107. LM Griffith and TD Pollard. Evidence for actin filament-microtubule interaction mediated by microtubule-associated proteins. *The Journal of Cell Biology*, 78(3): 958–965, 1978.

108. J Gruska. Descriptional complexity issues in quantum computing. *Journal of Automata, Languages and Combinatorics*, 5(3): 191–218, 2000.

109. HA Gutowitz and JD Victor. Local structure theory in more than one dimension. *Complex Systems*, 1(57): 8, 1987.

110. PR Halmos. The basic concepts of algebraic logic. *The American Mathematical Monthly*, 63(6): 363–387, 1956.

111. SR Hameroff. Quantum computation in brain microtubules? The penrose-hameroff 'orch or' model of consciousness. *Philosophical Transactions of the Royal Society of London A*, 356: 1869–1896, 1998.

112. SR Hameroff. Coherence in the cytoskeleton: Implications for biological information processing. In *Biological Coherence and Response to External Stimuli*, pp. 242–265. Springer, 1988.

113. SR Hameroff and S Rasmussen. Microtubule automata: Sub-neural information processing in biological neural networks. In *Molecular Electronics: Biosensors and Biocomputers*, pp. 243–257. Boston, MA: Springer US, 1989.

114. SR Hameroff, Judith E. JE Dayhoff, R Lahoz-Beltra, AV Samsonovich, and Steen Rasmussen. Models for molecular computation: Conformational automata in the cytoskeleton. *Computer*, 25(11): 30–39, 1992.

115. SR Hameroff and S Rasmussen. Information processing in microtubules: Biomolecular automata and nanocomputers. In *Molecular Electronics*, pp. 243–257. Springer, 1989.

116. S Harding, J Koutnik, K Greff, J Schmidhuber, and A Adamatzky. Discovering Boolean gates in slime mould. In *Inspired by Nature – Emergence, Complexity and Computation*, Vol. 28, S. Stepney and A. Adamatzky (Eds.) (2018), pp. 323–337.

117. S Harding, J Koutník, J Schmidhuber, and A Adamatzky. Discovering Boolean gates in slime mould. In *Inspired by Nature*, pp. 323–337. Springer, 2018.

118. S Harding and JF Miller. Evolution in materio: Evolving logic gates in liquid crystal. *International Journal of Unconventional Computing*, 3(4): 243–257, 2007.

119. JH Hartwig, J Tyler, and TP Stossel. Actin-binding protein promotes the bipolar and perpendicular branching of actin filaments. *The Journal of Cell Biology*, 87(3): 841–848, 1980.

120. Wei-Chiang Hong, Yucheng Dong, Wen Yu Zhang, Li-Yueh Chen, and BK Panigrahi. Cyclic electric load forecasting by seasonal SVR with chaotic genetic algorithm. *International Journal of Electrical Power & Energy Systems*, 44(1): 604–614, 2013.

121. JE Hopcroft and JD Ullman. *Introduction to Automata Theory, Languages and Computation*. Adison-Wesley, Reading, MA., 1979.

122. F Huber, J Schnauß, S Rönicke, P Rauch, K Müller, C Fütterer, and JA Käs. Emergent complexity of the cytoskeleton: From single filaments to tissue. *Advances in Physics*, 62(1): 1–112, 2013.

123. F Huber, D Strehle, and J Käs. Counterion-induced formation of regular actin bundle networks. *Soft Matter*, 8: 931–936, 2012.

124. F Huber, D Strehle, J Schnauß, and J Käs. Formation of regularly spaced networks as a general feature of actin bundle condensation by entropic forces. *New Journal of Physics*, 17(4): 043029, 2015.

125. WTS Huck. Responsive polymers for nanoscale actuation. *Materials Today*, 11(7-8): 24–32, 2008.

126. C Hunley, D Uribe, and M Marucho. A multi-scale approach to describe electrical impulses propagating along actin filaments in both intracellular and *in vitro* conditions. *RSC Advances*, 8(22): 12017–12028, 2018.

127. JD Hunter. Matplotlib: A 2d graphics environment. *Computing in Science & Engineering*, 9(3): 90–95, 2007.

128. Y Hwang and AI Barakat. Dynamics of mechanical signal transmission through prestressed stress fibers. *PLoS One*, 7(4): e35343, 2012.

129. S Inokuchi and Y Mizoguchi. Generalized partitioned quantum cellular automata and quantization of classical CA. *International Journal of Unconventional Computing*, 1: 149–160, 2005.

130. JA Tuszynski, MV Sataric, DL Sekulic, BM Sataric, and S Zdravkovic Nonlinear calcium ion waves along actin filaments control active hair–bundle motility. *bioxiv.org*, doi.org/10.1101/292292, 2018.

131. L Jaeken. A new list of functions of the cytoskeleton. *IUBMB Life*, 59(3): 127–133, 2007.

132. MH Jakubowski, K Steiglitz, and R Squier. Information transfer between solitary waves in the saturable schrödinger equation. *Physical Review E*, 56(6): 7267, 1997.

133. MH Jakubowski, K Steiglitz, and R Squier. State transformations of colliding optical solitons and possible application to computation in bulk media. *Physical Review E*, 58(5): 6752, 1998.

134. MH Jakubowski, K Steiglitz, and R Squier. Computing with solitons: A review and prospectus. In A Adamatzky (ed.), *Collision-based Computing*, pp. 277–297. Springer, 2002.

135. MH Jakubowski, K Steiglitz, and RK Squier. When can solitons compute? *Complex Systems*, 10(1): 1–22, 1996.

136. PA Janmey. The cytoskeleton and cell signalling: Component localization and mechanical coupling. *Physiological Reviews*, 78: 763–781, 1998.

137. PA Janmey, JX Tang, and CF Schmidt. Actin filaments. *Supramolecular Assemblies*, 81, 2001.

138. F Jelínek, M Cifra, J Pokorný, J Vaniš, J Šimša, J Hašek, and I Frýdlová. Measurement of electrical oscillations and mechanical vibrations of yeast cells membrane around 1 khz. *Electromagnetic Biology and Medicine*, 28(2): 223–232, 2009.

139. M Jibu, S Hagan, SR Hameroff, KH Pribram, and K Yasue. Quantum optical coherence in cytoskeletal microtubules: Implications for brain function. *Biosystems*, 32(3): 195–209, 1994.

140. W. Jin, F Chen, and C Yang. Topological chaos of cellular automata rules. In *2009 International Workshop on Chaos-Fractals Theories and Applications*, pp. 216–220. IEEE, 2009.

141. C Joachim, JK Gimzewski, and A Aviram. Electronics using hybrid-molecular and mono-molecular devices. *Nature*, 408(6812): 541–548, 2000.

142. L Kavitha, A Muniyappan, A Prabhu, S Zdravković, S Jayanthi, and D Gopi. Nano breathers and molecular dynamics simulations in hydrogen-bonded chains. *Journal of Biological Physics*, 39(1): 15–35, 2013.

143. C-H Kim and JE Lisman. A role of actin filament in synaptic transmission and long-term potentiation. *The Journal of Neuroscience*, 19(11): 4314–4324, 1999.

144. OG Kisselev, J Kao, JW Ponder, YC Fann, N Gautam, and GR Marshall. Light-activated rhodopsin induces structural binding motif in g protein α subunit. *Proceedings of the National Academy of Sciences*, 95(8): 4270–4275, 1998.

145. N Kojima and T Shirao. Synaptic dysfunction and disruption of postsynaptic drebrin–actin complex: A study of neurological disorders accompanied by cognitive deficits. *Neuroscience Research*, 58(1): 1–5, 2007.

146. A Kondacs and J Watrous. On the power of quantum finite state automata. In *Foundations of Computer Science, 1997. Proceedings 38th Annual Symposium on*, pp. 66–75. IEEE, 1997.

147. F Korobova and T Svitkina. Molecular architecture of synaptic actin cytoskeleton in hippocampal neurons reveals a mechanism of dendritic spine morphogenesis. *Molecular Biology of the Cell*, 21(1): 165–176, 2010.

148. A Kotsialos, MK Massey, F Qaiser, DA Zeze, C Pearson, and MC Petty. Logic gate and circuit training on randomly dispersed carbon nanotubes. *International Journal of Unconventional Computing*, 10(5-6): 473–497, 2014.

149. N Kovaleva and L Manevitch. Analytical study of discrete optical breathers in spiral polymer chain. *Macromolecular Theory and Simulations*, 21(8): 516–528, 2012.

150. M Krész. Soliton automata with constant external edges. *Information and Computation*, 206(9): 1126–1141, 2008.

151. O Kučera, D Havelka, and M Cifra. Vibrations of microtubules: Physics that has not met biology yet. *Wave Motion*, 72: 13–22, 2017.

152. R Lahoz-Beltra, SR Hameroff, and JE Dayhoff. Cytoskeletal logic: A model for molecular computation via Boolean operations in microtubules and microtubule-associated proteins. *Biosystems*, 29(1): 1–23, 1993.

153. J Lechleiter, S Girard, E Peralta, and D Clapham. Spiral calcium wave propagation and annihilation in xenopus laevis oocytes. *Science*, 252(5002): 123–126, 1991.

154. EC Lin and HF Cantiello. A novel method to study the electrodynamic behavior of actin filaments. Evidence for cable-like properties of actin. *Biophysical Journal*, 65(4): 1371, 1993.

155. AV Liopo, MP Stewart, J Hudson, JM Tour, and TC Pappas. Biocompatibility of native and functionalized single-walled carbon nanotubes for neuronal interface. *Journal of Nanoscience and Nanotechnology*, 6(5): 1365–1374, 2006.

156. JS Lorenz, J Schnauß, M Glaser, M Sajfutdinow, C Schuldt, JA Käs, and DM Smith. Synthetic transient crosslinks program the mechanics of soft, biopolymer' based materials. *Advanced Materials*, 30(13): 1706092, 2018.

157. René Lozi. Emergence of randomness from chaos. *International Journal of Bifurcation and Chaos*, 22(02): 1250021, 2012.

158. B Ludin and A Matus. The neuronal cytoskeleton and its role in axonal and dendritic plasticity. *Hippocampus*, 3(S1): 61–71, 1993.

159. Z-S Ma, J Wang, and H Guo. Weakly nonlinear ac response: Theory and application. *Physical Review B*, 59(11): 7575, 1999.

160. GS Manning. The molecular theory of polyelectrolyte solutions with applications to the electrostatic properties of polynucleotides. *Quarterly Reviews of Biophysics*, 11(2): 179–246, 1978.

161. S Marcovitch and B Reznik. Entanglement of solitons in the frenkel-kontorova model. *arxiv: 0809.1857*, 2009.

162. M Markus and B Hess. Isotropic cellular automaton for modelling excitable media. *Nature*, 347(6288): 56, 1990.

163. GJ Martinez, A Adamatzky, and R Alonso-Sanz. Complex dynamics of elementary cellular automata emerging from chaotic rules. *International Journal of Bifurcation and Chaos*, 22(2): 1250023, 2012.

164. GJ Martinez, A Adamatzky, and R Alonso-Sanz. Designing complex dynamics in cellular automata with memory. *International Journal of Bifurcation and Chaos*, 23(10): 1330035, 2013.

165. GJ Martínez, A Adamatzky, B Chen, F Chen, and JC Seck-Tuoh-Mora. Simple networks on complex cellular automata: From de bruijn diagrams to jump-graphs. In *Evolutionary Algorithms, Swarm Dynamics and Complex Networks: Methodology, Perspectives and Implementation*, pp. 241–264, 2018.

166. GJ Martinez, A Adamatzky, F Chen, and L Chua. On soliton collisions between localizations in complex elementary cellular automata: Rules 54 and 110 and beyond. http://www.complex-systems.com/abstracts/v23_i0 3_a04.html, 2013.

167. GJ Martínez, A Adamatzky, and HV McIntosh. Computing with virtual cellular automata collider. In *Science and Information Conference (SAI), 2015*, pp. 62–68. IEEE, 2015.

168. GJ Martínez, A Adamatzky, JC Seck-Tuoh-Mora, and R Alonso-Sanz. How to make dull cellular automata complex by adding memory: Rule 126 case study. *Complexity*, 15(6): 34–49, 2010.

169. GJ Martínez, A Adamatzky, CR Stephens, and AF Hoeflich. Cellular automaton supercolliders. *International Journal of Modern Physics C*, 22(04): 419–439, 2011.

170. GJ Martínez, A Adamatzky, and R Alonso-Sanz. On the dynamics of cellular automata with memory. *Fundamenta Informaticae*, 138(1-2): 1–16, 2015.

171. GJ Martínez, A Adamatzky, and HV McIntosh. Phenomenology of glider collisions in cellular automaton rule 54 and associated logical gates. *Chaos, Solitons & Fractals*, 28(1): 100–111, 2006.

172. M Matsumoto and T Nishimura. Mersenne twister: A 623-dimensionally equidistributed uniform pseudo-random number generator. *ACM Transactions on Modeling and Computer Simulation (TOMACS)*, 8(1): 3–30, 1998.

173. R Mayne and A Adamatzky. The physarum polycephalum actin network: Formalisation, topology and morphological correlates with computational ability. In *Proceedings of the 8th International Conference on Bioinspired Information and Communications Technologies*, pp. 87–94. ICST (Institute for Computer Sciences, Social-Informatics and Telecommunications Engineering), 2014.

174. R Mayne and A Adamatzky. On the computing potential of intracellular vesicles. *PloS one*, 10(10): e0139617, 2015.

175. R Mayne, A Adamatzky, and J Jones. On the role of the plasmodial cytoskeleton in facilitating intelligent behaviour in slime mould physarum polycephalum. *Communicative and Integrative Biology*, 7(1), 2014.

176. AB McGeachie, LA Cingolani, and Y Goda. A stabilising influence: Integrins in regulation of synaptic plasticity. *Neuroscience Research*, 70(1): 24–29, 2011.

177. HV McIntosh. Wolfram's class iv automata and a good life. *Physica D: Nonlinear Phenomena*, 45(1–3): 105–121, 1990.

178. HV McIntosh. *Linear Cellular Automata via de Bruijn Diagrams*, 1991.

179. HV McIntosh. *One Dimensional Cellular Automata*. Luniver Press, 2009.

180. B Mehrafrooz and A Shamloo. Mechanical differences between ATP and ADP actin states: A molecular dynamics study. *Journal of Theoretical Biology*, 448: 94–103, 2018.

181. MA Mikati, EE Grintsevich, and E Reisler. Drebrin-induced stabilization of actin filaments. *Journal of Biological Chemistry*, 288(27): 19926–19938, 2013.

182. JF Miller, SL Harding, and G Tufte. Evolution-in-materio: Evolving computation in materials. *Evolutionary Intelligence*, 7(1): 49–67, 2014.

183. PB Moore, HE Huxley, and DJ DeRosier. Three-dimensional reconstruction of F-actin, thin filaments and decorated thin filaments. *Journal of Molecular Biology*, 50(2): 279–292, 1970.

184. RD Mullins, JA Heuser, and TD Pollard. The interaction of arp2/3 complex with actin: Nucleation, high affinity pointed end capping, and formation of branching networks of filaments. *Proceedings of the National Academy of Sciences*, 95(11): 6181–6186, 1998.

185. Y Nabekawa, T Okino, and K Midorikawa. Probing attosecond dynamics of molecules by an intense a-few-pulse attosecond pulse train. In *31st International Congress on High-Speed Imaging and Photonics*, pp. 103280B–103280B. International Society for Optics and Photonics, 2017.

186. J Nagumo, S Arimoto, and S Yoshizawa. An active pulse transmission line simulating nerve axon. *Proceedings of the IRE*, 50(10): 2061–2070, 1962.

187. S Ninagawa and A Adamatzky. Classifying elementary cellular automata using compressibility, diversity and sensitivity measures. *International Journal of Modern Physics C*, 25(03): 1350098, 2014.

188. T Oda, M Iwasa, T Aihara, Y Maéda, and A Narita. The nature of the globular-to fibrous-actin transition. *Nature*, 457(7228): 441–445, 2009.

189. K Shankar, S Paladini, and JA Tuszynski. Measurement and characterization of the electrical properties of actin filaments. 2023.

190. K Palczewski, T Kumasaka, T Hori, CA Behnke, H Motoshima, BA Fox, IL Trong, DC Teller, T Okada, RE Stenkamp, *et al.* Crystal structure of rhodopsin: Ag protein-coupled receptor. *Science*, 289(5480): 739–745, 2000.

191. CT Pappas, PA Krieg, and CC Gregorio. Nebulin regulates actin filament lengths by a stabilization mechanism. *The Journal of Cell Biology*, 189(5): 859–870, 2010.

192. JK Park, K Steiglitz, and WP Thurston. Soliton-like behavior in automata. *Physica D: Nonlinear Phenomena*, 19(3): 423–432, 1986.

193. AM Persico and T Bourgeron. Searching for ways out of the autism maze: Genetic, epigenetic and environmental clues. *Trends in Neurosciences*, 29(7): 349–358, 2006.

194. KJ Persohn and Richard J Povinelli. Analyzing logistic map pseudorandom number generators for periodicity induced by finite precision floating-point representation. *Chaos, Solitons & Fractals*, 45(3): 238–245, 2012.

195. AM Pertsov, JM Davidenko, R Salomonsz, WT Baxter, and J Jalife. Spiral waves of excitation underlie reentrant activity in isolated cardiac muscle. *Circulation Research*, 72(3): 631–650, 1993.

196. DS Peterka, H Takahashi, and R Yuste. Imaging voltage in neurons. *Neuron*, 69(1): 9–21, 2011.

197. V Petit and J-P Thiery. Focal adhesions: Structure and dynamics. *Biology of the Cell*, 92(7): 477–494, 2000.

198. M Pluhacek, R Senkerik, I Zelinka, and D Davendra. Chaos pso algorithm driven alternately by two different chaotic maps—an initial study. In *2013 IEEE Congress on Evolutionary Computation*, pp. 2444–2449. IEEE, 2013.

199. J Pokorný. Excitation of vibrations in microtubules in living cells. *Bioelectrochemistry*, 63(1-2): 321–326, 2004.

200. J Pokorný, F Jelínek, V Trkal, I Lamprecht, and R Hölzel. Vibrations in microtubules. *Journal of Biological Physics*, 23(3): 171–179, 1997.

201. TD Pollard and MS Mooseker. Direct measurement of actin polymerization rate constants by electron microscopy of actin filaments nucleated by isolated microvillus cores. *The Journal of Cell Biology*, 88(3): 654–659, 1981.

202. PP Pompa, S Sabella, R Rinaldi, R Cingolani, and F Calabi. Method and a microdevice for the identification and/or quantification of an analyte in a biological sample, December 10 2009. US Patent App. 12/518,816.

203. A Priel and JA Tuszyński. A nonlinear cable-like model of amplified ionic wave propagation along microtubules. *EPL (Europhysics Letters)*, 83(6): 68004, 2008.

204. A Priel, AJ Ramos, JA Tuszynski, and HF Cantiello. A biopolymer transistor: Electrical amplification by microtubules. *Biophysical Journal*, 90(12): 4639–4643, 2006.

205. A Priel, JA Tuszynski, and HF Cantiello. Electrodynamic signaling by the dendritic cytoskeleton: Toward an intracellular information processing model. *Electromagnetic Biology and Medicine*, 24(3): 221–231, 2005.

206. A Priel, JA Tuszynski, and HF Cantiello. Ionic waves propagation along the dendritic cytoskeleton as a signaling mechanism. *Advances in Molecular and Cell Biology*, 37: 163–180, 2006.

207. A Priel, JA Tuszynski, and HF Cantiello. The dendritic cytoskeleton as a computational device: An hypothesis. In *The Emerging Physics of Consciousness*, pp. 293–325. Springer, 2006.

208. A Priel, JA Tuszynski, and NJ Woolf. Neural cytoskeleton capabilities for learning and memory. *Journal of Biological Physics*, 36(1): 3–21, 2010.

209. D Rand, K Steiglitz, and Paul R Prucnal. Signal standardization in collision-based soliton computing. *International Journal of Unconventional Computing*, 1: 31, 2005.

210. S Rasmussen, H Karampurwala, R Vaidyanath, KS Jensen, and S Hameroff. Computational connectionism within neurons: A model of cytoskeletal automata subserving neural networks. *Physica D: Nonlinear Phenomena*, 42(1-3): 428–449, 1990.

211. M Redeker, A Adamatzky, and GJ Martìnez. Expressiveness of elementary cellular automata. *International Journal of Modern Physics C*, 24(3): 1350010, 2013.

212. F Rodríguez-Henríquez, NA Saqib, AD Pérez, and CK Koc. *Cryptographic Algorithms on Reconfigurable Hardware*. Springer Science & Business Media, 2007.

213. JM Rogers and AD McCulloch. A collocation-Galerkin finite element model of cardiac action potential propagation. *IEEE Transactions on Biomedical Engineering*, 41(8): 743–757, 1994.

214. LJ Sampson, ML Leyland, and C Dart. Direct interaction between the actin-binding protein filamin-a and the inwardly rectifying potassium channel, kir2. 1. *Journal of Biological Chemistry*, 278(43): 41988–41997, 2003.

215. E Sapin, L Bull, and A Adamatzky. Genetic approaches to search for computing patterns in cellular automata. *Computational Intelligence Magazine, IEEE*, 4(3): 20–28, 2009.

216. MV Satarić, N Bednar, BM Satarić, and G Stojanović. Actin filaments as nonlinear RLC transmission lines. *International Journal of Modern Physics B*, 23(22): 4697–4711, 2009.

217. MV Satarić, DI Ilić, N Ralević, and Jack Adam Tuszynski. A nonlinear model of ionic wave propagation along microtubules. *European Biophysics Journal*, 38(5): 637–647, 2009.

218. MV Satarić and BM Satarić. Ionic pulses along cytoskeletal protophilaments. In *Journal of Physics: Conference Series*, 329, 012009, 2011.

219. MV Satarić, D Sekulić, and M Živanov. Solitonic ionic currents along microtubules. *Journal of Computational and Theoretical Nanoscience*, 7(11): 2281–2290, 2010.

220. BEH Saxberg and RJ Cohen. Cellular automata models of cardiac conduction. In *Theory of Heart*, pp. 437–476. Springer, 1991.

221. K Schmidt-Nielsen. *Animal Physiology: Adaptation and Environment*. Cambridge University Press, 1997.

222. J Schnauß, T Golde, C Schuldt, BUS Schmidt, M Glaser, D Strehle, T Händler, C Heussinger, and JA. Käs. Transition from a linear to a harmonic potential in collective dynamics of a multifilament actin bundle. *Physical Review Letters*, 116(10): 108102, 2016.

223. J Schnauß, T Händler, and JA Käs. Semiflexible biopolymers in bundled arrangements. *Polymers*, 8(8): 274, 2016.

224. T Schubert and A Akopian. Actin filaments regulate voltage-gated ion channels in salamander retinal ganglion cells. *Neuroscience*, 125(3): 583–590, 2004.

225. B Schumacher and RF Werner. Reversible quantum cellular automata. *arxiv*, quant-ph(0405174), 2004.

226. JC Seck-Tuoh-Mora, J Medina-Marin, N Hernandez-Romero, GJ Martinez, and I Barragan-Vite. Welch sets for random generation and representation of reversible one-dimensional cellular automata. *Information Sciences*, 382: 81–95, 2017.

227. Y Sekino, N Kojima, and T Shirao. Role of actin cytoskeleton in dendritic spine morphogenesis. *Neurochemistry International*, 51(2): 92–104, 2007.

228. DL Sekulić, BM Satarić, JA Tuszynski, and MV Satarić. Nonlinear ionic pulses along microtubules. *The European Physical Journal E*, 34(5): 49, 2011.

229. PM Shaibani, K Jiang, G Haghighat, M Hassanpourfard, H Etayash, S Naicker, and T Thundat. The detection of *Escherichia coli (E. coli)* with the pH sensitive hydrogel nanofiber-light addressable potentiometric sensor (nf-laps). *Sensors and Actuators B: Chemical*, 226: 176–183, 2016.

230. NWS Kam, TC Jessop, PA Wender, and H Dai. Nanotube molecular transporters: Internalization of carbon nanotube-protein conjugates into mammalian cells. *Journal of the American Chemical Society*, 126(22): 6850–6851, 2004.

231. S Siccardi and A Adamatzky. Actin quantum automata: Communication and computation in molecular networks. *Nano Communication Networks*, 6(1): 15–27, 2015.

232. S Siccardi and A Adamatzky. Logical gates implemented by solitons at the junctions between one-dimensional lattices. *International Journal of Bifurcation and Chaos*, 26(6): 1650107, 2016.

233. S Siccardi and A Adamatzky. Quantum actin automata and three-valued logics. *IEEE Journal on Emerging and Selected Topics in Circuits and Systems*, 6(1): 53–61, 2016.

234. S Siccardi and A Adamatzky. Models of computing on actin filaments. In *Advances in Unconventional Computing*, pp. 309–346. Springer, 2017.

235. S Siccardi, JA Tuszynski, and A Adamatzky. Boolean gates on actin filaments. *arXiv preprint arXiv:1506.09044*, 2015.

236. W Sierpiński. On a curve every point of which is a point of ramification. *Prace Matematyczno-Fizyczne (In Polish)*, 27: 77–86, 1916.

237. S Sinha. Phase transitions in the computational complexity of "elementary" cellular automata. In *Unifying Themes in Complex Systems: New Research Volume IIIB. Proceedings from the Third International Conference on Complex Systems*, pp. 337–348. Springer, 2006.

238. P Siregar, JP Sinteff, N Julen, and P Le Beux. An interactive 3D anisotropic cellular automata model of the heart. *Computers and Biomedical Research*, 31(5): 323–347, 1998.

239. Yuri M Sirenko, Michael A Stroscio, and KW Kim. Dynamics of cytoskeletal filaments. *Physical Review E*, 54(2): 1816, 1996.

240. J Southgate, P Harnden, and LK Trejdosiewicz. A review of the biological and diagnostic implications. *Histol Histopathol*, 14: 657–664, 1999.

241. Micha E Spira and Aviad Hai. Multi-electrode array technologies for neuroscience and cardiology. *Nature Nanotechnology*, 8(2): 83, 2013.

242. JA Spudich, HE Huxley, and JT Finch. Regulation of skeletal muscle contraction: II. Structural studies of the interaction of the tropomyosin-troponin complex with actin. *Journal of Molecular Biology*, 72(3): 619–632, 1972.

243. FF Spukti and J Schnauß. Large and stable: Actin aster networks formed via entropic forces. *Frontiers in Chemistry*, 951, 2022.

244. RK Squier and K Steiglitz. Programmable parallel arithmetic in cellular automata using a particle model. *Complex Systems*, 8(5): 311–324, 1994.

245. K Steiglitz. Time-gated manakov spatial solitons are computationally universal. *Physical Review E*, 63(1): 016608, 2000.

246. K Steiglitz, I Kamal, and A Watson. Embedding computation in one-dimensional automata by phase coding solitons. *Computers, IEEE Transactions on*, 37(2): 138–145, 1988.

247. D Strehle, J Schnauß, C Heussinger, J Alvarado, M Bathe, JA Käs, and B Gentry. Transiently crosslinked F-actin bundles. *European Biophysics Journal*, 40(1): 93–101, 2011.

248. H Stuart. Quantum computation in brain microtubules? The penrose–hameroff 'orch or 'model of consciousness. *Philosophical Transactions of the Royal Society of London. Series A: Mathematical, Physical and Engineering Sciences*, 356(1743): 1869–1896, 1998.

249. Yang Sun, Lingbo Zhang, and Xingsheng Gu. A hybrid co-evolutionary cultural algorithm based on particle swarm optimization for solving global optimization problems. *Neurocomputing*, 98: 76–89, 2012.

250. D Takahashi and J Satsuma. A soliton cellular automaton. *Journal of the Physical Society of Japan*, 59(10): 3514–3519, 1990.

251. T Toffoli. Reversible computing. In *International Colloquium on Automata, Languages, and Programming*, pp. 632–644. Springer, 1980.

252. R Toth, C Stone, A Adamatzky, BDL Costello, and L Bull. Dynamic control and information processing in the Belousov–Zhabotinsky reaction using a coevolutionary algorithm. *The Journal of Chemical Physics*, 129(18): 184708, 2008.

253. R Toth, C Stone, BDL Costello, A Adamatzky, and L Bull. Simple collision-based chemical logic gates with adaptive computing. *International Journal of Nanotechnology and Molecular Computation (IJNMC)*, 1(3): 1–16, 2009.

254. JA Tuszyński, JA Brown, and P Hawrylak. Dielectric polarization, electrical conduction, information processing and quantum computation in microtubules. Are they plausible? *Philosophical Transactions of the Royal Society of London. Series A: Mathematical, Physical and Engineering Sciences*, 356(1743): 1897–1926, 1998.

255. JA Tuszyński, S Portet, JM Dixon, C Luxford, and HF Cantiello. Ionic wave propagation along actin filaments. *Biophysical Journal*, 86(4): 1890–1903, 2004.

256. JA Tuszynski, A Priel, JA Brown, HF Cantiello, and JM Dixon. Electronic and ionic conductivities of microtubules and actin filaments, their consequences for cell signaling and applications to bioelectronics. *Nano and Molecular Electronics Handbook*, 9, 2007.

257. J Tuszyński, S Portet, and J Dixon. Nonlinear assembly kinetics and mechanical properties of biopolymers. *Nonlinear Analysis: Theory, Methods & Applications*, 63(5-7): 915–925, 2005.

258. PB Umbanhowar, F Melo, and HL Swinney. Localized excitations in a vertically vibrated granular layer. *Nature*, 382(6594): 793–796, 1996.

259. M van Spronsen and CC Hoogenraad. Synapse pathology in psychiatric and neurologic disease. *Current Neurology and Neuroscience Reports*, 10(3): 207–214, 2010.

260. AE Van Woerkom. The major hallucinogens and the central cytoskeleton: An association beyond coincidence? Towards sub-cellular mechanisms in schizophrenia. *Medical Hypotheses*, 31(1): 7–15, 1990.

261. MG Velarde, W Ebeling, and AP Chetverikov. On the possibility of electric conduction mediated by dissipative solitons. *International Journal of Bifurcation and Chaos*, 15(1): 245–251, 2005.

262. MG Velarde, W Ebeling, AP Chetverikov, and D Hennig. Electron trapping by solutions. Classical versus quantum mechanical approach. *International Journal of Bifurcation and Chaos*, 18(2): 521–526, 2008.

263. MG Velarde, W Ebeling, D Hennig, and C Neißner. On soliton-mediated fast electric conduction in a nonlinear lattice with morse interactions. *International Journal of Bifurcation and Chaos*, 16(4): 1035–1039, 2006.

264. P Virtanen, R Gommers, TE Oliphant, M Haberland, T Reddy, D Cournapeau, E Burovski, P Peterson, W Weckesser, J Bright, SJ van der Walt, M Brett, J Wilson, KJ Millman, N Mayorov, ARJ Nelson, E Jones, R Kern, E Larson, CJ Carey, İ Polat, Y Feng, EW Moore, JV erPlas, D Laxalde, J Perktold, R Cimrman, I Henriksen, EA Quintero, CR Harris, AM Archibald, AH Ribeiro, F Pedregosa, P van Mulbregt, and SciPy 1. 0 Contributors. SciPy 1.0–fundamental algorithms for scientific computing in python. *arXiv e-prints*, arXiv:1907.10121, Jul 2019.

265. J von Neumann. Re-evaluation of the problems of complicated automata problems of hierarchy and evolution. *Theory of Selfreproducing Automata*, pp. 74–87, 1949.

266. Burton H Voorhees. *Computational Analysis of One-dimensional Cellular Automata*, Vol. 15. World Scientific, 1996.

267. A Waksman. An optimum solution to the firing squad synchronization problem. *Information and Control*, 9(1): 66–78, 1966.

268. BG Wang, X Zhao, J Wang, and H Guo. Nonlinear quantum capacitance. *Applied Physics Letters*, 74(19): 2887–2889, 1999.

269. B Wang, X Zhao, J Wang, and H Guo. Nonlinear quantum capacitance. *Applied Physics Letters*, 74(19): 2887–2889, 1999.

270. J Watrous. On one-dimensional quantum cellular automata. In *Foundations of Computer Science, 1995. Proceedings, 36th Annual Symposium on*, pp. 528–537. IEEE, 1995.

271. AM Weaver, AV Karginov, AW Kinley, SA Weed, Y Li, JT Parsons, and JA Cooper. Cortactin promotes and stabilizes arp2/3-induced actin filament network formation. *Current Biology*, 11(5): 370–374, 2001.

272. JR Weimar, JJ Tyson, and LT Watson. Diffusion and wave propagation in cellular automaton models of excitable media. *Physica D: Nonlinear Phenomena*, 55(3-4): 309–327, 1992.

273. SJ Winder and KR Ayscough. Actin-binding proteins. *Journal of Cell Science*, 118(4): 651–654, 2005.

274. S Wolfram. Statistical mechanics of cellular automata. *Reviews of Modern Physics*, 55(3): 601, 1983.

275. S Wolfram. Cellular automata as models of complexity. *Nature*, 311(5985): 419–424, 1984.

276. S Wolfram. Twenty problems in the theory of cellular automata. *Physica Scripta*, 1985(T9): 170, 1985.

277. A Wuensche, M Lesser, *et al*. *Global Dynamics of Cellular Automata: An Atlas of Basin of Attraction Fields of One-dimensional Cellular Automata*, Vol. 1., 1992.

278. J Xu, Y Tseng, and D Wirtz. Strain hardening of actin filament networks regulation by the dynamic cross-linking protein α-actinin. *Journal of Biological Chemistry*, 275(46): 35886–35892, 2000.

279. P Ye, E Entcheva, R Grosu, and SA Smolka. Efficient modeling of excitable cells using hybrid automata. In *Proceedings of CMSB*, Vol. 5, pp. 216–227, 2005.

280. JG Zabolitzky. Critical properties of rule 22 elementary cellular automata. *Journal of Statistical Physics*, 50: 1255–1262, 1988.

281. Ivan Zelinka, Sergej Celikovskỳ, Hendrik Richter, and Guanrong Chen. *Evolutionary algorithms and chaotic systems*, Vol. 267. Springer, 2010.

282. H Zenil and E Villarreal-Zapata. Asymptotic behavior and ratios of complexity in cellular automata. *International Journal of Bifurcation and Chaos*, 23(09): 1350159, 2013.

283. L Zhang and A Adamatzky. Collision-based implementation of a two-bit adder in excitable cellular automaton. *Chaos, Solitons & Fractals*, 41(3): 1191–1200, 2009.

284. P Zhong, M Martonosi, P Ashar, and S Malik. Solving Boolean satisfiability with dynamic hardware configurations. In *International Workshop on Field Programmable Logic and Applications*, pp. 326–335. Springer, 1998.

285. D Ciucci and D Dubois. Three-valued logics for incomplete information and epistemic logic. In *Logics in Artificial Intelligence*, pages 147–159. Springer, 2012.

286. D Ciucci and D Dubois. From paraconsistent three-valued logics to multiple-source epistemic logic. In *8th Conference European Society for Fuzzy Logic and Technology (EUSFLAT-13)*. Atlantis Press, 2013.

287. S Stefano and A Andrew. Quantum actin automata and three-valued logics. *IEEE Journal on Emerging and Selected Topics in Circuits and Systems*, IEEE, 6(1): 53–61, 2016.

288. J Watrous. On one-dimensional quantum cellular automata. In *Foundations of Computer Science, 1995. Proceedings, 36th Annual Symposium on*, pages 528–537. IEEE, 1995.

289. A Kondacs and J Watrous. On the power of quantum finite state automata. In *Foundations of Computer Science, 1997. Proceedings, 38th Annual Symposium on*, pages 66–75. IEEE, 1997.

290. J Gruska. *Quantum Computing* (Springer, 1999).

291. C Boutilier. Modal logics for qualitative possibility and beliefs. In *Proceedings of the 8th International Conference on Uncertainty in Artificial Intelligence*, pages 17–24. Morgan Kaufmann Publishers Inc., 1992.

292. D Dubois and H Prade. Possibilistic logic: a retrospective and prospective view. *Fuzzy Sets and Systems*, 144(1):3–23, 2004.

293. G Benenti, G Casati and G Strini. Principles of Quantum Computation and Information. Vol. I-II. World Scientific, 2007.

294. M Fisch and A Turquette. Peirce's triadic logic. *Trans of the Charles S. Peirce Society*, pages 71–85, 1966.

295. J Łukasiewicz. On three-valued logic. *The Polish Review*, pages 43–44, 1968.

296. AI Arruda. On the imaginary logic of N. A. Vasil'ev. *Studies in Logic and the Foundations of Mathematics*, 89:3–24, 1977.

297. B Russell and H MacColl. The existential import of propositions. *Mind*, pages 398–402, 1905.

298. DA Bochvar and M Bergmann. On a three-valued logical calculus and its application to the analysis of the paradoxes of the classical extended functional calculus. *History and Philosophy of Logic*, 2(1–2): 87–112, 1981.

299. SC Kleene. *Introduction to Metamathematics*. North-Holland, 1952.

300. ND Belnap Jr. A useful four-valued logic. In *Modern Uses of Multiple-Valued Logic*, pages 5–37. Springer, 1977.

301. WS McCulloch and W. Pitts. A logical calculus of the ideas immanent in nervous activity. *The Bull Math Bioph*, 5(4):115–133, 1943.

302. SA Kauffman. Metabolic stability and epigenesis in randomly constructed genetic nets. *Journal Theor Biology*, 22(3):437–467, 1969.

303. RV Solé and B. Luque. Phase transitions and antichaos in generalized Kauffman networks. *Physics Letters A*, 196(5):331–334, 1995.

304. A Wuensche. The ghost in the machine. In *Santa Fe Institute Studies in the Sciences of Complexity*, Vol. 17, pages 465–465. Addison-Wesley Publishing, 1994.

305. A Wuensche. Attractor basins of discrete networks. *Cognitive Science Research Paper*, 461, 1997.

306. W Kinsner. Complexity and its measures in cognitive and other complex systems. In *Cognitive Informatics, 2008. ICCI 2008. 7th IEEE International Conference on*, pages 13–29. IEEE, 2008.

307. A Muthukrishnan and CR Stroud Jr. Multivalued logic gates for quantum computation. *Physical Review A*, 62(5):052309, 2000.

308. I Lebar Bajec, N Zimic, and M Mraz. Towards the bottom-up concept: Extended quantum-dot cellular automata. *Microelectronic Engineering*, 83(4):1826–1829, 2006.

309. IL Bajec and M Mraz. Multi-valued logic based on quantum-dot cellular automata. *Int J of Unconv Comput*, 3(4):311, 2007.

310. IL Bajec and M Mraz. Towards multi-state based computing using quantum-dot cellular automata. *Unconv Comput*, pages 105–116, 2005.

311. J Watrous. On one-dimensional quantum cellular automata. In *Foundations of Computer Science, 1995. Proceedings, 36th Annual Symposium on*, pages 528–537. IEEE, 1995.
312. A Kondacs and J Watrous. On the power of quantum finite state automata. In *Foundations of Computer Science, 1997. Proceedings, 38th Annual Symposium on*, pages 66–75. IEEE, 1997.
313. J Gruska. *Quantum Computing* (Springer, 1999).
314. D. Eppstein. Wolfram's classification of cellular automata. https://www. ics.uci.edu/~eppstein/ca/wolfram.html.
315. A Adamatzky. Computing in verotoxin. *ChemPhysChem*, 18(13):1822–1830, 2017.
316. A Adamatzky and L Bull. Are complex systems hard to evolve? *Complexity*, 14(6):15–20, 2009.
317. AC Yao. Theory and application of trapdoor functions, SFCS' 1982, *23rd Annual Symposium on Foundations of Computer Science*, IEEE: 80–91, 1982.
318. R Impagliazzo and A Wigderson. P = BPP unless E has sub-exponential circuits: Derandomizing the XOR Lemma, *Proceedings of the 29th Symposium on Theory of Computing STOC*: 220–229, 1997.
319. Z Zhang and Y. Shi. Communication complexities of symmetric XOR functions. *Quantum Information & Computation*, 9(3):255–263, 2009.
320. A Montanaro and T. Osborne. On the communication complexity of XOR functions. *Computational Complexity*: 1–18, 2010, arXiv:0909.3392.

Index

www.ingramcontent.com/pod-product-compliance
Lightning Source LLC
Chambersburg PA
CBHW072256210326
41458CB00074B/1844